T0296755

Machine Learning Guide for Oil and Gas Using Python

Machine Learning Guide for Oil and Gas Using Python

A Step-by-Step Breakdown with Data, Algorithms, Codes, and Applications

Hoss Belyadi
Obsertelligence, LLC

Alireza Haghighat
IHS Markit

Gulf Professional Publishing
An imprint of Elsevier

Gulf Professional Publishing is an imprint of Elsevier
50 Hampshire Street, 5th Floor, Cambridge, MA 02139, United States
The Boulevard, Langford Lane, Kidlington, Oxford, OX5 1GB, United Kingdom

Library of Congress Cataloging-in-Publication Data
A catalog record for this book is available from the Library of Congress

British Library Cataloguing-in-Publication Data
A catalogue record for this book is available from the British Library

ISBN: 978-0-12-821929-4

For information on all Gulf Professional Publishing publications visit our website at
https://www.elsevier.com/books-and-journals

Publisher: Joe Hayton
Senior Acquisitions Editor: Katie Hammon
Editorial Project Manager: Hillary Carr
Production Project Manager: Poulouse Joseph
Cover Designer: Christian Bilbow

Typeset by TNQ Technologies

Working together
to grow libraries in
developing countries

www.elsevier.com • www.bookaid.org

Contents

5. Supervised learning

8. Fuzzy logic

9. Evolutionary optimization

For additional information on the topics covered in the book, visit the companion site:
https://www.elsevier.com/books-and-journals/book-companion/9780128219294

Biography

Hoss Belyadi is the founder and CEO of Obsertelligence, LLC, focused on providing artificial intelligence (AI) in-house training and solutions. As an adjunct faculty member at multiple universities, including West Virginia University, Marietta College, and Saint Francis University, Mr. Belyadi taught data analytics, natural gas engineering, enhanced oil recovery, and hydraulic fracture stimulation design. With over 10 years of experience working in various conventional and unconventional reservoirs across the world, he works on diverse machine learning projects and holds short courses across various universities, organizations, and the department of energy (DOE). Mr. Belyadi is the primary author of *Hydraulic Fracturing in Unconventional Reservoirs* (first and second editions) and is the author of *Machine Learning Guide for Oil and Gas Using Python*. Hoss earned his BS and MS, both in petroleum and natural gas engineering from West Virginia University.

Dr. Alireza Haghighat is a senior technical advisor and instructor for Engineering Solutions at IHS Markit, focusing on reservoir/production engineering and data analytics. Prior to joining IHS, he was a senior reservoir engineer at Eclipse/Montage resources for nearly 5 years. As a reservoir engineer, he was involved in well performance evaluation with data analytics, rate transient analysis of unconventional assets (Utica and Marcellus), asset development, hydraulic fracture/reservoir simulation, DFIT analysis, and reserve evaluation. He was an adjunct faculty at Pennsylvania State University (PSU) for 5 years, teaching courses in Petroleum Engineering/Energy, Business and Finance departments. Dr. Haghighat has published several technical papers and book chapters on machine learning applications in smart wells, CO_2 sequestration modeling, and production analysis of unconventional reservoirs. He has received his PhD in petroleum and natural gas engineering from West Virginia University and a master's degree in petroleum engineering from Delft University of Technology.

Acknowledgment

We would like to thank the whole Elsevier team including Katie Hammon, Hilary Carr, and Poulouse Joseph for their continued support in making the publication process a success. I, Hoss Belyadi, would like to thank two individuals who have truly helped with the grammar and technical review of this book. First, I would like to thank my beautiful wife, Samantha Walstra, for her continuous support and encouragement during the past 2 years of writing this book. I would also like to express my deepest appreciation to Dr. Neda Nasiriani for her technical review of the book.

I, Alireza Haghighat, want to acknowledge Dr. Shahab D. Mohaghegh, who was my PhD advisor. He, a pioneer of AI & ML applications in the oil and gas industry, has guided me in my journey to learn petroleum data analytics. I would like to thank my wife, Dr. Neda Nasiriani, who has been incredibly supportive throughout the process of writing this book. She encouraged me to write, made recommendations that resulted in improvements, and reviewed every chapter of the book from a computer science point of view. I also want to thank Samantha Walstra for reviewing the technical writing of this book.

Chapter 1

Introduction to machine learning and Python

Introduction

Artificial Intelligence (AI) and machine learning (ML) have grown in popularity throughout various industries. Corporations, universities, government, and research groups have noticed the true potential of various applications of AI and ML to automate various processes while increasing predicting capabilities. The potential of AI and ML is a remarkable game changer in various industries. The technological AI advancements of self-driving cars, fraud detection, speech recognition, spam filtering, Amazon and Facebook's product and content recommendations, etc., have generated massive amounts of net asset value for various corporations. The energy industry is at the beginning phase of applying AI to different applications. The rise in popularity in the energy industry is due to new technologies such as sensors and high-performance computing services (e.g., Apache Hadoop, NoSQL, etc.) that enable big data acquisition and storage in different fields of study. Big data refers to a quantity of data that is too large to be handled (i.e., gathered, stored, and analyzed) using common tools and techniques, e.g., terabytes of data. The number of publications in this domain has exponentially increased over the past few years. A quick search on the number of publications in the oil and gas industry with Society of Petroleum Engineer's OnePetro or American Association of Petroleum Geologists (AAPG) in the past few years attests to this fact. As more companies realize the value added through incorporating AI into daily operations, more creative ideas will foster. The intent of this book is to provide a step-by-step, easy-to-follow workflow on various applications of AI within the energy industry using Python, a free open source programming language. As one continues through this book, one will notice the incredible work that the Python community has accomplished by providing various libraries to perform ML algorithms easily and efficiently. Therefore, our main goal is to share our knowledge of various ML applications within the energy industry with this step-by-step guide. Whether you are new to data science/programming language or at an advanced level, this book is written in a manner suitable for anyone. We will use many examples throughout the book

Machine Learning Guide for Oil and Gas Using Python. https://doi.org/10.1016/B978-0-12-821929-4.00006-8

that can be followed using Python. The primary user interface that we will use in this book is "Jupyter Notebook" and the download process of Anaconda package is explained in detail in the following sections.

Artificial intelligence

Terminologies such as AI, ML, big data, and data mining are used interchangeably across different organizations. Therefore, it is crucial to understand the true meaning of each terminology before diving deeper into various applications. AI is simply the use of machine or computer intelligence rather than human or animal intelligence. It is a branch of computer science that studies the simulation of human intelligence processes such as learning, reasoning, problem-solving, and self-correction by computers. Creating intelligent machines that work, react, and mimic cognitive functions of humans is the primary goal of AI. Examples of AI include email classification (categorization), smart personal assistants such as Siri, Alexa, and Google, automated respondents, process automation, security surveillance, fraud detection and prevention, pattern and image recognition, product recommendation and purchase prediction, smart searches, sales, volumes, and business forecasting, advertisement targeting, news feed personalization, terrorist activity detection, self-driving cars, health diagnostics, mortgage default prediction, house pricing prediction, robo-advisors (automated portfolio manager), and virtual travel assistant. As shown, the field of AI is only growing with extraordinary potential for decades to come. In addition, the demand for data science jobs has also exponentially grown in the past few years where companies search desperately for computer scientists, mathematicians, data scientists, and engineers that have postgraduate and preferably PhD degrees from accredited universities.

Data mining

Data mining is a terminology used in computer science and is defined as the process of extracting specific information from a database that was hidden and not explicitly available for the user, using a set of different techniques such as ML. It is also called **knowledge discovery in databases (KDD)**. Teaching someone how to play basketball is ML; however, using someone to find the best basketball centers is data mining. Data mining is used by ML algorithms to find links between various linear and nonlinear relationships. Data mining is often used to help **collect data** on various aspects of the business such as nonproductive time, sales trend, production key performance indicators, drilling data, completions data, stock market key indicators and information, etc. Data mining can also be used to go through websites, online platforms, and social media to collect and compile information (Belyadi et al., 2019).

Machine learning

ML is a subset of *AI*. It is defined as the collection of using various algorithms to teach computers to find patterns in data to be used for future prediction and forecasting or as a quality check for performance optimization. ML provides computers the ability to learn without being explicitly programmed. Some of the patterns may be hidden and therefore, finding those hidden patterns can add significant shareholder value to any organization. Please note that data mining deals with searching specific information while ML focuses on performing a certain task. In Chapter 2 of this book, various types of ML algorithms will be discussed. Also note that deep learning is a subset of machine learning in which multi-layer neural networks are used for various purposes including but not limited to image and facial recognition, time series forecasting, autonomous cars, language translation, etc. Examples of deep learning algorithms are convolution neural network (CNN) and recurrent neural network (RNN) that will be discussed with various O&G applications in Chapter 6.

Python crash course

Before covering the essentials of all the algorithms as well as the codes in Python, it is imperative to understand the fundamentals of Python. Therefore, this chapter will be used to illustrate the fundamentals before diving deeper into various workflow and examples.

Anaconda introduction

It is highly recommended to download Anaconda, the standard platform for Python data science which includes many of the necessary libraries with its installation. Most libraries used in this book are already preinstalled with Anaconda, so they don't need to be downloaded individually. The libraries that are not preinstalled in Anaconda will be mentioned throughout the chapters.

Anaconda installation

To install Anaconda, go on Anaconda's website (www.anaconda.com) and click on "Get Started." Afterward, click on "Download Anaconda Installers" and download the latest version of Anaconda either using Windows or Mac. Anaconda distribution will have over 250 packages some of which will be used throughout this book. If you do not download Anaconda, most libraries must be installed separately using the command prompt window. Therefore, it is highly advisable to download Anaconda to avoid downloading majority of the libraries that will be used in this book. Please note that while majority of the libraries will be installed by installing Anaconda, there will be some libraries where they would have to separately get installed using the command prompt or Anaconda prompt window. For those libraries that have not been preinstalled, simply open "Anaconda prompt" from the "start" menu, and type in "pip install (library name)" where "library name" is the name of the library that would like to be installed. Once the Anaconda has been successfully installed, search for "Jupyter Notebook" under start menu. Jupyter Notebook is a web-based, interactive computing notebook environment. Jupyter Notebook loads quickly, is user-friendly, and will be used throughout this book. There are other user interfaces such as Spyder, JupyterLab, etc. Fig. 1.1 shows the Jupyter Notebook's window after opening. Simply go into "Desktop" and create a folder called "ML Using Python." Afterward, go to the created folder ("ML Using Python") and click on "New" on the top right-hand corner as illustrated in Fig. 1.2.

You now have officially launched a new Jupyter Notebook and are ready to start coding as shown in Fig. 1.3.

Displayed in Fig. 1.4, the top left-hand corner indicates the Notebook is "Untitled." Simply click on "Untitled" and name the Jupyter Notebook "Python Fundamentals."

FIGURE 1.1   Jupyter Notebook window.

FIGURE 1.2   Opening a new Jupyter Notebook.

FIGURE 1.3   A blank Jupyter Notebook.

FIGURE 1.4   Python Fundamentals.

Jupyter Notebook interface options

To run a line in Jupyter Notebook, the "Run" button or preferably "SHIFT + ENTER" can be used. To add a line in Jupyter Notebook, hit "ALT + ENTER" or simply use "Insert → Cell Below." "Insert Cell Above can also be used for inserting a cell above where the current cell is. To delete a line in Jupyter Notebook, while that line is selected, hit "DD" (in other words, hit the "D" word key button twice in a row). If at any point when coding, the Jupyter Notebook would like to be stopped, select "Kernel → Interrupt" or "Kernel → Restart" to restart the cell. "Kernel → Restart and Run All" is another handy tool in Jupyter Notebook that can be used to run the whole notebook from top to bottom as opposed to using "SHIFT + ENTER" to run each line of code manually which can reduce productivity. Below are some of the handy shortcuts that are recommended to be used in Jupyter Notebook:

Shift + Enter → Run the current cell, select below
Ctrl + Enter → Run selected cells
Alt + Enter → Run the current cell, insert below
Ctrl + S → Save and checkpoint
Enter → Takes you into an edit mode
When in command mode "ESC" will get you out of edit mode.
H → Shows all shortcuts (use H when in command mode)
Up → Select cell above
Down → Select cell below
Shift + Up → Extends selected cells above (use when in command mode)

Shift + Down → Extends selected cells below (use when in command mode)

A → Inserts cell above (use when in command mode)

B → Inserts cell below (use when in command mode)

X → Cuts selected cells (use when in command mode)

C → Copy selected cells (use when in command mode)

V → Paste cells below (use when in command mode)

Shift + V → Paste cells above (use when in command mode)

DD (press the "D" keyword twice) → Deletes selected cells (use when in command mode)

Z → Undo cell deletion (use when in command mode)

Ctrl + A → Selects all (use when in command mode)

Ctrl + Z → Undo (use when in command mode)

Please spend some time using these key shortcuts to get comfortable with the Jupyter Notebook user interface. Other important notes throughout this book:

- Jupyter Notebook is extremely helpful when it comes to autocompleting some codes. The keyword to remember is the "tab" keyword which will help in autocompleting and faster coding. For example, if one wants to import the matplotlib library, simply type in "mat" and hit tab. Two available options such as "math" and "matplotlib" will be populated. This important feature enables one to obtain a library more quickly. In addition, it helps with remembering the syntax, library, or command names when coding. Therefore, for faster coding habits, feel free to use the "tab" keyword for autopopulating and autocompleting.

- Another especially useful shortcut is "shift + tab." Pressing this keyword inside a library's parenthesis one time will open all the features associated within that library. For example, if after importing "import numpy as np" np.linspace() is typed and "shift + tab" is hit once, it will populate a window that will show all the arguments that can be passed inside. Pressing "shift + tab" two, three, and four times will keep expanding the argument window until it occupies half of the page.

Basic math operations

Next, let's go over the basic operations in Python:

```
4*4
Python output=16
```

Please note that if a mathematical operation below is typed in a cell, Python will use **order of operations** to solve this; therefore, the answer is 42 for the example below.

```
4*4+2*4+9*2
Python output=42
```

```
(4*2)+(8*7)
Python output=64
```

To raise a variable or number to power, "**" can be used. For example,

```
10**3
Python output=1000
```

The remainder of a division can also be found using "%" sign. For example, remainder of 13 divided by 2 is 1.

```
13%2
Python output=1
```

Assigning a variable name

To assign a variable name in Python, it is important to note that a variable name cannot start with a number

```
x=100
y=200
z=300
d=x+y+z
d
Python output=600
```

To separate variable names, it is recommended to use underscore. If a variable name such as "critical rate" is defined with a space in between, Python will give an "invalid syntax" error. Therefore, underscore (_) must be used between critical and rate and "critical_rate" would be used as the variable name.

Creating a string

To create a string, single or double quotes can be used.

```
x='I love Python'
Python output='I love Python'
```

```
x="I love Python"
Python output='I love Python'
```

To index a string, bracket ([]) notation along with the element number can be used. It is crucial to remember that indexing in Python starts with 0. Let's assume that variable name y is defined as a string "Oil_Gas." Y[0] means the first element in Oil_Gas, while a y[5] means the sixth element in Oil_Gas since indexing starts with 0.

```
y="Oil_Gas"
y[0]
```
Python output='O'

```
y[5]
```
Python output='a'

To get the first 4 letters, y [0:4] can be used.

```
y[0:4]
```
Python output='Oil_'

To obtain the whole string, it is sufficient to use y[:]. Therefore,

```
y[:]
```
Python output='Oil_Gas'

Another way of indexing everything until an n-element is as follows:

```
y[:6]
```
Python output='Oil_Ga'

y[:6] essentially indicates indexing everything **up until the sixth element** (excluding the sixth element). y[2:] can also be used to index the second element and thereafter.

```
y[2:]
```
Python output='l_Gas'

To obtain the length of a string, "len()" can simply be used. For instance, to obtain the length of the string "The optimum well spacing is 950 ft" that is defined in variable z, len() can be used to do so.

```
z='The optimum well spacing is 950 ft'
len(z)
```
Python output=34

Defining a list

A list can be defined as follows:

```
list=['Land','Geology','Drilling']
list
```
Python output=['Land', 'Geology', 'Drilling']

```
list.append('Frac')
list
```
Python output=['Land', 'Geology', 'Drilling', 'Frac']

To append a number such as 100 to the list above, the following line of code can be performed:

```
list.append(100)
list
```
Python output=['Land', 'Geology', 'Drilling', 'Frac', 100]

To index from a list, the same bracket notation as **string indexing** can be used. For example, using the print syntax to print a different element in the defined list above would be as follows:

```
print (list[0])
print (list[1])
print (list[2])
print (list[3])
print (list[4])
Python output=Land
Geology
Drilling
Frac
100
```

To get elements 1 to 4 (excluding the fourth element), the following line can be used:

```
list[0:4]
Python output=['Land', 'Geology','Drilling', 'Frac']
```

Notice that list[0,4] excludes the fourth element which is 100 in this example.

To **replace** the first element with "Title_Search," the following line can be used:

```
list[0]='Title_Search'
list
Python output=['Title_Search', 'Geology', 'Drilling', 'Frac', 100]
```

To replace more elements and keeping the last element of 100, the following lines can be used:

```
list[0]='Reservoir_Engineer'
list[1]='Data_Engineer'
list[2]='Data_Scientist'
list[3]='Data Enthusiast'
list
Python output=['Reservoir_Engineer', 'Data_Engineer',
   'Data_Scientist', 'Data Enthusiast', 100]
```

Creating a nested list

A nested list is when there are list(s) inside of another list.

```
nested_list=[10,20, [30,40,50,60]]
nested_list
Python output=[10, 20, [30, 40, 50, 60]]
```

To grab number 30 from the nested_list above, the following line can be used:

```
nested_list[2][0]
Python output=30
```

Nested lists can become confusing, especially as the number of nested lists inside the bracket increases. Let's examine the example below called **nested_list2**. If the objective is to get **number 3** from the nested_list2 shown below, three brackets will be needed to accomplish this task. First, use **nested_list2[2]** to obtain [30, 40, 50, 60, [4, 3, 1]]. Afterward, use **nested_list2[2] [4]** to get [4, 3, 1], and finally to get 3, use **nested_list2[2][4][1]**.

```
nested_list2=[10,20, [30,40,50,60, [4,3,1]]]
nested_list2[2][4][1]
Python output=3
```

Creating a dictionary

Thus far, we have covered strings and lists and the next section is to talk about dictionary. When using dictionary, wiggly brackets ({}) are used. Below, a dictionary was created and named "a" for various ML models and their respective scores.

```
a={'ML_Models':['ANN','SVM','RF','GB','XGB'],
  'Score':[90,85,95,90,100]}
a
Python output={'ML_Models': ['ANN', 'SVM', 'RF', 'GB', 'XGB'],
  'Score': [90, 85, 95, 90, 100]}
```

To index off this dictionary, the below command can be used. As shown, calling a['ML_Models'] lists the name of the ML models that are under ML_Models.

```
a['ML_Models']
Python output=['ANN', 'SVM', 'RF', 'GB', 'XGB']
```

To yield the name of the ML models including ANN, SVM, and RF and excluding the rest, the following command can be used.

```
a['ML_Models'][0:3]
Python output=['ANN', 'SVM', 'RF']
```

Another nested dictionary example is listed below:

```
d={'a':{'inner_a':[1,2,3]}}
d
{'a': {'inner_a': [1, 2, 3]}}
```

If the objective is to show number 2 in the above dictionary, the following indexing can be used to accomplish this task. The first step is to use d['a'] which would result in showing {'inner_a': [1, 2, 3]}. Next, using d['a'] ['inner_a'] would result in [1, 2, 3]. Finally, to pick number 2, we simply need to add index 1 to yield d['a'] ['inner_a'][1].

```
d['a']['inner_a'][1]
Python output=2
```

Creating a tuple

As opposed to lists that use **brackets**, tuples use **parentheses** to define a sequence of elements. One of the advantages of using a list is that items can be assigned; however, tuples do not support item assignments which means they are immutable. For instance, let's create a list and replace one of its elements and examine the same concept with tuples. As shown below, a list with 4 elements of 100, 200, 300, and 400 was created. The first index of 100 was replaced with "New" and the new list is as follows:

```
list=[100,200,300,400]
list[0]='New'
list
Python output=['New', 200, 300, 400]
```

Next, let's create a tuple with the same numbers as follows:

```
t= (100,200,300,400)
t
Python output= (100, 200, 300, 400)
```

As shown, Python generated a list of tuples. Now, let's assign "New" to the first element in the generated tuple above. However, after running this, Python will return an error indicating that **"tuple's object does not support item assignment."** This is essentially the primary difference between lists and tuples.

```
t[0]='New'
```

Creating a set

Set is defined by unique elements which means defining the same numbers multiple times will only return the unique numbers and will not show the repetitive numbers. The wiggly brackets ({}) can be used to generate a set as follows. As displayed, the generated output only has 100,200,300 since each number was repeated twice.

```
set={100,200,300,100,200,300}
set
Python output={100, 200, 300}
```

The add() syntax can be attached to a set to add a number at the end of the set as shown below:

```
set.add(400)
set
Python output={100, 200, 300, 400}
```

If statements

If statements are perhaps one of the most important concepts in any programming language. Let's start with a simple example and define if 100 is equal to 200, print good job, otherwise, print not good. Make sure the print statements following "if 100 == 200:" and "else:" are indented, otherwise, an error will be received. The "tab" keyword can be used to indent in Jupyter Notebook. Please note that indenting in Python means **4 spaces**.

```
if 100= =200:
    print('Good Job!')
else:
    print('Not Good!')
Python output=Not Good!
```

Now, let's define X, Y, and Z variables and write another if statement referencing these variables. As shown below, if X is bigger than Y, print "GOOD" which is not the case. Therefore, the term "elif" can be used to define multiple conditions. The next condition is if Z < Y to print "SO SO" which is again not the case and therefore, the term "else" is used to define all other cases, and the output would be "BAD."

```
X=100
Y=200
Z=300
if X>Y:
    print('Good')
elif Z<Y:
    print('SO SO')
else:
    print('BAD')
Python output=BAD
```

The if statement above can also be written as follows to obtain numeric output as opposed to string.

```
X=100
Y=200
Z=300
if X>Y:
```

```
    A=X+Y
elif Z<Y:
    B=X+Y+Z
else:
    C=2*(X+Y+Z)
C
```
Python output=1200

Let's do another if statement example. First, let's define n as an input number that the user can enter. If n is equal to 0, print "ZERO." If n is less than 0, print "NEGATIVE Number," and finally if n is bigger than 0, print "POSITIVE Number." When the code below is run, enter a number and click on enter. Next, depending on the number that was entered, an appropriate statement will be printed.

```
n=float(input("Enter any number"))
if n==0:
    print('ZERO')
elif n>0:
    print('POSITIVE Number')
else:
    print('NEGATIVE Number')
```

For loop

For loop is another very useful tool in any programming language and allows for iterating through a sequence. Let's define i to be a range between 0 and 5 (excluding 5). A for loop is then written to result in writing 0 to 4. As shown below, **"for x in i"** is the same as **"for x in range(0,5)"**

```
i=range(0,5)
for x in i:
    print(x)
```
Python output=
```
0
1
2
3
4
```

Another for loop example can be written as follows:

```
for x in range(0,3):
    print('Edge computing in the O&G industry is very valuable')
```
Python output=Edge computing in the O&G industry is very valuable
Edge computing in the O&G industry is very valuable
Edge computing in the O&G industry is very valuable

The "break" function allows stopping through the loop before looping through all the items. Below is an example of using an if statement and break function within the for loop. As displayed below, if the for loop sees "Frac_Crew_2," it will break and not finish the for-loop iteration.

```
Frac_Crews=['Frac_Crew_1', 'Frac_Crew_2', 'Frac_Crew_3',
'Frac_Crew_4']
for x in Frac_Crews:
    print(x)
    if x=='Frac_Crew_2':
        break
```
Python output=Frac_Crew_1
Frac_Crew_2

With the "continue" statement, it is possible to stop the current iteration of the loop and continue with the next. For example, if it is desirable to skip "Frac_Crew_2" and move to the next name, the continue statement can be used as follows:

```
Frac_Crews=['Frac_Crew_1', 'Frac_Crew_2', 'Frac_Crew_3',
'Frac_Crew_4']
for x in Frac_Crews:
    if x=='Frac_Crew_2':
        continue
    print(x)
```
Python output=Frac_Crew_1
Frac_Crew_3
Frac_Crew_4

The "range" function can also be used in different increments. By default, the range function uses the following sequence to generate the numbers: start, stop, increments. For example, if the number that is desirable to start with is 10 with the final number of 18, and an increment of 4, the following lines can be written:

```
for x in range(10, 19, 4):
    print(x)
```
Python output=10
14
18

Nested loops

A nested loop refers to a loop inside another loop. In various ML optimization programs such as grid search (which will be discussed), nested loops are very common to optimize the ML hyperparameters. Below is a simple example of a nested loop:

```
Performance=["Low_Performing", "Medium_Performing",
   'High_Performing']
Drilling_Crews=["Drilling_Crew_1", "Drilling_Crew_2",
   "Drilling_Crew_3"]
for x in Performance:
    for y in Drilling_Crews:
        print(x, y)
```
Python output=Low_Performing Drilling_Crew_1
Low_Performing Drilling_Crew_2
Low_Performing Drilling_Crew_3
Medium_Performing Drilling_Crew_1
Medium_Performing Drilling_Crew_2
Medium_Performing Drilling_Crew_3
High_Performing Drilling_Crew_1
High_Performing Drilling_Crew_2
High_Performing Drilling_Crew_3

Let's try another for loop example except for creating an empty list first, followed by performing basic algebra on a list of numbers as follows:

```
i=[10,20,30,40]
out=[]
for x in i:
    out.append(x**2+200)
print(out)
```
Python output= [300, 600, 1100, 1800]

List comprehension

List comprehension is another powerful way of performing calculations quickly. The calculations listed above could have been simplified in the following list comprehension:

```
[x**2+200 for x in i]
```
Python output= [300, 600, 1100, 1800]

As shown in the list comprehension example above, first write down the mathematical calculations that is desired followed by "for x in i" inside a bracket. Below is another example:

```
y=[5,6,7,8,9,10]
[x**2+3*x+369 for x in y]
```
Python output= [409, 423, 439, 457, 477, 499]

As demonstrated above, each element in y goes through the quadratic equation of "$x^2+3x+369$." This is a more advanced way of writing the same code above. At first glance, it might not be intuitive, but over time, list comprehension becomes much easier to understand and apply.

Defining a function

The next concept in Python is defining a function. A function can be defined by using "def" followed by "return" to perform various mathematical equations:

```
def linear_function(n):
    return 2*n+20
linear_function(20)
Python output=60
```

As shown above, first use the syntax "def" to define any name that is desirable. Afterward, return the equation that is desirable. Finally, call the defined name followed by the number that is desired to use to run the calculations.

Below is another example:

```
def Turner_rate(x):
    return x**2+50
Turner_rate(20)
Python output=450
```

Introduction to pandas

Pandas is one of the most famous libraries in Python, and it is essentially designed to replicate the excel sheet formats in Python. The primary role of pandas is data manipulation and analysis, and it is heavily used for data preprocessing before implementing ML models. Building various ML models becomes much easier after learning the fundamentals of pandas and numpy (which will be discussed next) libraries. To start off, let's create a dictionary and covert that dictionary into a pandas table format as follows:

```
dictionary={'Column_1':[10,20,30],'Columns_2':[40,50,60],
    'Column_3':[70,80,90]}
dictionary
Python output={'Column_1': [10, 20, 30], 'Columns_2': [40, 50, 60],
    'Column_3': [70, 80, 90]}
```

To convert this dictionary into a data frame, the **pd.DataFrame** command line can be used. It is also important to "import pandas as pd" prior to the command line as follows:

```
import pandas as pd
dictionary=pd.DataFrame(dictionary)
dictionary
Python output=Fig. 1.5
```

	Column_1	Columns_2	Column_3
0	10	40	70
1	20	50	80
2	30	60	90

FIGURE 1.5   Converted dictionary to data frame.

The above command could have been performed in one step as follows:

```
dictionary=pd.DataFrame({'Column_1':[10,20,30],'Columns_2':
    [40,50,60],'Column_3':[70,80,90]})
dictionary
```

For practice, let's create another data frame with 3 columns including drilling_CAPEX, completions_CAPEX, and production CAPEX but also have well_1, well_2, and well_3 as indices in lieu of 0, 1, and 2. As shown below, the "index" can be added to define the name of each row or index. Note that backslash (\) symbol was added twice in to the code below. Adding backslash will essentially tell Python that the remaining portion of the code is being continued on the next line. Otherwise, Python will return an error when a portion of the code is abruptly located in the next line.

```
CAPEX=pd.DataFrame({'Drilling_CAPEX':[2000000,2200000,2100000],
    'Completions_CAPEX':[5000000,5200000,5150000], 'Production_\
    CAPEX':\
    [500000,550000,450000]}, index=['Well_1','Well_2','Well_3'])
CAPEX
Python output=Fig. 1.6
```

	Drilling_CAPEX	Completions_CAPEX	Production_CAPEX
Well_1	2000000	5000000	500000
Well_2	2200000	5200000	550000
Well_3	2100000	5150000	450000

FIGURE 1.6   CAPEX data frame table.

Next, let's create an 8×5 matrix using "np.random.seed" (with a seed number of 100) and transfer this matrix array to a data frame named "life_-cycle" as shown below. Note that instead of defining index and column names inside single or double quotes, ".split" function can be used as shown below to save typing. Please note that numpy will be covered in detail in this chapter. For now, simply make sure to import the numpy library using "import numpy as np" as illustrated below:

```
from numpy.random import randn
import numpy as np
seed=100
np.random.seed(seed)
life_cycle=pd.DataFrame(randn(8,5),index='Land Seismic Geology\
    Drilling Completions Production Facilities Midstream'.split(),
    columns='Cycle_1 Cycle_2 Cycle_3 Cycle_4 Cycle_5'.split())
life_cycle
```
Python output=Fig. 1.7

	Cycle_1	Cycle_2	Cycle_3	Cycle_4	Cycle_5
Land	-1.749765	0.342680	1.153036	-0.252436	0.981321
Seismic	0.514219	0.221180	-1.070043	-0.189496	0.255001
Geology	-0.458027	0.435163	-0.583595	0.816847	0.672721
Drilling	-0.104411	-0.531280	1.029733	-0.438136	-1.118318
Completions	1.618982	1.541605	-0.251879	-0.842436	0.184519
Production	0.937082	0.731000	1.361556	-0.326238	0.055676
Facilities	0.222400	-1.443217	-0.756352	0.816454	0.750445
Midstream	-0.455947	1.189622	-1.690617	-1.356399	-1.232435

FIGURE 1.7 life_cycle data frame.

As shown in the Python output above, a random **8×5 matrix data frame** with a seed number of 100 was generated with index names of Land, Seismic, etc., and column names of Cycle_1, Cycle_2, etc. This data frame was called "life_cycle." Data frames look like excel tables and are easy to read and perform various calculations. Now let's only visualize the **first and last two rows of data** by using ".head" and ".tail" pandas functions:

```
life_cycle.head(2)
```
Python output=Fig. 1.8

	Cycle_1	Cycle_2	Cycle_3	Cycle_4	Cycle_5
Land	-1.749765	0.34268	1.153036	-0.252436	0.981321
Seismic	0.514219	0.22118	-1.070043	-0.189496	0.255001

FIGURE 1.8 life_cycle head.

```
life_cycle.tail(2)
```
Python output=Fig. 1.9

	Cycle_1	Cycle_2	Cycle_3	Cycle_4	Cycle_5
Facilities	0.222400	-1.443217	-0.756352	0.816454	0.750445
Midstream	-0.455947	1.189622	-1.690617	-1.356399	-1.232435

FIGURE 1.9   life_cycle tail.

".describe" function can also be used to provide basic statistics (count, mean, standard deviation, min, 25%, 50%, 75%, and max) of each column. Let's also get the basic statistics of the life_cycle data frame as follows:

```
life_cycle.describe()
Python output=Fig. 1.10
```

	Cycle_1	Cycle_2	Cycle_3	Cycle_4	Cycle_5
count	8.000000	8.000000	8.000000	8.000000	8.000000
mean	0.065566	0.310844	-0.101020	-0.221480	0.068616
std	1.019031	0.946721	1.142760	0.745362	0.829178
min	-1.749765	-1.443217	-1.690617	-1.356399	-1.232435
25%	-0.456467	0.033065	-0.834775	-0.539211	-0.237823
50%	0.058994	0.388922	-0.417737	-0.289337	0.219760
75%	0.619935	0.845656	1.060558	0.061992	0.692152
max	1.618982	1.541605	1.361556	0.816847	0.981321

FIGURE 1.10   life_cycle description.

Another very useful function is the ".cumsum()" function which would result in the CUM values in each column. This is specifically useful when trying to add a column for cumulative production volumes over time having daily or monthly production rates. Let's apply the ".cumsum()" function to the life_cycle data frame as follows:

```
life_cycle.cumsum()
Python output=Fig. 1.11
```

To only select a column from the created "life_cycle" data frame, it suffices to pass in the column name inside a bracket as follows:

```
life_cycle['Cycle_2']
Python output=Fig. 1.12
```

	Cycle_1	Cycle_2	Cycle_3	Cycle_4	Cycle_5
Land	-1.749765	0.342680	1.153036	-0.252436	0.981321
Seismic	-1.235547	0.563860	0.082992	-0.441932	1.236322
Geology	-1.693574	0.999024	-0.500603	0.374915	1.909043
Drilling	-1.797985	0.467743	0.529130	-0.063220	0.790725
Completions	-0.179003	2.009348	0.277251	-0.905656	0.975243
Production	0.758079	2.740349	1.638807	-1.231894	1.030919
Facilities	0.980479	1.297132	0.882455	-0.415440	1.781364
Midstream	0.524532	2.486754	-0.808162	-1.771839	0.548930

FIGURE 1.11   life_cycle CUM.

```
Land          0.342680
Seismic       0.221180
Geology       0.435163
Drilling     -0.531280
Completions   1.541605
Production    0.731000
Facilities   -1.443217
Midstream     1.189622
Name: Cycle_2, dtype: float64
```

FIGURE 1.12   Cycle_2 column.

The recommended method of obtaining a column is to pass in the column name inside a bracket. However, another methodology is to use ".column name" as shown below to obtain the same information:

```
life_cycle.Cycle_2
Python output=same as before
```

Please note that the second methodology is not recommended because, if there is any space in the column name, Python will yield an error message. For instance, if the name of the "Cycle_2" column was defined as "Cycle 2" instead, and "life_cycle.Cycle 2" is called to obtain this column, Python will yield an invalid syntax error. **Therefore, it is strongly recommended to pass in the column names inside a bracket.**

To select multiple columns, use **double bracket** notation as follows:

```
life_cycle[['Cycle_2','Cycle_3','Cycle_4']]
Python output=Fig. 1.13
```

	Cycle_2	Cycle_3	Cycle_4
Land	0.342680	1.153036	-0.252436
Seismic	0.221180	-1.070043	-0.189496
Geology	0.435163	-0.583595	0.816847
Drilling	-0.531280	1.029733	-0.438136
Completions	1.541605	-0.251879	-0.842436
Production	0.731000	1.361556	-0.326238
Facilities	-1.443217	-0.756352	0.816454
Midstream	1.189622	-1.690617	-1.356399

FIGURE 1.13   Cycle_2, Cycle_3, and 4 columns.

It is also easy to perform quick calculations using pandas. For example, if all the cycles in the life_cycle data frame must be added to create a new life_cycle data frame called "life_cycle['Cycle_Total']", the following lines can be used:

```
life_cycle['Cycle_Total']=life_cycle['Cycle_1']+life_cycle\
   ['Cycle_2']+life_cycle['Cycle_3']+life_cycle['Cycle_4']+\
life_cycle['Cycle_5']
life_cycle['Cycle_Total']
Python output=Fig. 1.14
```

```
Land              0.474835
Seismic          -0.269139
Geology           0.883109
Drilling         -1.162413
Completions       2.250791
Production        2.759077
Facilities       -0.410271
Midstream        -3.545775
Name: Cycle_Total, dtype: float64
```

FIGURE 1.14   Cycle_Total.

Please note to always use backslash (\) when coding in Jupyter Note-book when going to a new line in the middle of performing calculations; otherwise, Python will not recognize that the calculation is being continued in the next line. Let's create another column called "life_cycle['Cycles_1_2_Mult']" as follows:

```
life_cycle['Cycle_1_2_Mult']=life_cycle['Cycle_1']*life_cycle\
    ['Cycle_2']
life_cycle['Cycle_1_2_Mult']
Python output=Fig. 1.15
```

```
Land         -0.599610
Seismic       0.113735
Geology      -0.199317
Drilling      0.055472
Completions   2.495831
Production    0.685007
Facilities   -0.320971
Midstream    -0.542405
Name: Cycle_1_2_Mult, dtype: float64
```

FIGURE 1.15   Cycle_1 and 2 multiplication.

If the life_cycle data frame is called now, the entire data frame including the two new added columns will be shown:

```
life_cycle
Python output=Fig. 1.16
```

	Cycle_1	Cycle_2	Cycle_3	Cycle_4	Cycle_5	Cycle_Total	Cycle_1_2_Mult
Land	-1.749765	0.342680	1.153036	-0.252436	0.981321	0.474835	-0.599610
Seismic	0.514219	0.221180	-1.070043	-0.189496	0.255001	-0.269139	0.113735
Geology	-0.458027	0.435163	-0.583595	0.816847	0.672721	0.883109	-0.199317
Drilling	-0.104411	-0.531280	1.029733	-0.438136	-1.118318	-1.162413	0.055472
Completions	1.618982	1.541605	-0.251879	-0.842436	0.184519	2.250791	2.495831
Production	0.937082	0.731000	1.361556	-0.326238	0.055676	2.759077	0.685007
Facilities	0.222400	-1.443217	-0.756352	0.816454	0.750445	-0.410271	-0.320971
Midstream	-0.455947	1.189622	-1.690617	-1.356399	-1.232435	-3.545775	-0.542405

FIGURE 1.16   life_cycle data frame with new calculated columns.

Dropping rows or columns in a data frame

To drop rows or columns in a data frame, ".drop()" syntax can be used. If this drop is going to be permanent, please make sure to include "inplace = True" because the default inplace is False in Python's pandas. To drop rows, use axis = 0 (which is the default in Python's pandas) and to drop columns, use axis = 1 as shown below. Below is an example of dropping the last two new created columns (Cycle_Total and Cycle_1_2_Mult). Axis = 1 represents dropping columns and inplace = True represents that this change is permanent.

```
life_cycle.drop(labels=['Cycle_Total','Cycle_1_2_Mult'], axis=1,
    inplace=True)
life_cycle
Python output=The original life_cycle data frame without the added
    calculated columns.
```

To drop the first two rows named Land and Seismic, the following command can be used:

```
life_cycle.drop(labels=['Land','Seismic'], axis=0, inplace=True)
life_cycle
Python output=Fig. 1.17
```

	Cycle_1	Cycle_2	Cycle_3	Cycle_4	Cycle_5
Geology	-0.458027	0.435163	-0.583595	0.816847	0.672721
Drilling	-0.104411	-0.531280	1.029733	-0.438136	-1.118318
Completions	1.618982	1.541605	-0.251879	-0.842436	0.184519
Production	0.937082	0.731000	1.361556	-0.326238	0.055676
Facilities	0.222400	-1.443217	-0.756352	0.816454	0.750445
Midstream	-0.455947	1.189622	-1.690617	-1.356399	-1.232435

FIGURE 1.17   life_cycle without the first two rows.

loc and iloc

loc is for selecting rows and columns by **label**. The format for loc is the rows of interest followed by columns of interest. Let's create a 5×5 random number matrix as follows:

```
from numpy.random import randn
seed=200
np.random.seed(seed)
matrix=pd.DataFrame(randn(5,5), columns='Cycle_1 Cycle_2 Cycle_3\
    Cycle_4 Cycle_5'.split())
matrix
Python output=Fig. 1.18
```

	Cycle_1	Cycle_2	Cycle_3	Cycle_4	Cycle_5
0	-1.450948	1.910953	0.711879	-0.247738	0.361466
1	-0.032950	-0.221347	0.477257	-0.691939	0.792006
2	0.073249	1.303286	0.213481	1.017349	1.911712
3	-0.529672	1.842135	-1.057235	-0.862916	0.237631
4	-1.154182	1.214984	-1.293759	0.822723	-0.332151

FIGURE 1.18 matrix data frame.

From the 5×5 matrix above, let's select rows 0 through 2 and all columns as follows:

```
matrix.loc[0:2,:]
Python output=Fig. 1.19
```

	Cycle_1	Cycle_2	Cycle_3	Cycle_4	Cycle_5
0	-1.450948	1.910953	0.711879	-0.247738	0.361466
1	-0.032950	-0.221347	0.477257	-0.691939	0.792006
2	0.073249	1.303286	0.213481	1.017349	1.911712

FIGURE 1.19 First three rows of the matrix data frame.

As demonstrated, bracket notation must be used with ".loc[]." In addition, ":" can be used to represent either all rows or all columns. The code above can also be rewritten as follows with the same outcome:

```
matrix.loc[[0,1,2],:]
Python output=same as above
```

Please note instead of specifying rows of 0, 1, and 2, it is recommended to use [0:2] if the rows of interest are continuous. If the rows of interests are not continuous, the above command can be used instead.

To select all rows and the first two columns, the code below can be used:

```
matrix.loc[:,['Cycle_1','Cycle_2']]
Python output=Fig. 1.20
```

	Cycle_1	Cycle_2
0	-1.450948	1.910953
1	-0.032950	-0.221347
2	0.073249	1.303286
3	-0.529672	1.842135
4	-1.154182	1.214984

FIGURE 1.20   Cycle_1 and 2 columns.

To select the first four rows and cycles 3 and 5, the command can be written as follows:

```
matrix.loc[0:3,['Cycle_3','Cycle_5']]
Python output=Fig. 1.21
```

	Cycle_3	Cycle_5
0	0.711879	0.361466
1	0.477257	0.792006
2	0.213481	1.911712
3	-1.057235	0.237631

FIGURE 1.21   First four rows and Cycle_3 and 5 columns.

Please note that loc is inclusive on both sides when selecting a range. For example, when selecting matrix.loc[0:3,:], it will return rows 0, 1, 2, and 3. However, when using iloc, it will be **inclusive** of the first number but **exclusive** of the second number. Therefore, when selecting matrix.iloc[0:3,:], it will return rows 0, 1, and 2 only and will **exclude** row 3. "iloc" is used for selecting rows and columns by **integer** position. That's why it is referred to as iloc ("i" stands for integer and loc stands for location). Let's use the iloc to get all rows of the matrix data frame and only the first two columns:

```
matrix.iloc[:,0:2]
Python output=the output of this is the same as "matrix.loc
   [:,['Cycle_1','Cycle_2']]"
```

If it is desirable to obtain the number 1.017349 from the matrix data frame, iloc can be used as follows:

```
matrix.iloc[2,3]
Python output=1.0173489526279178
```

The code above will simply select the number in row 3 (since indexing starts with 0 in Python) and column 4. To select certain rows and columns, the notation below can be used:

```
matrix.iloc[[0,2,4],[0,2,4]]
Python output=Fig. 1.22
```

	Cycle_1	Cycle_3	Cycle_5
0	-1.450948	0.711879	0.361466
2	0.073249	0.213481	1.911712
4	-1.154182	-1.293759	-0.332151

FIGURE 1.22   iloc selection for various rows and columns.

Therefore, loc is primarily used for labels and iloc is primarily used for integer selection. loc is inclusive on both sides while iloc is inclusive on one side and exclusive on another.

Please note that if it is desirable to obtain the first two columns from the matrix data frame, the following two methodologies will yield the same result:

```
matrix[['Cycle_1','Cycle_2']]
matrix.loc[:,['Cycle_1','Cycle_2']]
Python output for both lines of codes=Fig. 1.20
```

"ix" allows the mix of labels and integers when performing selection. It is therefore a blend between loc and iloc. Note that ix is deprecated, and it is generally recommended to use loc and iloc for any coding. Therefore, it will not be discussed in this book because an attribute error will be received when using "ix" in newer Python versions.

Conditional selection

Pandas allows for conditional selection for filtering the desired data. To illustrate an example, let's create an 8×4 matrix data frame as follows:

```
from numpy.random import randn
seed=400
np.random.seed(seed)
Decision=pd.DataFrame(randn(8,4),columns='Drill No_Drill Frac\
    No_Frac'.split())
Decision
```
Python output=Fig. 1.23

	Drill	No_Drill	Frac	No_Frac
0	-1.130571	0.696200	-0.432293	0.741020
1	-0.478137	1.386040	0.125180	1.148860
2	-2.350259	0.183293	-0.311386	-0.294066
3	0.400061	1.005500	0.501643	-0.000668
4	1.303972	1.073324	-0.515345	1.662900
5	0.551817	-0.579494	-0.848800	-0.410862
6	2.160625	-1.557699	0.134852	0.005454
7	0.646549	-0.592073	0.551924	1.378349

FIGURE 1.23   Decision data frame.

If it is desired to return a Boolean value for a certain condition, the data frame can be passed on with the condition as follows:

```
Decision>0
```
Python output=Fig. 1.24

To return the Boolean value for a column, the following can be done:

```
Decision['Drill']>0
```
Python output=Fig. 1.25

To select the rows greater than 0 of the "Drill" column, the following can be executed.

```
Decision[Decision['Drill']>0]
```
Python output=Fig. 1.26

	Drill	No_Drill	Frac	No_Frac
0	False	True	False	True
1	False	True	True	True
2	False	True	False	False
3	True	True	True	False
4	True	True	False	True
5	True	False	False	False
6	True	False	True	True
7	True	False	True	True

FIGURE 1.24 Boolean return when decision>0.

```
0      False
1      False
2      False
3       True
4       True
5       True
6       True
7       True
Name: Drill, dtype: bool
```

FIGURE 1.25 Boolean return when decision[Drill]>0.

	Drill	No_Drill	Frac	No_Frac
3	0.400061	1.005500	0.501643	-0.000668
4	1.303972	1.073324	-0.515345	1.662900
5	0.551817	-0.579494	-0.848800	-0.410862
6	2.160625	-1.557699	0.134852	0.005454
7	0.646549	-0.592073	0.551924	1.378349

FIGURE 1.26 Filtered decision data frame.

As noted above, all the rows that have values greater than 0 in the "Drill" column have been selected only.

Multiconditional filtering can also be applied as follows:

```
Decision[(Decision['Drill']>0) & (Decision['No_Drill']>0) &
    (Decision['Frac']>0) & (Decision['No_Frac']<0)]
Python output=Fig. 1.27
```

	Drill	No_Drill	Frac	No_Frac
3	0.400061	1.0055	0.501643	-0.000668

FIGURE 1.27   Filtered decision data frame using multiconditions.

Please note that "&" is used in lieu of "and" in Python. If "&" symbol is replaced with "and," Python will return an error. Therefore, the conditional statement above indicates that Drill, No_Drill, and Frac columns should all be greater than 0 and No_Frac should be less than 0. This condition is only satisfied in index 3 (or row 4) of the Decision data frame.

Another conditional statement is to use "|" for "or" as follows. For example, if the Drill column of greater than 0 **OR** No_Drill column of less than 0 is desirable, Python will filter out remaining rows that fail to meet these criteria.

```
Decision[(Decision['Drill']>0) | (Decision['No_Drill']<0)]
Python output=Fig. 1.26
```

Conditional filtering is very useful for automating various processes prior to implementing ML models. As the amount of data increases in the source file, it is much more convenient to perform all the filtering processes in Python prior to performing any analysis.

Pandas groupby

Groupby is a very useful way of grouping rows together and performing an aggregate function on them. To illustrate the concept, let's create a data frame with 3 different corporation names and sales amount per corporation and person as follows:

```
df=pd.DataFrame({'Corporation_Name':['CVX','EXXON', 'CVX',
    'EXXON', 'GE','GE'],'Sales_Person':['Adam','Alex','Bruce',
    'Jessica', 'Natalie','Rachel'],'Sales_Amount':[2000,5000,
    10000,45000,60000, 20000]})
df
Python output=Fig. 1.28
```

	Corporation_Name	Sales_Person	Sales_Amount
0	CVX	Adam	2000
1	EXXON	Alex	5000
2	CVX	Bruce	10000
3	EXXON	Jessica	45000
4	GE	Natalie	60000
5	GE	Rachel	20000

FIGURE 1.28 df data frame.

To obtain the mean of sales amount by corporation, the groupby function can be used as follows:

```
df_group_by=df.groupby('Corporation_Name')
df_group_by.mean()
Python output=Fig. 1.29
```

	Sales_Amount
Corporation_Name	
CVX	6000
EXXON	25000
GE	40000

FIGURE 1.29 Mean of sales amount by corporation.

The following line would have also resulted in the same outcome:

```
df.groupby('Corporation_Name').mean()
df_group_by.std()
Python output=Fig. 1.30
```

	Sales_Amount
Corporation_Name	
CVX	5656.854249
EXXON	28284.271247
GE	28284.271247

FIGURE 1.30 Standard deviation of sales amount by corporation.

Groupby can also be performed on multiple columns at one time. For example, to group by two columns named "col_name1" and "col_name_2" on a data frame called "df" and perform an average aggregate function, use the following format:

```
df.groubpy(['col_name_1', 'col_name_2']).mean()
```

Other useful built-in pandas aggregation functions are listed below:

count() = total number of items
mean(), median() = mean and median
first(), last() = first and last item
std(), var() = standard deviation and variance
mad() = mean absolute deviation
prod() = product of all items
size() = calculate group sizes
sum() = calculate sum of group values

Another useful way of utilizing the groupby function is to apply the describe function to create a summary by a column name.

```
df.groupby('Corporation_Name').describe().transpose()
Python output=Fig. 1.31
```

Corporation_Name		CVX	EXXON	GE
Sales_Amount	count	2.000000	2.000000	2.000000
	mean	6000.000000	25000.000000	40000.000000
	std	5656.854249	28284.271247	28284.271247
	min	2000.000000	5000.000000	20000.000000
	25%	4000.000000	15000.000000	30000.000000
	50%	6000.000000	25000.000000	40000.000000
	75%	8000.000000	35000.000000	50000.000000
	max	10000.000000	45000.000000	60000.000000

FIGURE 1.31   Group by description.

Pandas data frame concatenation

Concatenation is another commonly used pandas function to append various columns vertically or horizontally. Concatenation can also be done using multiple files. Later in this section, importing various files into Python will be discussed. To illustrate the concept, let's create 3 data frames called df1, df2, and df3 as follows:

```
df1=pd.DataFrame({'A':['H1','H2','H3','H4'],
   'B':['I1','I2','I3','I4'],'C':['J1','J2','J3','J4'],
   'D':['K1','K2','K3','K4']}, index=[0,1,2,3])
df1
```
Python output=Fig. 1.32

	A	B	C	D
0	H1	I1	J1	K1
1	H2	I2	J2	K2
2	H3	I3	J3	K3
3	H4	I4	J4	K4

FIGURE 1.32 df1 data frame.

```
df2=pd.DataFrame({'A':['H5','H6','H7','H8'],
   'B':['I5','I6','I7','I8'], 'C':['J5','J6','J7','J8'],
   'D':['K5','K6','K7','K8']}, index=[4,5,6,7])
df2
```
Python output=Fig. 1.33

	A	B	C	D
4	H5	I5	J5	K5
5	H6	I6	J6	K6
6	H7	I7	J7	K7
7	H8	I8	J8	K8

FIGURE 1.33 df2 data frame.

```
df3=pd.DataFrame({'A':['H9','H10','H11','H12'],
  'B':['I9','I10','I11','I12'],'C':['J9','J10','J11','J12'],
  'D':['K9','K10','K11','K12'],'E':['L1','L2','L3','L4']},
  index=[8,9,10,11])
df3
Python output=Fig. 1.34
```

	A	B	C	D	E
8	H9	I9	J9	K9	L1
9	H10	I10	J10	K10	L2
10	H11	I11	J11	K11	L3
11	H12	I12	J12	K12	L4

FIGURE 1.34   df3 data frame.

Please note that df3 has an additional column called "E." Therefore, pay attention to the resulting concatenated data frame when combining df1, df2, and df3. "pd.concat()" function can be used to glue one data frame right below one another as follows:

```
pd.concat([df1,df2,df3])
Python output=Fig. 1.35
```

Note that each data frame is glued right below one another. The last data frame had an additional column. Therefore, column E shows "NaN" for df1 and df2 but shows the actual elements in df3. To concatenate data frames horizontally, axis = 1 can be placed inside pd.concat() function as shown below:

```
pd.concat([df1,df2,df3],axis=1)
Python output=Fig. 1.36
```

	A	B	C	D	E
0	H1	I1	J1	K1	NaN
1	H2	I2	J2	K2	NaN
2	H3	I3	J3	K3	NaN
3	H4	I4	J4	K4	NaN
4	H5	I5	J5	K5	NaN
5	H6	I6	J6	K6	NaN
6	H7	I7	J7	K7	NaN
7	H8	I8	J8	K8	NaN
8	H9	I9	J9	K9	L1
9	H10	I10	J10	K10	L2
10	H11	I11	J11	K11	L3
11	H12	I12	J12	K12	L4

FIGURE 1.35 Concatenated df1, df2, and df3 vertically.

Pandas merging

Aside from concatenation, it might be necessary to merge tables with unique properties together. For example, if a geologic properties table should be merged with a well data table with the unique identifier of "API number," merging can easily take place in pandas. Before going through the coding example, let's define some of the key terms that are used for merging in Python. The definitions below are from the official pandas documentation

	A	B	C	D	A	B	C	D	A	B	C	D	E
0	H1	I1	J1	K1	NaN	NaN	NaN	NaN	NaN	NaN	NaN	NaN	NaN
1	H2	I2	J2	K2	NaN	NaN	NaN	NaN	NaN	NaN	NaN	NaN	NaN
2	H3	I3	J3	K3	NaN	NaN	NaN	NaN	NaN	NaN	NaN	NaN	NaN
3	H4	I4	J4	K4	NaN	NaN	NaN	NaN	NaN	NaN	NaN	NaN	NaN
4	NaN	NaN	NaN	NaN	H5	I5	J5	K5	NaN	NaN	NaN	NaN	NaN
5	NaN	NaN	NaN	NaN	H6	I6	J6	K6	NaN	NaN	NaN	NaN	NaN
6	NaN	NaN	NaN	NaN	H7	I7	J7	K7	NaN	NaN	NaN	NaN	NaN
7	NaN	NaN	NaN	NaN	H8	I8	J8	K8	NaN	NaN	NaN	NaN	NaN
8	NaN	NaN	NaN	NaN	NaN	NaN	NaN	NaN	H9	I9	J9	K9	L1
9	NaN	NaN	NaN	NaN	NaN	NaN	NaN	NaN	H10	I10	J10	K10	L2
10	NaN	NaN	NaN	NaN	NaN	NaN	NaN	NaN	H11	I11	J11	K11	L3
11	NaN	NaN	NaN	NaN	NaN	NaN	NaN	NaN	H12	I12	J12	K12	L4

FIGURE 1.36   Concatenated df1, df2, and df3 horizontally.

(link = https://pandas.pydata.org/pandas-docs/stable/reference/api/pandas.DataFrame.merge.html):

left=use only keys from left frame, similar to a SQL left outer join

right=use only keys from right frame, similar to a SQL right outer join

outer=use union of keys from both frames, similar to a SQL full outer join

inner=use intersection of keys from both frames, similar to a SQL inner join; preserve the order of the left keys.

Fig. 1.37 shows the difference between inner, left, right, and outer joins.

Let's create two data frames and join them based on their "unique_ID" as follows:

```
test1=pd.DataFrame({'A':['X1','X2','X3','X4'],
   'B':['Y1','Y2','Y3','Y4'], 'unique_ID':['Z1','Z2','Z3',
   'Z5'],})
test1
```
Python output=Fig. 1.38

```
test2=pd.DataFrame({'C':['L1','L2','L3','L4'],
   'D':['M1','M2','M3','M4'], 'unique_ID':['Z1','Z2','Z3','Z6']})
test2
```
Python output=Fig. 1.39

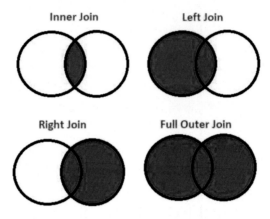

FIGURE 1.37 Merging concepts.

	A	B	unique_ID
0	X1	Y1	Z1
1	X2	Y2	Z2
2	X3	Y3	Z3
3	X4	Y4	Z5

FIGURE 1.38 test1 data frame.

	C	D	unique_ID
0	L1	M1	Z1
1	L2	M2	Z2
2	L3	M3	Z3
3	L4	M4	Z6

FIGURE 1.39 test2 data frame.

Please note that Z1, Z2, and Z3 exist in both data frames; however, Z5 and Z6 do not exist in both data frames. Let's join these two data frames on their "unique_ID" column and using inner join, explained and illustrated in Fig 1.37

```
pd.merge(test1,test2, on='unique_ID', how='inner')
Python output=Fig. 1.40
```

	A	B	unique_ID	C	D
0	X1	Y1	Z1	L1	M1
1	X2	Y2	Z2	L2	M2
2	X3	Y3	Z3	L3	M3

FIGURE 1.40   Merged test1 and test2 data frames using "inner".

Please note that only the first three rows with identical unique_IDs were matched and shown as the resulting merged data frame. Now, let's repeat the same merging except using "outer" in lieu of "inner" as follows:

```
pd.merge(test1,test2, on='unique_ID', how='outer')
Python output=Fig. 1.41
```

	A	B	unique_ID	C	D
0	X1	Y1	Z1	L1	M1
1	X2	Y2	Z2	L2	M2
2	X3	Y3	Z3	L3	M3
3	X4	Y4	Z5	NaN	NaN
4	NaN	NaN	Z6	L4	M4

FIGURE 1.41   Merged test1 and test2 data frames using "outer".

As is illustrated above, "outer" will result in showing all the rows in both data frames.

It is also possible to perform these calculations using multiple columns. If it is desired to merge on additional columns, just pass the names of all the columns inside a bracket. Let's create two more data frames called test3 and test4 and perform merging on two unique columns:

```
test3=\
pd.DataFrame({'A':['X1','X2','X3','X4'],'B':['Y1','Y2','Y3','Y4'],
    'unique_ID_1':['Z1','Z2','Z5','Z6'],
    'unique_ID_2':['N1','N2','N6','N7']})

test4=\
pd.DataFrame({'C':['L1','L2','L3','L4'],'D':['M1','M2',
    'M3','M4'],'unique_ID_1':['Z1','Z2','Z9','Z10'],
    'unique_ID_2':['N1','N2','N11','N12']})
```

test3 and test4 data frames will look as follows:

	A	B	unique_ID_1	unique_ID_2		C	D	unique_ID_1	unique_ID_2
0	X1	Y1	Z1	N1	0	L1	M1	Z1	N1
1	X2	Y2	Z2	N2	1	L2	M2	Z2	N2
2	X3	Y3	Z5	N6	2	L3	M3	Z9	N11
3	X4	Y4	Z6	N7	3	L4	M4	Z10	N12

test 3 and 4 data frames

Let's merge the two data frames on unique_ID_1 and unique_ID_2 using inner join as follows:

```
pd.merge(test3,test4, on=['unique_ID_1','unique_ID_2'],
    how='inner')
Python output=Fig. 1.42
```

	A	B	unique_ID_1	unique_ID_2	C	D
0	X1	Y1	Z1	N1	L1	M1
1	X2	Y2	Z2	N2	L2	M2

FIGURE 1.42 Merged test3 and 4 data frames using "inner".

Pandas joining

In addition to merging, joining can be performed in pandas. Joining is a convenient method for combining two columns of two differently indexed data frames into a single resulting data frame. It is the same as pd.merge() except that indexing (rows) is used for joining. Let's create two more data frames called test5 and test6 and use the pd.join() function to join the two tables:

```
test5=pd.DataFrame({'A':['X1','X2','X3','X4'],
  'B':['Y1','Y2','Y3','Y4']}, index=['H1','H2','H3','H4'])
test6=pd.DataFrame({'C':['X1','X2','X3','X4'],
  'D':['Y1','Y2','Y3','Y4']}, index=['H2','H3','H1','H0'])
```

Test5 and test6 data frames will look as follows:

	A	B		C	D
H1	X1	Y1	H2	X1	Y1
H2	X2	Y2	H3	X2	Y2
H3	X3	Y3	H1	X3	Y3
H4	X4	Y4	H0	X4	Y4

test 5 and 6 data frames.

pd.join() function can be used as follows to merge the two data frames based on their index:

```
test5.join(test6, how='inner')
Python output=Fig. 1.43
```

	A	B	C	D
H1	X1	Y1	X3	Y3
H2	X2	Y2	X1	Y1
H3	X3	Y3	X2	Y2

FIGURE 1.43 Joined test5 and test6 using "inner".

```
test5.join(test6, how='outer')
Python output=Fig. 1.44
```

Please note that if indices were simply integer values as opposed to H0, H1, etc. shown above, the same joining can be performed.

	A	B	C	D
H0	NaN	NaN	X4	Y4
H1	X1	Y1	X3	Y3
H2	X2	Y2	X1	Y1
H3	X3	Y3	X2	Y2
H4	X4	Y4	NaN	NaN

FIGURE 1.44   Joined test5 and test6 using "outer".

Pandas operation

In this section, important and useful pandas operations will be examined and illustrated. Let's create a data frame called df_ops as follows:

```
df_ops=pd.DataFrame({'A':[24,21,74,21],
    'B':[32,31,65,54],'C':['a','b','c','d']})
df_ops
Python output=Fig. 1.45
```

	A	B	C
0	24	32	a
1	21	31	b
2	74	65	c
3	21	54	d

FIGURE 1.45   df_ops data frame.

To obtain **a list** of unique values, ".unique()" function can be used as follows:

```
df_ops['A'].unique()
Python output=array([24, 21, 74], dtype=int64)
```

To obtain the **number** of unique values, ".nunique()" function can be used instead. This will return the number of elements in a data frame that is unique.

```
df_ops['A'].nunique()
Python output=3
```

We can also simply obtain the number of unique elements in each column as follows:

```
df_ops.nunique()
Python output=Fig. 1.46
```

A   3
B   4
C   4
dtype: int64

FIGURE 1.46   Number of unique elements in df_ops.

To obtain the number of times unique values have been repeated in each column, ".value_counts()" can be used:

```
df_ops['B'].value_counts()
Python output=Fig. 1.47
```

31   1
54   1
65   1
32   1
Name: B, dtype: int64

FIGURE 1.47   value_counts() for df_ops.

To return values with the criteria listed below, the following line of code can be written:

- Column A greater or equal to 20
- Column B greater or equal to 35
- Column C equal to "c"

```
df_ops[(df_ops['A']>=20) & (df_ops['B']>=35) & (df_ops['C']== 'c')]
Python output=Fig. 1.48
```

A	B	C
2 74	65	c

FIGURE 1.48   Filtered df_ops.

Pandas' "apply" function can be used to perform quick calculations. Below "economics" is first defined using the def function. This economic calculation was then applied to column A of the df_ops data frame as follows:

```
def economics(x):
    return x**2
df_ops['A'].apply(economics)
Python output=Fig. 1.49
```

```
0      576
1      441
2     5476
3      441
Name: A, dtype: int64
```

FIGURE 1.49   Apply function example.

This might be a lengthy version of writing the code. Therefore, pandas "lambda expression" can be used to simplify the code and perform the same calculation faster and more easily.

Pandas lambda expressions

As discussed, pandas' lambda expression can be combined with the apply function to perform quick calculations without defining the function first. For instance, if the values in column B of the df_ops data frame are desired to be multiplied by 4, the following lambda expression can be combined with the "apply" function to perform such calculation:

```
df_ops['B'].apply(lambda x: x*4)
Python output=Fig. 1.50
```

```
0    128
1    124
2    260
3    216
Name: B, dtype: int64
```

FIGURE 1.50   Lambda expression example.

Let's create 5 more columns and attach it to the df_ops data frame as follows:

```
df_ops['D']=df_ops['A'].apply(lambda x: x**2+1)
df_ops['E']=df_ops['B'].apply(lambda x: x**3+x**2+1)
df_ops['F']=df_ops['A'].apply(lambda x: x**4+x**2+5)
df_ops['G']=df_ops['B'].apply(lambda x: x**2+10)
df_ops['ALL']=df_ops['A'].apply(lambda x: x**2+10)+df_ops\
    ['A'].apply(lambda x: x**2+10)
df_ops
Python output=Fig. 1.51
```

	A	B	C	D	E	F	G	ALL
0	24	32	a	577	33793	332357	1034	1172
1	21	31	b	442	30753	194927	971	902
2	74	65	c	5477	278851	29992057	4235	10972
3	21	54	d	442	160381	194927	2926	902

FIGURE 1.51   Lambda expressions.

To obtain the list of all columns, ".columns" function can be used as follows:

```
df_ops.columns
Python output=Index(['A', 'B', 'C', 'D', 'E', 'F', 'G', 'ALL'],
    dtype='object')
```

To sort values in ascending or descending order, the following can be performed:

```
df_ops.sort_values('ALL', ascending=True)
Python output=Fig. 1.52
```

	A	B	C	D	E	F	G	ALL
1	21	31	b	442	30753	194927	971	902
3	21	54	d	442	160381	194927	2926	902
0	24	32	a	577	33793	332357	1034	1172
2	74	65	c	5477	278851	29992057	4235	10972

FIGURE 1.52 Sorted values using ALL column.

Dealing with missing values in pandas

Dealing with missing values is another crucial preprocessing step that must be performed to get the data ready for any ML analysis. The main methods for dealing with NA values are as follows:

- Dropping NA
- Filling NA with basic mathematical functions
- Using various algorithms that will be discussed in the supervised learning chapter of this book

Dropping NAs

Dropping NAs can be performed by using ".dropna()" as shown below. Let's first create a data frame with some missing values. To perform this, pandas and numpy libraries must be imported. Please note that numpy is a linear algebra library and is a fundamental package for scientific computing in Python. Almost all libraries in the PyData Ecosystem heavily rely on numpy as one of their fundamental building blocks. After creating the df_missing data frame, let's visualize the top 5 rows of data.

```
import pandas as pd
import numpy as np
df_missing\
=pd.DataFrame({'A':[1,20,31,43,59,np.nan,6,4,5,2,5,6,7,3,np.nan],
    'B':[1,20,np.nan,43,52,32,6,9,5,2,90,6,np.nan,3,100],
    'C':[1,30,68,43,np.nan,32,6,4,np.nan,2,5,6,888,3,100]})
df_missing.head(5)
Python output=Fig. 1.53
```

	A	B	C
0	1.0	1.0	1.0
1	20.0	20.0	30.0
2	31.0	NaN	68.0
3	43.0	43.0	43.0
4	59.0	52.0	NaN

FIGURE 1.53   df_missing data frame.

As shown, each column has several NA values. To drop rows with missing value, ".dropna()" can be applied:

```
df_missing.dropna(how='any')
Python output=Fig. 1.54
```

If it is desired to drop NA values when all columns have missing values, "how = 'all'" can be used instead to perform this. In addition, to drop columns with missing values as opposed to rows, use "axis = 1" inside the parenthesis of the dropna function as follows:

```
df_missing.dropna(how='any', axis=1)
```

Please note that "thresh" can be used inside the parenthesis, requiring the number of non-NA values to be entered for the rows or columns to get dropped. To make these changes permanent, it is very important to include "inplace = True" inside the parenthesis. This is because pandas' default for "inplace = False."

	A	B	C
0	1.0	1.0	1.0
1	20.0	20.0	30.0
3	43.0	43.0	43.0
6	6.0	6.0	6.0
7	4.0	9.0	4.0
9	2.0	2.0	2.0
10	5.0	90.0	5.0
11	6.0	6.0	6.0
13	3.0	3.0	3.0

FIGURE 1.54   df_missing data frame after dropping the NA cells.

Filling NAs

The next useful function is referred to as ".fillna()" to fill in the missing data with numeric values or strings. Let's create another data frame and call it df_filling as follows:

```
df_filling=pd.DataFrame({'A':[43,59,np.nan,6,3,np.nan],
   'B':[1,20,np.nan,43,np.nan,32],'C':[np.nan,32,6,4,np.nan,2]})
df_filling
Python output=Fig. 1.55
```

To fill in the missing values permanently with "Hydraulic Fracturing," the following line of code is entered:

```
df_filling.fillna(value='Hydraulic Fracturing', inplace=True)
df_filling
Python output=Fig. 1.56
```

	A	B	C
0	43.0	1.0	NaN
1	59.0	20.0	32.0
2	NaN	NaN	6.0
3	6.0	43.0	4.0
4	3.0	NaN	NaN
5	NaN	32.0	2.0

FIGURE 1.55 df_filling data frame.

	A	B	C
0	43	1	Hydraulic Fracturing
1	59	20	32
2	Hydraulic Fracturing	Hydraulic Fracturing	6
3	6	43	4
4	3	Hydraulic Fracturing	Hydraulic Fracturing
5	Hydraulic Fracturing	32	2

FIGURE 1.56   Filled df_filling data frame using "Hydraulic Fracturing."

To fill in the missing values with some statistical aggregate functions, let's import the same data frame and apply the mean of each column in lieu of missing values as follows:

```
df_filling_mean=pd.DataFrame({'A':[43,59,np.nan,6,3,np.nan],
    'B':[1,20,np.nan,43,np.nan,32],'C':[np.nan,32,6,4,np.nan,2]})
df_filling_mean.fillna(value=df_filling_mean.mean(), inplace=True)
Python output=Fig. 1.57
```

As demonstrated, column A's missing values were filled with 27.75 which is the arithmetic average of 43, 59, 6, and 3. Other statistical functions such as standard deviation (.std()), median (.median()), variance (.var()), skewness (.skew()), kurtosis (.kurt()), minimum (.min()), maximum (.max()), count (.count()), etc., could also be applied (this is not to recommend using these

	A	B	C
0	43.00	1.0	11.0
1	59.00	20.0	32.0
2	27.75	24.0	6.0
3	6.00	43.0	4.0
4	3.00	24.0	11.0
5	27.75	32.0	2.0

FIGURE 1.57   Filled df_filling data frame using the mean.

statistical measures to fill in the missing values but to illustrate the availability of these options).

Other convenient functions such as "ffill" and "bfill" can be used to fill in the missing values as follows:

ffill = will place the last valid observation forward to fill in the next observation

bfill = uses the next valid observation to fill in the previous value

Let's use the bfill method by creating a data frame called df_filling_backfill and applying "bfill" as the method to **temporarily** fill in the missing values. If this change is desired to be permanent, simply add "inplace = True".

```
df_filling_backfill=pd.DataFrame({'A':[43,59,np.nan,6,3,np.nan],
   'B':[1,20,np.nan,43,np.nan,32],'C':[np.nan,32,6,4,np.nan,2]})
df_filling_backfill.fillna(method='bfill')
Python output=Fig. 1.58
```

Numpy introduction

Numpy is a linear algebra library in Python and one of the foundational libraries in Python that almost all libraries build upon. It is also considered very fast and can simply be imported as one of the main libraries when doing any type of ML analysis. It is highly recommended to install Anaconda distribution to make sure all the underlying libraries and dependencies get installed. Numpy array will be used throughout this book and manifests in two distinct formats which are vectors and matrices. The primary difference between

	A	B	C
0	43.0	1.0	32.0
1	59.0	20.0	32.0
2	6.0	43.0	6.0
3	6.0	43.0	4.0
4	3.0	32.0	2.0
5	NaN	32.0	2.0

FIGURE 1.58   Filled df_filling data frame using the bfill method.

vectors and matrices is that vectors are 1d array while matrices are 2d arrays. Let's implement the following code:

Generate a list of numbers and place them under a variable called "A" as shown below. The list can then be converted to an array using np.array function as follows:

```
import numpy as np
A=[1,2,3,4,5,6,7,8,9,10]
np.array(A)
Python output=array([ 1, 2, 3, 4, 5, 6, 7, 8, 9, 10])
```

A two-dimensional array can also be generated as follows:

```
B=[[4,5,6],[7,8,9],[10,11,12]]
np.array(B)
Python output=Fig. 1.59
```

```
array([[ 4,  5,  6],
       [ 7,  8,  9],
       [10, 11, 12]])
```

FIGURE 1.59   B array.

There are various built-in functions such as np.arange that can be used to generate an array. For instance, to generate an array starting with 0, ending with 40, and every 5 increments, the following lines of code can be applied:

```
np.arange(0,40,5)
Python output=array([ 0, 5, 10, 15, 20, 25, 30, 35])
```

After typing np.arange(), pressing shift + tab **inside the parenthesis** would populate a window that shows the arguments that must be passed on. Those arguments in this example are start, stop, and step. Numpy library can also be used to create 1d or 2d arrays using np.zeros and no.ones as follows:

```
np.zeros(5)
Python output=array([0., 0., 0., 0., 0.])

np.zeros((5,5))
Python output=Fig. 1.60
```

```
array([[0., 0., 0., 0., 0.],
       [0., 0., 0., 0., 0.],
       [0., 0., 0., 0., 0.],
       [0., 0., 0., 0., 0.],
       [0., 0., 0., 0., 0.]])
```

FIGURE 1.60 5×5 zero matrix.

```
np.ones(5)
Python output=array([1., 1., 1., 1., 1.])
np.ones((5,5))
Python output=Fig. 1.61
```

```
array([[1., 1., 1., 1., 1.],
       [1., 1., 1., 1., 1.],
       [1., 1., 1., 1., 1.],
       [1., 1., 1., 1., 1.],
       [1., 1., 1., 1., 1.]])
```

FIGURE 1.61 5×5 matrix of one.

Another useful numpy function is np.linspace which can be used to create an np array starting with a certain value, ending with another value, and having a number of increments in between. For example, let's create an array from 0 to 20 and dividing it into 10 equal increments as follows:

```
np.linspace(0,20,10)
Python output=array([0., 2.22222222, 4.44444444, 6.66666667,
    8.88888889, 11.11111111, 13.33333333, 15.55555556, 17.77777778,
    20.])
```

"np.linspace" is a useful function that can be used to generate an array for sensitivity analysis after training a ML model. This function will be used extensively throughout this book.

In linear algebra, identity matrix (also called unit matrix) is an n by n square matrix with 1 in the diagonal and 0s elsewhere. The numpy library can be used to generate an identity matrix as follows:

```
np.eye(10)
```
Python output=Fig. 1.62

```
array([[1., 0., 0., 0., 0., 0., 0., 0., 0., 0.],
       [0., 1., 0., 0., 0., 0., 0., 0., 0., 0.],
       [0., 0., 1., 0., 0., 0., 0., 0., 0., 0.],
       [0., 0., 0., 1., 0., 0., 0., 0., 0., 0.],
       [0., 0., 0., 0., 1., 0., 0., 0., 0., 0.],
       [0., 0., 0., 0., 0., 1., 0., 0., 0., 0.],
       [0., 0., 0., 0., 0., 0., 1., 0., 0., 0.],
       [0., 0., 0., 0., 0., 0., 0., 1., 0., 0.],
       [0., 0., 0., 0., 0., 0., 0., 0., 1., 0.],
       [0., 0., 0., 0., 0., 0., 0., 0., 0., 1.]]])
```

FIGURE 1.62   10 × 10 identity matrix.

As illustrated, the above identity matrix is 10 by 10 matrix with 1s on the diagonal and 0s elsewhere.

Random number generation using numpy

Numpy has extensive functionalities when it comes to generating random numbers. For instance, to generate a random sample from a uniform distribution, the following command lines can be used:

```
seed=100
np.random.seed(seed)
np.random.rand(5)
```
**Python output=array([0.54340494, 0.27836939, 0.42451759,
    0.84477613, 0.00471886])**

To generate a 5 by 5 matrix of random number, the following lines of code can be used:

```
seed=100
np.random.seed(seed)
np.random.rand(5,5)
```
Python output=Fig. 1.63

```
array([[0.54340494, 0.27836939, 0.42451759, 0.84477613, 0.00471886],
       [0.12156912, 0.67074908, 0.82585276, 0.13670659, 0.57509333],
       [0.89132195, 0.20920212, 0.18532822, 0.10837689, 0.21969749],
       [0.97862378, 0.81168315, 0.17194101, 0.81622475, 0.27407375],
       [0.43170418, 0.94002982, 0.81764938, 0.33611195, 0.17541045]]])
```

FIGURE 1.63   5 by 5 matrix of random number.

Please note that the random numbers that are generated will be the same since the same seed number is selected every time random number is generated. Seed numbers can be fixed or simply remove the seed number to get a different random number every time a line of code is executed.

If the generated random number is desired to have a normal or Gaussian distribution, just add "n" at the end of "np.random.rand" as follows: "np.random.rand**n**":

```
seed=100
np.random.seed(seed)
np.random.randn(5)
Python output=array([-1.74976547, 0.3426804, 1.1530358,
  -0.25243604, 0.98132079])
seed=100
np.random.seed(seed)
np.random.randn(5,5)
Python output=
```

```
array([[-1.74976547,  0.3426804 ,  1.1530358 , -0.25243604,  0.98132079],
       [ 0.51421884,  0.22117967, -1.07004333, -0.18949583,  0.25500144],
       [-0.45802699,  0.43516349, -0.58359505,  0.81684707,  0.67272081],
       [-0.10441114, -0.53128038,  1.02973269, -0.43813562, -1.11831825],
       [ 1.61898166,  1.54160517, -0.25187914, -0.84243574,  0.18451869]])
```

An integer random number can also be generated using np.random.randint(). To create 10 integer random numbers between 1 and 500, the following line of code can be used without using a fixed seed number:

```
np.random.randint(1,500,10)
```

Now, let's create a numpy array between 20 and 30 (excluding 30) and reshape the array to a 2d matrix of 5 by 2. The reshape function is a useful term for converting various numpy arrays to different shapes.

```
A=np.arange(20,30)
A.reshape(5,2)
Python output=Fig. 1.64
```

```
array([[20, 21],
       [22, 23],
       [24, 25],
       [26, 27],
       [28, 29]])
```

FIGURE 1.64 Reshaped array.

In addition, maximum, minimum, location of maximum and minimum, mean, and standard deviation of the above array can also be determined using the following keywords:

A.min() = to obtain the minimum value in the A array
A.max() = to obtain the maximum value in the A array
A.argmin() = to obtain the location of the minimum value in the A array
A.argmax() = to obtain the location of the maximum value in the A array
A.std() = to obtain the standard deviation in the A array
A.mean() = to obtain the mean in the A array

Numpy indexing and selection

Numpy indexing and selection is another important tool that must be comprehended. Let's create an array between 30 and 40 (including 40) and call it a variable name X. If obtaining indices 0 to 7 is desired, it can be written as "X [0:7]" (as shown below). Please note that "X[:7]" can also be used as opposed to "X[0:7]" to get the same output. Python by default recognizes that "X[:7]" indicates obtaining all indices from index 0 to index 7 (not including index 7). Starters will usually start off by using "X[0:7]" and switch over to "X[:7]" as progression is made with Python.

```
X=np.arange(30,41)
X
Python output=array([30, 31, 32, 33, 34, 35, 36, 37, 38, 39, 40])

X[0:7]
Python output=array([30, 31, 32, 33, 34, 35, 36])
```

A code of "X[:]" will return all the indices in the created variable as follows:

```
X[:]
Python output=array([30, 31, 32, 33, 34, 35, 36, 37, 38, 39, 40])
```

If index 2 and beyond are the desired indices, "X[2:]" can be used to denote that as follows:

```
X[2:]
Python output=array([32, 33, 34, 35, 36, 37, 38, 39, 40])
```

Let's create a copy of array X and multiply it by 2 and call it Y as follows:

```
Y=X.copy()*2
Y
Python output=array([60, 62, 64, 66, 68, 70, 72, 74, 76, 78, 80])
```

In the example below, indices 60, 62, 64, 66, and 68 (first 5 indices in the Y array) are replaced with a generic number of 1000.

```
Y[0:5]=1000
Y
```
Python output=array([1000, 1000, 1000, 1000, 1000, 70, 72, 74, 76, 78, 80])

Indexing can also be applied on 2d arrays. Let's create a matrix with 4 rows and 4 columns (4d matrix) as follows:

```
array_2d=np.array([[50,52,54],[56,58,60],[62,64,66],
[68,70,72]])
array_2d
```
Python output=Fig. 1.65

```
array([[50, 52, 54],
       [56, 58, 60],
       [62, 64, 66],
       [68, 70, 72]])
```
FIGURE 1.65 array_2d.

If number 64 in the 4d matrix above is desired, it can be attained by using the notation "array_2d[2,1]". 2 represents the row number and 1 represents the column number. Please note that as previously discussed, indexing in Python starts with 0. Therefore, 2 means the third row and 1 indicates the second column in this example.

```
array_2d[2,1]
```
Python output=64

To obtain numbers 64, 66, 70, and 72, "array_2d[2:,1:]" can be used to obtain the third row and beyond and second column and beyond. In this example, "2:" indicates the third row and beyond, while "1:" indicates the second column and beyond.

```
array_2d[2:,1:]
```
Python output=Fig. 1.66

```
array([[64, 66],
       [70, 72]])
```
FIGURE 1.66 array_2d[2;1:].

To obtain numbers 68 and 70, the below can be performed:

```
array_2d[3,0:2]
```
Python output=array([68, 70])

Please spend some time creating various n-dimensional matrices and obtain various sets of rows and columns to get used to the concept.

Reference

Belyadi, H., Fathi, E., & Belyadi, F. (2019). *Hydraulic fracturing in unconventional reservoirs* (2nd ed.). Elsevier.

Chapter 2

Data import and visualization

Data import and export using pandas

In this chapter, data visualization with various Python libraries will be discussed in detail. Prior to discussing data visualization, let's address data imports using pandas. To import a csv or excel file, "pd.read" syntax can be used as follows. As shown in Fig. 2.1, the "tab" key can be hit on your keyboard to see the list of available options. To import an excel file, simply choose "pd.read_excel."

```
import pandas as pd
df=pd.read_csv('Chapter2_Shale_Gas_Wells_DataSet.csv')
```

To write a data frame to a csv or an excel file, the following lines of code can be executed:

```
df=df.describe()
df.to_csv('Shale Gas Wells Description.csv', index=False)
```
Python output = this will write "Shale Gas Wells Description" csv file to the same folder where the original "ipynb" Python script was written.

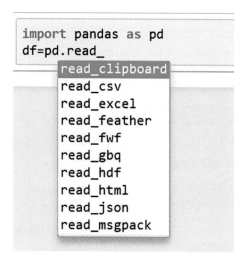

FIGURE 2.1   pd.read options in pandas.

Machine Learning Guide for Oil and Gas Using Python. https://doi.org/10.1016/B978-0-12-821929-4.00007-X
57

To obtain the exact directory of where the file is saved, simply execute "pwd" as shown below:

```
Pwd
Python output='C:\\Users\\hossb\\Desktop\\Files\\Machine Learning (ML)
\\Chapters\\Chapter 2\\Codes'
```

When "pwd" is executed, the directory at which Python ".ipynb" is stored will be displayed. Please note that by default when writing a data frame to a file, "index = True" which means row numbers will be shown in the exported file; however, to avoid showing the index numbers, simply put "index -= False" as was displayed above. To read the same file in an excel format, let's create another data frame and name it df2:

```
df2=pd.read_excel('Chapter2_Shale_Gas_Wells_DataSet.xlsx',
sheet_name='Shale Gas Wells')
```

Please note that "sheet_name" can be specified to import the proper excel sheet into Jupyter Notebook (in the event there are multiple sheets within an excel document). To import the description of df2 data frame into an excel file with a sheet name of "New_Sheet" and a file name of "Excel_Output," the following two lines of code can be executed:

```
df2=df2.describe()
df2.to_excel('Excel_Output.xlsx', sheet_name='New_Sheet')
```

To read an HTML table into a "list" of "DataFrame" objects for natural gas pricing, the code below can be executed:

```
df3=pd.read_html('https://www.eia.gov/dnav/ng/hist/n9190us3m.htm')
df3
Python output=Fig. 2.2 illustrates the natural gas pricing since 1973
using U.S. energy information administration data.
```

Please note that reading HTML from the internet is most effective when the entire page is in a table format. To create a simple SQL engine memory followed by a temporarily SQL light engine, the following can be executed:

```
from sqlalchemy import create_engine
engine=create_engine('sqlite:///:memory:')
```

To write the df data frame to a SQL file following by reading the file using "pd.read_sql," the following lines can be executed:

```
df.to_sql('my_table', engine)
sqldf=pd.read_sql('my_table',con=engine)
sqldf
Python output=Fig. 2.3 illustrates the output which is a sqldf data
frame description.
```

```
[                                                        0
0   View History: Monthly Annual Download Data (XL...
1   U.S. Natural Gas Wellhead Price (Dollars per T...,
                            0                           1
0   View History: Monthly Annual  Download Data (XLS File),
                            0
0   View History: Monthly Annual,
                    0    1      2    3      4
0   View History: NaN  Monthly NaN  Annual,
        Year    Jan    Feb    Mar    Apr    May    Jun    Jul    Aug    Sep    Oct
0     1973.0    NaN    NaN    NaN    NaN    NaN    NaN    NaN    NaN    NaN    NaN
1     1974.0    NaN    NaN    NaN    NaN    NaN    NaN    NaN    NaN    NaN    NaN
2        NaN    NaN    NaN    NaN    NaN    NaN    NaN    NaN    NaN    NaN    NaN
3     1975.0    NaN    NaN    NaN    NaN    NaN    NaN    NaN    NaN    NaN    NaN
4     1976.0   0.54   0.54   0.54   0.55   0.55   0.58   0.58   0.60   0.60   0.62
5     1977.0   0.67   0.71   0.75   0.77   0.77   0.82   0.83   0.82   0.83   0.84
6     1978.0   0.87   0.88   0.89   0.88   0.91   0.91   0.89   0.91   0.92   0.92
7     1979.0   1.02   1.05   1.10   1.11   1.15   1.17   1.20   1.24   1.24   1.28
8        NaN    NaN    NaN    NaN    NaN    NaN    NaN    NaN    NaN    NaN    NaN
9     1980.0   1.37   1.42   1.46   1.51   1.56   1.57   1.64   1.64   1.69   1.71
10    1981.0   1.77   1.81   1.86   1.93   1.95   1.95   2.01   2.02   2.08   2.11
11    1982.0   2.23   2.30   2.35   2.40   2.45   2.45   2.47   2.53   2.56   2.60
12    1983.0   2.66   2.66   2.58   2.53   2.53   2.59   2.52   2.58   2.67   2.58
13    1984.0   2.67   2.71   2.67   2.64   2.67   2.70   2.68   2.69   2.62   2.63
14       NaN    NaN    NaN    NaN    NaN    NaN    NaN    NaN    NaN    NaN    NaN
15    1985.0   2.64   2.71   2.62   2.64   2.53   2.58   2.51   2.47   2.42   2.37
```

FIGURE 2.2   HTML reading using pd.read_html.

Index		Stage Spacing	bbl/ft	Well Spacing	Dip	Thickness	Lateral Length	Injection Rate	Porosity	ISIP	Water Saturation	Percentage of LG	Pressure Gradient
0	count	506.000000	506.000000	506.000000	506.000000	506.000000	506.000000	506.000000	506.000000	506.000000	506.000000	506.000000	506.000000
1	mean	147.640316	35.134387	820.158103	0.069170	162.365613	8153.086957	63.079051	7.337549	7010.490119	19.213439	64.845455	0.930257
2	std	18.392128	10.533197	135.736986	0.253994	15.471044	942.393981	7.250106	0.749451	1211.452205	3.198579	18.427813	0.046507
3	min	140.000000	30.000000	650.000000	0.000000	120.000000	4500.000000	55.000000	5.500000	5000.000000	15.000000	15.000000	0.750000
4	25%	140.000000	30.000000	700.000000	0.000000	153.000000	7617.750000	57.000000	6.600000	5000.000000	16.800000	55.900000	0.940000
5	50%	141.000000	30.000000	800.000000	0.000000	165.000000	8051.000000	61.000000	7.500000	7643.000000	17.700000	69.900000	0.950000
6	75%	148.000000	36.000000	900.000000	0.000000	176.000000	8608.000000	69.000000	8.000000	7783.000000	24.100000	79.700000	0.950000
7	max	330.000000	75.000000	1350.000000	1.000000	185.000000	11500.000000	80.000000	8.500000	8200.000000	25.000000	95.000000	0.950000

FIGURE 2.3   sqldf data frame description.

Data visualization

Data visualization is one of the most important aspects of data preparation and analysis. In this step, various plots such as distribution plots, box plots, scatter plots, etc., are used to visualize the data and identify the outliers. Outlier can be easily identified with basic plots discussed throughout this chapter. Using data visualization can help with understanding the underlying nature of the data and is a crucial step before applying the appropriate ML algorithms.

Matplotlib library

Matplotlib is one of the most important and common visualization libraries in Python. It provides great flexibility when plotting any aspect of a figure. This library was designed to have a similar user interaction as Matlab's graphical plotting. One of the main advantages of this library is that it is versatile, and anything can be plotted. However, it might be challenging to do more complex plots. Those complex plots will require more coding. Please refer to Matplotlib's website if some more complex plots that have not been covered in this chapter would like to be explored (Matplotlib's website: https://matplotlib.org/tutorials/introductory/sample_plots.html). Please note that matplotlib creates static image files such as JPEG or PNG files and does not change once created.

To start creating some plots, let's import pandas, numpy, and matplotlib libraries. "%matplotlib inline" is also added (only in Jupyter Notebook), so all the plots are shown within Jupyter Notebook. Variables x and y can also be created as follows:

```
import numpy as np
import pandas as pd
import matplotlib.pyplot as plt
%matplotlib inline
x=np.linspace(0,10,100)
y=x**3
```

Now let's plot the created x and y variables above using "plt.plot." Please note that any color can be simply entered as shown below. In this plot, red color, line width of 4, and line style of "–" are used. In addition, "plt.xlabel" and "plt.ylabel" are used to show the x and y labels, and "plt.title" is used to show the title of the plot.

```
plt.plot(x,y,'red',linewidth=4, linestyle='--')
plt.xlabel('Time')
plt.ylabel('Number of iterations')
plt.title('Number of Iterations Vs. Time')
Python output=Fig. 2.4
```

Next, let's use the matplotlib library to create a subplot of three plots as illustrated below. The first line of code is to define the figure size. In this example, a figsize of 10 by 5 is being used. 10 represents the length of the plot and 5 represents the width. Please feel free to adjust these numbers until your desired plot size is achieved. In addition, "plt.subplot(1,3,1)" represents a plot with 1 row, three columns, and plot #1 of the subplot. In other words, "plt.subplot(1,3,1)" can be considered as plt.subplot(row,column,plot#).

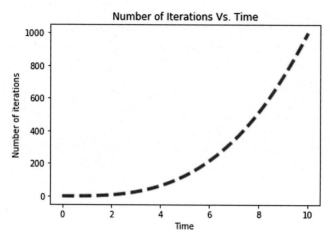

FIGURE 2.4 plt plot of x and y variables.

```
fig=plt.figure(figsize=(10,5))
plt.subplot(1,3,1)
plt.plot(x,y,'b')
plt.title('Plot 1')
plt.subplot(1,3,2)
plt.plot(y*2,x**2,'r')
plt.title('Plot 2')
plt.subplot(1,3,3)
plt.plot(x,y,'purple')
plt.title('Plot 3')
```
Python output=Fig. 2.5

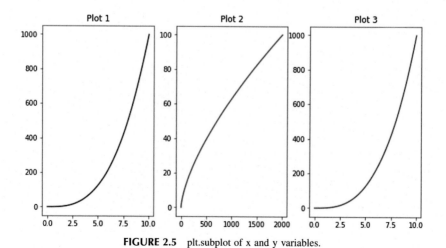

FIGURE 2.5 plt.subplot of x and y variables.

Next, let's create two plots inside of one large plot. This can be achieved by creating an empty figure first. Then, use "fig.add_axes" to add the dimensions of each plot as revealed below. Afterward, each plot along with their titles and labels can simply be added as illustrated.

```python
# plt.figure() creates an empty figure
fig=plt.figure()
# figsize can be defined
fig=plt.figure(figsize=(10,6))
# The dimensions can then be added (left, bottom, width, height) for
each figure
fig1=fig.add_axes([0.1,0.1,0.8,0.8])
fig2=fig.add_axes([0.25,0.5,0.3,0.2])
# Each figure can be plotted and axes and tiles can be added
fig1.plot(x,y)
fig1.set_title('Large Plot')
fig1.set_xlabel('Time')
fig1.set_ylabel('Number of iteration')
fig2.plot(x*2,y)
fig2.set_title('Small Plot')
fig2.set_xlabel('Time')
fig2.set_ylabel('Number of iteration')
Python output=Fig. 2.6
```

Another **method** to draw a subplot is using "fig,axis" command as shown below:

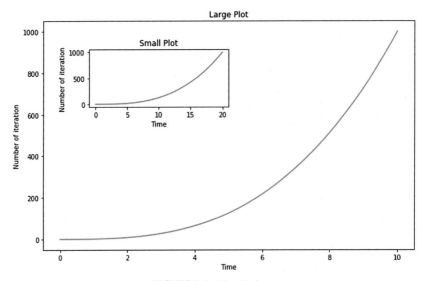

FIGURE 2.6 Nested plots.

```
fig,axes=plt.subplots(nrows=2, ncols=2,figsize=(5,5),dpi=100)
# Plot #1 in the first row and first column
axes[0,0].plot(x,y)
axes[0,0].set_title('Plot 1')
axes[0,0].set_xlabel('Time')
axes[0,0].set_ylabel('Production 1')
# Plot #2 in the first row and second column
axes[0,1].plot(y,x)
axes[0,1].set_title('Plot 2')
axes[0,1].set_xlabel('Time')
axes[0,1].set_ylabel('Production 2')
# Plot #3 in the second row and first column
axes[1,0].plot(y,x**2)
axes[1,0].set_title('Plot 3')
axes[1,0].set_xlabel('Time')
axes[1,0].set_ylabel('Production 3')
# Plot #4 in the second row and second column
axes[1,1].plot(x,y**2)
axes[1,1].set_title('Plot 4')
axes[1,1].set_xlabel('Time')
axes[1,1].set_ylabel('Production 4')
plt.tight_layout()
```
Python output=Fig. 2.7

As illustrated above, Fig. 2.7 subplot contains 2 rows and 2 columns (nrows = 2, ncols = 2). "figsize" and dpi of 100 are defined for plot resolution (feel free to adjust those accordingly). "axes[0,0].plot(x,y)" indicates to plot x and y variables (defined earlier) as x-axis and y-axis on the first plot located on the first row and first column. Title, x and y axes labels can also be set using "axes[row #, column #].set_title('NAME')," "axes[row #, column #].set_xlabel('NAME')," and "axes[row #, column #].set_ylabel ('NAME')." Please note that "plt.tight_layout()" is used to avoid overlapping. This is a very useful function in matplotlib.

Let's import natural gas pricing using a csv file and call it "df_ng" as follows:

```
df_ng=pd.read_csv('Chapter2_Natural_Gas_Pricing_DataSet.csv',
parse_dates=['Date'],index_col=['Date'])
df_ng.head()
```
Python output=Fig. 2.8

Please note that "parse_dates = ['Date']" and "index_col = ['Date']" are used when importing the csv file to make sure the date column when plotting is formatted/read properly and would not overlap. Next, let's plot date versus gas pricing over time. To call the gas pricing column (which is called Value in this data frame), df_ng['Value'] can be used and to call the date column in the csv file, "df_ng.index.values" can be used.

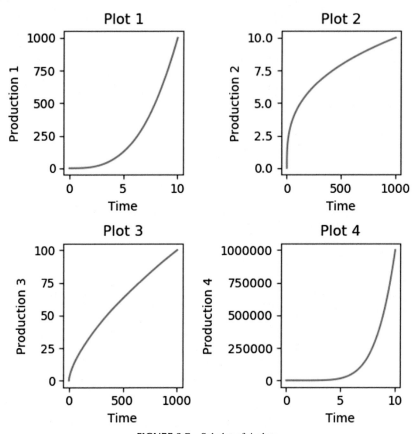

FIGURE 2.7 Subplot of 4 plots.

Date	Value
1997-01-01	5.25
1997-02-01	2.86
1997-03-01	2.95
1997-04-01	3.45
1997-05-01	3.57

FIGURE 2.8 df_ng data frame.

```
fig=plt.figure(figsize=(10,10), dpi=90)
plt.plot(df_ng.index.values,df_ng['Value'], color='red')
plt.xlabel('Date')
plt.ylabel('Gas Pricing, $/MMBTU')
plt.title('Gas Pricing Vs. Date')
Python output=Fig. 2.9
```

Let's create another two variables called g and h and plot those variables except this time setting x and y axes limits using "ax.set_xlim([0,8])" which limits the x-axis to between 0 and 8 and "ax.set_ylim([0,80])" which limits the y-axis to between 0 and 80. This example is illustrated to **show** the customization flexibility in matplotlib library, including but not limited to colors, line width, line style, marker size, marker face color, marker edge width, marker edge color, font sizes in various parts of the plot, and adding text within the plot which is shown below under "ax.text(3,3, "Commodity Pricing Plot", fontsize = 30, color = "red")." Please explore various customization codes

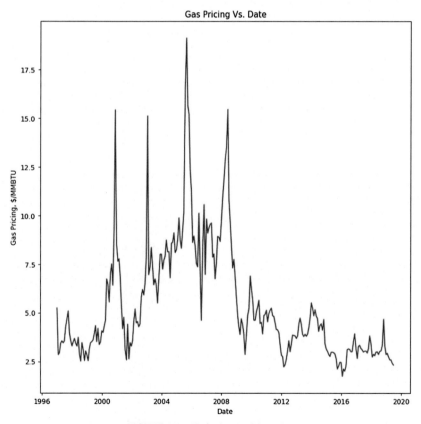

FIGURE 2.9 Natural gas pricing plot.

and numbers listed below to understand the impact of each size and variable on the result. Our recommendation is to copy and paste the codes below and start exploring various parameters and sizes listed below.

```
g=np.linspace(1,10,5)
h=g**2
fig=plt.figure(figsize=(10,10))
ax=fig.add_axes([0,0,1,1])
#RGB Hex codes are used for custom colors (use the link https://www.
rapidtables.com/web/color/RGB_Color.html to find your custom color)
# Line styles such as '--', '-.', 'steps' can be used
# markers such as 'o' and '+' can be used to show markers on a plot
ax.plot(g,h, color='#FF8C00', linewidth=3, linestyle='-.', alpha=1,
marker='o', markersize=10, markerfacecolor='green',
markeredgewidth=3, markeredgecolor='red')
ax.set_xlabel('Time')
ax.set_ylabel('Pricing')
ax.set_title('Time Vs. Pricing')
ax.set_xlim([0,8])
ax.set_ylim([0,80])
plt.rc('font', size=18) # Controls default text sizes
plt.rc('axes', titlesize=18) # Fontsize of the axes title
plt.rc('axes', labelsize=18) # Fontsize of the x and y labels
plt.rc('xtick', labelsize=18) # Fontsize of the tick labels
plt.rc('ytick', labelsize=18) # Fontsize of the tick labels
plt.rc('legend', fontsize=18) # Legend fontsize
plt.rc('figure', titlesize=18) # Fontsize of the figure title
ax.text(3,3, 'Commodity Pricing Plot', fontsize=30, color='red')
```
Python output=Fig. 2.10

To illustrate plotting two separate columns on primary and secondary axes using matplotlib, let's discuss another example by importing "Chapter2_Production_DataSet.**csv**", removing NA values along the rows (as illustrated in Chapter 1) and reading the head of the data which will show the first 5 rows as follows:

```
df=pd.read_csv('Chapter2_Production_DataSet.csv')
df.dropna(axis=0).head()
```
Python output=Fig. 2.11

To plot gas rate, casing pressure, and tubing pressure all on the same axis, the following lines of code are run (this was discussed earlier in the chapter):

```
fig=plt.figure(figsize=(10,7))
plt.plot(df['Prod Date'],df['Gas Prod'], color='red')
plt.plot(df['Prod Date'],df['Casing'], color='blue')
plt.plot(df['Prod Date'],df['Tubing'], color='green')
plt.title('Production Rate Vs. Time')
plt.xlabel('Time (days)')
plt.ylabel('Gas Production Rate (MSCF/D), Casing Pressure (psi),
Tubing Pressure (psi)')
```
Python output=Fig. 2.12

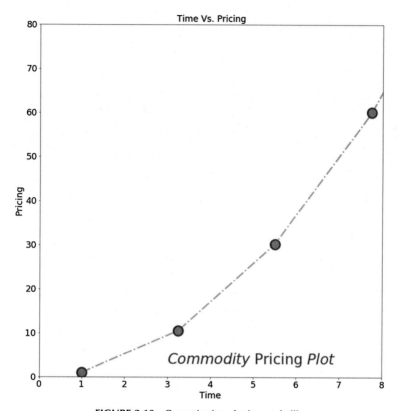

FIGURE 2.10   Customization plot in matplotlib.

	Well Name	Prod Date	Gas Prod	Water Prod	Tubing	Casing	Line
73	MIP 4H	74	2614.0	33.0	363.6	837.5	197.0
74	MIP 4H	75	2601.0	32.0	358.8	835.8	197.0
75	MIP 4H	76	2590.0	33.0	357.1	833.1	197.0
76	MIP 4H	77	2555.0	31.0	358.2	831.4	197.0
77	MIP 4H	78	2575.0	29.0	353.6	836.5	197.0

FIGURE 2.11   Production data head.

As opposed to placing all three columns on the primary axis, let's plot gas rate on the primary axis and casing pressure on the secondary axis as **shown** below. ax1 is used to plot production date versus gas production on the primary axis (red color), and ax2 is used to plot production date versus casing pressure on the secondary axis (blue color). Please note that "ax2 = ax1.twinx()" is used to instantiate a second axis that shares the same x-axis.

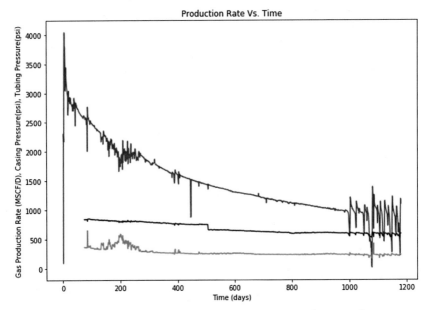

FIGURE 2.12 Gas rate, casing, and tubing pressures on the same axis.

```
fig, ax1=plt.subplots()
ax1.plot(df['Prod Date'],df['Gas Prod'], color='red')
ax1.set_xlabel('time (days)')
ax1.set_ylabel('Gas Production Rate', color='red')
ax1.tick_params(axis='y', labelcolor='red')
ax2=ax1.twinx() # instantiate a second axes that shares the same
x-axis
ax2.plot(df['Prod Date'],df['Casing'], color='blue')
ax2.set_ylabel('Casing Pressure', color='blue')
ax2.tick_params(axis='y', labelcolor='blue')
fig.tight_layout()
Python output=Fig. 2.13
```

Well log plotting using matplotlib

Another useful advantage of the matplotlib library is plotting the log data. Let's use MIP_4H well sonic log data from the MSEEL project that is publicly available at the following link:

http://www.mseel.org/data/Wells_Datasets/MIP/MIP_4H/GandG_and_Reservoir_Engineering/

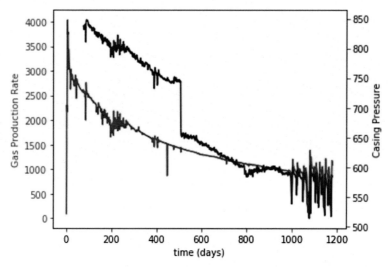

FIGURE 2.13   Gas rate and casing pressure on primary and secondary axes.

Once downloaded, save the excel as a csv file. Let's import the csv file and call it "df_log" as follows:

```
df_log=pd.read_csv('Chapter2_Sonic_Log_MIP_4H_DataSet.csv')
df_log.columns
Python output=Index(['Well_Name', 'DEPT', 'DPHZ', 'DT', 'GR_EDTC',
'HCAL', 'HDRA', 'NPHI', 'PEFZ', 'RHOZ', 'RLA3', 'RLA4', 'RLA5',
'RM_HRLT', 'SPHI', 'SVEL', 'TENS', 'TT1', 'TT2', 'TT3', 'TT4'],
dtype='object')
```

As shown above, this log contains many different properties. Let's plot depth (DEPT) on the y-axis and gamma ray (GR_EDTC), neutron porosity (NPHI), bulk density (RHOZ), and photoelectric (PEFZ) on the x axis on 4 different tracks (tracks 1 through 4) side by side using the subplot concept discussed above. Once the code for the first track is written, the remaining tracks can simply get copied and change "User#", name of the track, column number of the track, and color of the property. Please note that this log has not been quality checked and controlled for washout and the log is simply being plotted as is based on publicly available data.

```
# This will create 1 row and 4 columns with the title "MIP_4H Sonic log"
fig,ax=plt.subplots(nrows=1, ncols=4,figsize=(16,12),sharey=True)
fig.suptitle("MIP_4H Sonic Log", fontsize=30)
Name='MIP_4H'
# Track #1 which will assign "Use1" to be GR and "Use2" to be the depth
User1=df_log[df_log['Well_Name']==Name]['GR_EDTC']
User2=df_log[df_log['Well_Name']==Name]['DEPT']
```

```
# "ax[0]" will plot the first column track and ".twiny()" is a method to
share the y-axis.
ax1=ax[0].twiny()
ax1.invert_yaxis()
# This will plot "User1" which is gamma ray on the x-axis and "User2"
which is depth on the y-axis
ax1.plot(User1,User2, color='black',linestyle='-')
# This will assign the x label and color.
ax1.set_xlabel('GR_EDTC',color='black')
# This will adjust tick label, color, and axis on which to apply the
tick label
ax1.tick_params(axis='x', color='black')
# This indicates data area boundary and adjusting location of label
ax1.spines['top'].set_position(('outward',1))
# Track #2:
User3=df_log[df_log['Well_Name']==Name]['NPHI']
User4=df_log[df_log['Well_Name']==Name]['DEPT']
ax2=ax[1].twiny() ax01.invert_yaxis()
ax2.plot(User3,User4, color='red',linestyle='-')
ax2.set_xlabel('NPHI',color='red')
ax2.tick_params(axis='x', color='red')
ax2.spines['top'].set_position(('outward',1))
# Track #3:
User5=df_log[df_log['Well_Name']==Name]['RHOZ']
User6=df_log[df_log['Well_Name']==Name]['DEPT']
ax3=ax[2].twiny() ax01.invert_yaxis()
ax3.plot(User5,User6, color='green',linestyle='-')
ax3.set_xlabel('RHOZ',color='green')
ax3.tick_params(axis='x', color='green')
ax3.spines['top'].set_position(('outward',1))
# Track #4:
User7=df_log[df_log['Well_Name']==Name]['PEFZ']
User8=df_log[df_log['Well_Name']==Name]['DEPT']
ax4=ax[3].twiny()
ax4.invert_yaxis()
ax4.plot(User7,User8, color='brown',linestyle='-')
ax4.set_xlabel('PEFZ',color='brown')
ax4.tick_params(axis='x', color='brown')
ax4.spines['top'].set_position(('outward',1))
```

Python output=Fig. 2.14 (if you would like to revert the y-axis, simply add "plt.gca().invert_yaxis()" at the end of the code above to accomplish this).

Seaborn library

Seaborn is a statistical plotting library and is built on top of matplotlib. Seaborn library has many default styles that can be used for quick plotting, and it is designed to work well with pandas library. As it will be evident shortly through various examples, one of the main advantages of the seaborn library is

MIP_4H Sonic Log

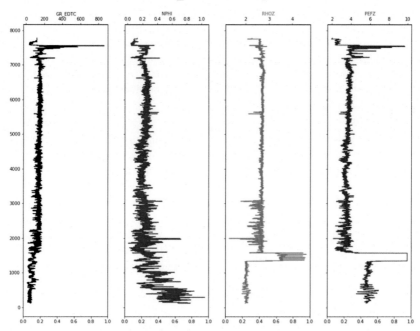

FIGURE 2.14   MIP_4H Sonic Log.

that it requires much less coding as compared to matplotlib. Various high-resolution, presentable, and sophisticated plots can be created rapidly with one or a few lines of code using this library. This library is specifically useful for presentation creation purposes. The disadvantage of the seaborn library is that although most of the important plots are available, it does not have as wide of a collection as the matplotlib library and can only create statistical plots that exist within the library. Some of the common useful plots in seaborn are distribution plot, joint plot, pair plots, rug plot, kernel density estimate (KDE) plot, bar plot, count plot, box plot, violin plot, and swarm plot. Let's go ahead and start importing the seaborn library (highlighted in bold) and "Shale Gas Wells." The link to this data set can be found below:

https://www.elsevier.com/books-and-journals/book-companion/9780128219294

```
import numpy as np
import pandas as pd
import matplotlib.pyplot as plt
import seaborn as sns
%matplotlib inline
df_Shale=pd.read_csv('Chapter2_Shale_Gas_Wells_DataSet.csv')
df_Shale.columns
```

```
Python output=Index(['Stage Spacing', 'bbl/ft', 'Well Spacing',
'Dip', 'Thickness', 'Lateral Length', 'Injection Rate', 'Porosity',
'ISIP', 'Water Saturation', 'Percentage of LG', 'Pressure Gradient',
'Proppant Loading', 'EUR', 'Category'], dtype='object')
Python output=Index(['Stage Spacing', 'bbl/ft', 'Well Spacing',
'Dip', 'Thickness', 'Lateral Length', 'Injection Rate', 'Porosity',
'ISIP', 'Water Saturation', 'Percentage of LG', 'Pressure Gradient',
'Proppant Loading', 'EUR', 'Category'], dtype='object')
```

Distribution plots

One of the most important plots to comprehend a data set prior to performing any type of ML analysis is the distribution plot. Here the distribution of each parameter is plotted for analysis. The main reason for using a distribution plot is to make sure the distribution of the input and output features are normal (Gaussian). This is because the majority of ML algorithms assume that the distribution of the parameters is normal. Therefore, when distribution of parameters is nonnormal, one should apply various techniques to normalize them. Seaborn library can be used to plot distribution plots. As illustrated below, "sns.distplot" method is used to plot the distribution of each parameter. In addition, color can be specified and bins can be set to a number. Otherwise, if "bins = None," Freedman–Diaconis rule will be used as the default parameter in the seaborn library.

Please note that Freedman–Diaconis rule is a popular method developed in 1981 to select the width of the bins used in a histogram. The bin width is calculated as shown below (Freedman & Diaconis, 1981):

$$Bin\ \ Width = 2\frac{IQR(x)}{\sqrt[3]{n}} \tag{2.1}$$

In this equation, IQR(x) is the interquartile range of the data and n is the number of observations in sample x. Interquartile range is defined as the difference between 75th and 25th percentile (IQR = Q3-Q1). IQR is also referred to as "middle 50," "H-spread," or "midspread" (Kokoska & Zwillinger, 2000).

Let's go over calculating the IQR on a sample assuming the following numbers:

1,3,5,2,6,12,17,8,15,18,20

Step 1) Place the numbers in order as follows:

1,2,3,5,6,**8**,12,15,17,18,20

Step 2) Find the median. The median for this example is 8.

Step 3) To make Q1 and Q3 easier to identify, let's place parentheses around the numbers above and below the median as follows:

(1,2,3,5,6),8,(12,15,17,18,20)

Step 4) Q1 is defined as the median in the **lower** half of the data and Q3 is defined as the median in the **upper** half of the data. Therefore, Q1 = 3 and Q3 = 17 in this example.

(1,2,**3**,5,6),**8**,(12,15,**17**,18,20)

Step 5) Finally, the last step is to subtract Q1 from Q3 as follows:

$$IQR = 17 - 3 = 14$$

To remove KDE line from the plot, simply add "KDE=False" as the default parameter is True. Below is a distribution plot of well spacing in blue with an x label of "Well Spacing," a y label of "Frequency," and a title of "Distribution Plot for Well Spacing." To plot the distribution of other features (parameters), simply replace "df_Shale['Well Spacing']" with "df_Shale ['Name of the desired column']" and adjust the size, label, and title accordingly.

```
fig=plt.figure(figsize=(5,5))
sns.distplot(df_Shale['Well Spacing'], color='blue', bins=None)
plt.xlabel('Well Spacing')
plt.ylabel('Frequency')
plt.title('Distribution Plot for Well Spacing')
Python output=Fig. 2.15
```

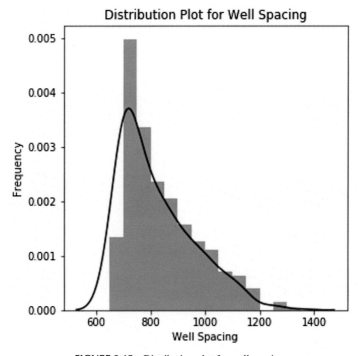

FIGURE 2.15   Distribution plot for well spacing.

Joint plots

A joint plot is another useful way of visualizing the relationship between two parameters as well as their distribution. To plot a joint plot, "sns.jointplot" method can be used. Please note that at any point during coding, if an exact method cannot be remembered, after typing in the first couple of words or so, hit the shift button and all the available options will be shown which is a very useful feature. Please note that auto-completion of different methods (or attributes) is a feature of the IDE (Integrated Development Environment) such as Jupyter. Let's plot stage spacing and proppant loading versus EUR in two separate joint plots. Pass in x-axis which is stage spacing and y-axis which is EUR and define "kind = 'scatter'" to plot these two variables as scatter plot. Please note that kind can be replaced with other options such as regular, kde, and hex. Also, to show the Pearson correlation coefficient and *P*-value on the plot, remember to import the scipy stats library and pass in "stat_func = -stats.pearsonr" as shown below:

```python
from scipy import stats
sns.jointplot(df_Shale['Stage Spacing'], df_Shale['EUR'], color='blue',
kind='scatter', stat_func=stats.pearsonr)
```
Python output=Fig. 2.16

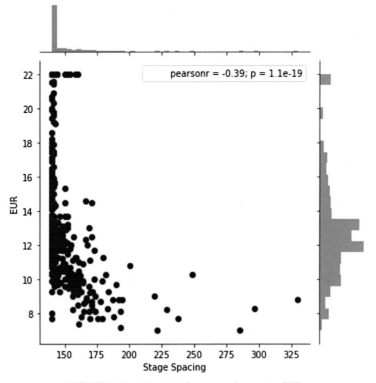

FIGURE 2.16 Joint plot of stage spacing versus EUR.

Let's also import proppant loading versus EUR using another joint plot. Please replace scatter with "kde" under "kind" as shown in bold below. KDE stands for kernel density estimate and plots a KDE in the margins by converting the interior into a shaded contour plot.

```
sns.jointplot(df_Shale['Proppant Loading'],df_Shale['EUR'],
color='blue',kind='kde',stat_func=stats.pearsonr)
Python output=Fig. 2.17
```

Pair plots

Another very useful plot showing the relationship between all the parameters in one plot is called a pairplot in seaborn library. As opposed to plotting scatter plots individually, a pairplot can be used to plot all variables against one another with one line of code (sns.pairplot()). If there is categorical data in the imported data frame, "hue = 'name of column'" can be specified to determine which column in the data frame should be used for color encoding. In the df_Shale data frame, there is a column called "Category" that has two

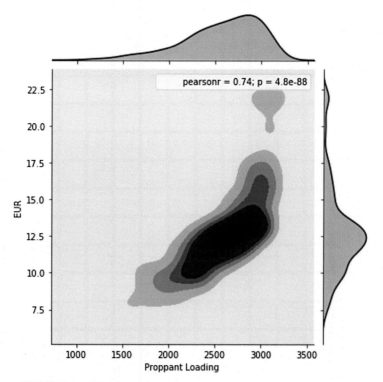

FIGURE 2.17   Joint plot of proppant loading versus EUR using "kind = kde".

categorical names of "Vintage" and "New_Design" and are used to color code the pairplot as shown below.

```
sns.pairplot(df_Shale, hue='Category', kind='scatter')
Python output=Fig. 2.18
```

lmplots

lmplots are basically scatter plots with overlaid regression lines. lmplot() combines linear regression model fit (regplot()) and FacetGrid(). The FacetGrid class aids in understanding the relationship between multiple variables separately within subsets of a data set as illustrated in Fig. 2.19. Please note that lmplots are computationally intensive as compared to regplot(). Let's use the lmplot() to plot "Lateral Length" on the x-axis, "EUR" on the y-axis, and set hue to be equal to "Category." Afterward, remove hue = 'Category' and use "sns.regplot()" method instead to compare the two methodologies. It will be evident that the resulting plots are identical but the figure shapes are

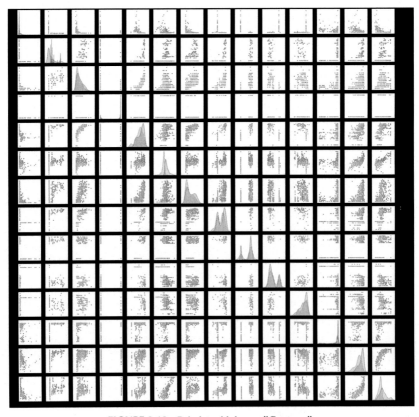

FIGURE 2.18 Pairplot with hue = "Category".

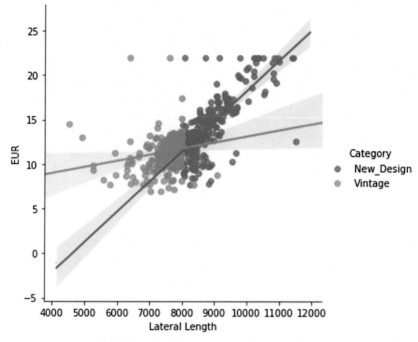

FIGURE 2.19 Implot of lateral length versus EUR.

different. Therefore, according to the seaborn library documentation, regplot() and lmplot() are closely related where regplot() is an axes-level function while lmplot() is a figure-level function that combines regplot() and FaceGrid() (Seaborn.lmplot, n.d.).

```
sns.lmplot(x='Lateral Length',y='EUR',data=df_Shale,hue='Category')
Python output=Fig. 2.19
```

Let's also use lmplot to visualize porosity versus water saturation and use "col = 'Category'" to show two side-by-side lmplots as illustrated below:

```
sns.lmplot(x='Porosity',y='Water Saturation',data=df_Shale,
col='Category',aspect=0.8,height=5)
Python output=Fig. 2.20
```

Bar plots

Bar plots can also be plotted using "sns.barplot()." Let's plot Category versus EUR on a bar plot using "estimator = mean" as shown below. Please note that other **estimators** such as std (standard deviation), median, and sum can also be passed on to produce these bar plots. The term "estimator" is used to set an aggregate function to use when plotting different categories in a bar plot. For

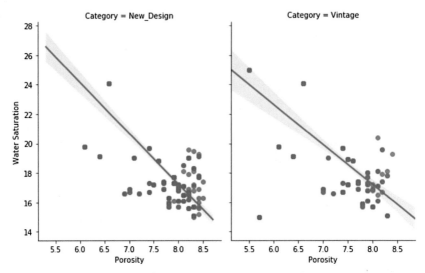

FIGURE 2.20 Implot using col = "Category".

example, when "estimator = np.mean" is used, the average of each category is taken and shown on the plot. In the example below, the average EUR for "Vintage" and "New_Design" categories are shown on the bar plot.

```
sns.barplot(df_Shale['Category'],df_Shale
['EUR'],estimator=np.mean)
```
Python output=Fig. 2.21

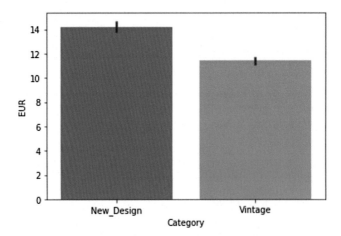

FIGURE 2.21 Bar plot of Category versus EUR using estimators = np.mean.

Count plots

Seaborn also has built-in count plot functionality that can be used. "sns.count-plot()" will show value count for each category. The example below shows the value count for each completions design category (Vintage vs. New_Design).

```
sns.countplot(df_Shale['Category'])
Python output=Fig. 2.22
```

Box plots

Another important visualization plot that is recommended to be paired with distribution plots is box plots. Box plots are great visualization tools to identify anomalous and outlier points. Please note that the following parameters can be obtained from a box plot:

- Median (50th percentile)
- First quartile (25th percentile)
- Third quartile (75th percentile)
- Interquartile range (25th to 75th percentile)
- Maximum: this is not the maximum value in a data set and it is defined as: Q3 + 1.5 * IQR
- Minimum: this is not the minimum value in a data set and it is defined as: Q1 - 1.5 * IQR
- Outliers are the points that lie outside of "Minimum" and "Maximum" terms defined above.

Fig. 2.23 illustrates the above definitions for simplicity of understanding box plots.

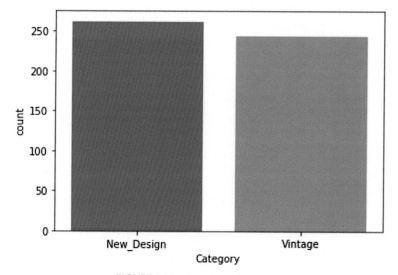

FIGURE 2.22   Count plot of category.

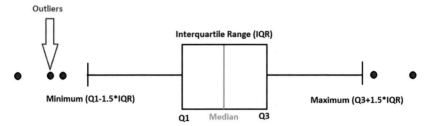

FIGURE 2.23 Box plot illustration.

Below is an example of porosity box plot using a green color:

```
sns.boxplot(df_Shale['Porosity'], color='green')
Python output=Fig. 2.24
```

Let's also use sns.boxplot to plot Category versus EUR. "Palette = 'rainbow'" can also be used to customize the color scheme.

```
sns.boxplot(df_Shale['Category'],df_Shale
['EUR'],palette='rainbow')
Python output=Fig. 2.25
```

Violin and swarm plots

The next advanced visualization plots that can be visualized using the seaborn library are called violin and swarm plots. A violin plot is another useful visualization tool that can be used in addition to a box plot. The main advantage of a violin plot as compared to a box plot is being able to show the

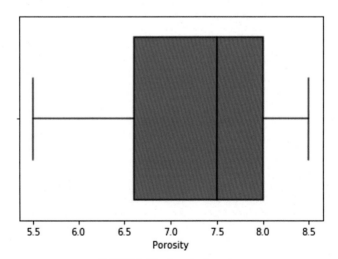

FIGURE 2.24 Porosity box plot.

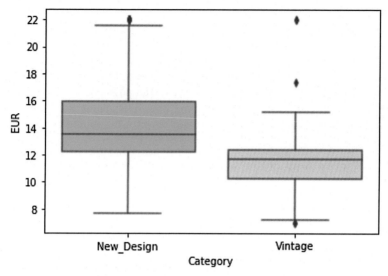

FIGURE 2.25 Box plot of category versus EUR.

full distribution of the data especially when a multimodal (more than one peak) distribution is observed within a data set. Please note that the **width** of a violin plot curve relates to the approximate frequency of data points in each region. Fig. 2.26 illustrates important statistics that can be obtained from a violin plot.

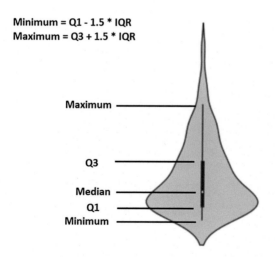

Anything outside of Minimum and Maximum are considered outlier points.

FIGURE 2.26 Violin plot illustration.

Let's plot category versus well spacing using both **violin** and **swarm** plots using "sns.violinplot()" and "sns.swarmplot()" as follows. Plot each one individually (as opposed to what is shown below) to make sure some of the important features of the violin plot such as **median** can be visually seen. Otherwise, median point in the violin plot is blocked by the swarm plot data points. Note that when a swarm plot is used, data points from the central line are offset to avoid overlaps. Spend some time overlapping box, violin, and swarm plots to make sure the concept is clear prior to proceeding to the next section.

```
sns.violinplot(df_Shale['Category'],df_Shale['Well Spacing'],palette=
'rainbow')
sns.swarmplot(df_Shale['Category'],df_Shale['Well Spacing'],
palette='rainbow')
Python output=Fig. 2.27
```

KDE plots

KDE stands for kernel density estimate and is used to visualize the probability density of a continuous variable and is used for nonparametric analysis. Please note that KDE plots can be useful in determining the shape of a distribution. When KDE is applied to a single parameter, the density of the observation is plotted on one axis and the height on the other axis (Visualizing the Distribution of a Dataset, n.d.). "sns.kdeplot()" method can be used to plot kde. Below is an example of plotting water saturation versus EUR (bivariate density

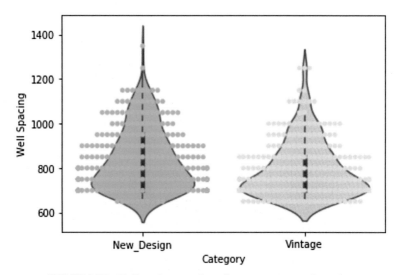

FIGURE 2.27 Violin and swarm plots of category versus well spacing.

plot). "cmap = 'Reds'" color maps the plot using a red color and "shade = True" provides shade in the area under the KDE curve.

```
sns.kdeplot(df_Shale['Water Saturation'],df_Shale['EUR'],cmap='Reds',
shade=True, shade_lowest=False)
Python output=Fig. 2.28
```

Heat maps

The next very important plot to identify relationship between parameters is called a heat map and "sns.heatmap()" can be used to perform such plotting. In addition to plotting the heat map, it is extremely important to add the Pearson correlation coefficient inside the heat map to identify collinear parameters. The Pearson correlation is calculated as follows:

$$P_{X,\ Y} = \frac{\text{cov}(X,\ Y)}{\sigma_X \sigma_Y} \quad (2.2)$$

where cov(X,Y) is covariance between parameters X and Y, σ_X is standard deviation of parameter X, and σ_Y is standard deviation of parameter Y. Pearson correlation coefficient ranges between -1 and 1. Positive values indicate the tendency of one variable to increase or decrease respective to another variable. Negative values, on the other hand, indicate the tendency that the increase in value of one variable is due to the decrease in the value of another variable and vice versa. The square of correlation coefficient (R) is referred to as the coefficient of determination (R^2). (Pearson's Correlation Coefficient, 2008). Please note that the general rule of thumb is to remove collinear input variables with $|\pm 90\%|$ as those variables provide the same information; hence are redundant and would increase the dimensionality of the input data set. Unless those collinear parameters happen to be collinear due to other factors such as timing, it is best to remove those parameters. For example, the oil and gas

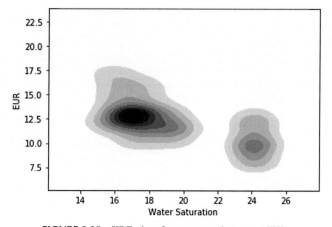

FIGURE 2.28   KDE plot of water saturation versus EUR.

industry switched to tighter stage spacing and higher proppant loading around the same time. Therefore, a Pearson correlation coefficient might indicate negative collinearity between these two parameters; however, in essence, this is due to timing rather than the inherent collinearity between these two variables. In this particular example, two separate models can be trained to understand the impact of each variable on the model. Let's use the "sns.heatmap()" to visualize this. As illustrated below, "df_Shale.corr()," used to define Pearson correlation coefficient in Python (the default method when calling ".corr()" is 'perason'), would be used to plot the heat map and "annot = True" would show the actual values on the heat map. With setting "annot = False," the Pearson correlation values will be shown on the heat map. Note that other methods such as "spearman" and "kendall" could also be passed using "df_Shale.corr(method='spearman')" instead of the "perason" method which is the default case.

```
fig=plt.figure(figsize=(17,8))
sns.heatmap(df_Shale.corr(), cmap='coolwarm', annot=True,
linewidths=4, linecolor='black')
```
Python output=Fig. 2.29

Cluster maps

Cluster maps can be used to plot a matrix data set as a hierarchically clustered heat map and seaborn has a powerful built-in function for that purpose. Hierarchical clustering concept is discussed in detail in unsupervised clustering chapter of this book. Therefore, in this section, an example with the code is illustrated without explaining the concept. To visualize, let's import the natural gas pricing csv file located below.

https://www.elsevier.com/books-and-journals/book-companion/9780128219294

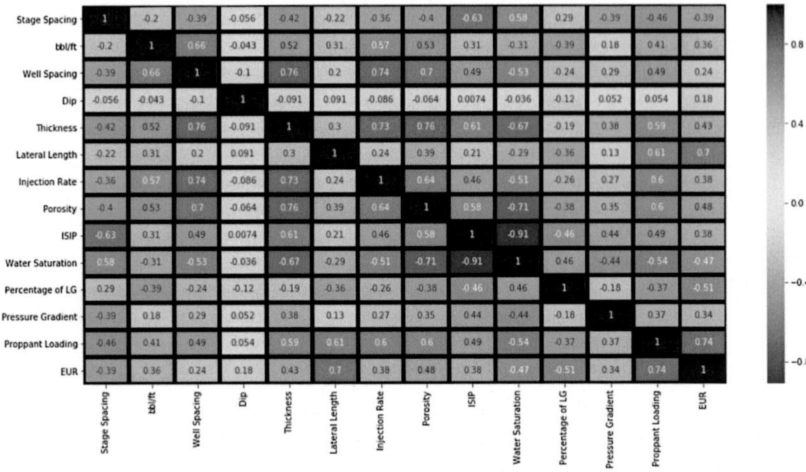

FIGURE 2.29 Pearson correlation coefficient heat map.

The code below illustrates importing a csv file called "Chapter2_Natural_ Gas_Pricing_Version2_DataSet," followed by pivoting the df_ngp data frame by using the "Month" column as rows, "Prior to 2000" column as columns, and using average of gas pricing per month as "values." For illustration purpose of this example, note that "Prior to 2000" column is an arbitrary categorical column that has 0 for each gas pricing before year 2000 and 1 after year 2000. Afterward, a heat map of gas pricing per month can simply be visualized.

```
df_ngp=pd.read_csv('Chapter2_Natural_Gas_Pricing_Version2_
DataSet.csv')
df_ngp_pivot=df_ngp.pivot_table(values='Value',index='Month',
columns='Prior to 2000')
fig=plt.figure(figsize=(10,5))
sns.heatmap(df_ngp_pivot, annot=True, linecolor='white',
linewidths=3)
```
Python output=Fig. 2.30

Next, let's use "sns.clustermap()" to use a hierarchically clustered heat map to visualize the data. "standard_scale = 1" standardizes the dimension in which for each row or column, the minimum is subtracted and divided by its maximum.

```
sns.clustermap(df_ngp_pivot,cmap='coolwarm',standard_scale=1)
```
Python output=Fig. 2.31

PairGrid plots

PairGrid plots are similar to pair plots in which all variables are plotted against one another (as was previously illustrated). PairGrid plots simply provide more flexibility to create various plots and customize the plots. In the example below, a PairGrid plot of df_Shale data frame is created. A histogram was used to plot all the diagonal plots, the upper portion above the diagonal is scatter plots, and the lower portion is KDE plots. Please note that "Category" was used to distinguish between "New Design" and "Vintage." Pay attention to the bold texts within the script.

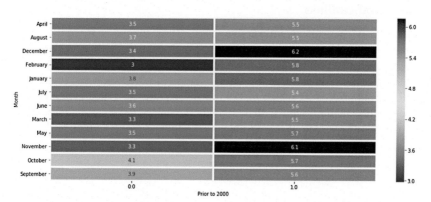

FIGURE 2.30   Heat map of gas pricing per month divided between before and after year 2000.

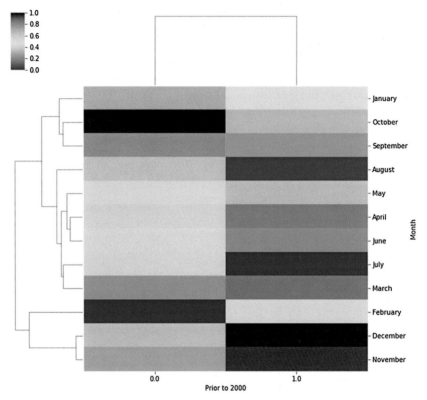

FIGURE 2.31 Hierarchically clustered heat map of gas pricing.

```
Shale_Properties=sns.PairGrid(df_Shale, hue='Category')
Shale_Properties.map_diag(plt.hist)
Shale_Properties.map_upper(plt.scatter)
Shale_Properties.map_lower(sns.kdeplot)
Python output=Fig. 2.32
```

Plotly and cufflinks

Plotly is an open-source **interactive** visualization library in Python, and the quality of the plots is higher than matplotlib and seaborn. Plotly provides many interactive features such as zooming in/out, panning, box and lasso selection, auto scaling, etc. In addition, hovering over the plot will show the underlying data behind each point which makes it a super powerful visualization library. The quality of the pictures makes plotly a perfect library for presentation purposes as well. One of the advantages of plotly is that it is like "R" in the event the user is a fan of plots created within the R programming language.

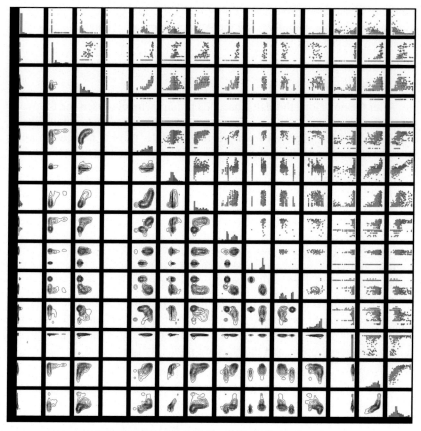

FIGURE 2.32   PairGrid plot of df_Shale data frame.

Please note that Plotly is both a company and an open-source interactive data visualization library. As it is inferred, the company Plotly specializes on data visualization for business intelligence such as creating dashboards, reporting, etc. The Plotly company released an open-source interactive data visualization library called plotly that primarily focuses on interactive data visualization as it will be illustrated shortly. Using the Plotly Python library by itself creates interactive plots as **.html files**. While these plots are high-quality interactive plots, they cannot be connected to changing data sources. Therefore, to create interactive real-time dashboards for changing data sources, the "Dash" library in conjunction with the plotly library can be used. Dash is also an open-source library from the plotly company that allows full dashboard creation with various interactive components and plots. The Dash library and creating dashboards are not discussed in this book because it is beyond the scope of this book. Cufflinks connects plotly with pandas. Please check out plotly.ly website for more info. Plotly and Cufflinks do not come with the Anaconda package; therefore, use the Anaconda prompt window to download these two packages as follows:

pip install plotly
pip install cufflinks

After the download through Anaconda prompt window, insert the following lines of code to work with this library offline. If pandas and numpy library have already been imported from previous module, there is no need to reimport these libraries as illustrated below.

```
import pandas as pd
import numpy as np
%matplotlib inline
from plotly.offline import download_plotlyjs, init_notebook_mode,
plot, iplot
import cufflinks as cf
init_notebook_mode(connected=True) # For Notebooks
cf.go_offline() # For offline use
```

Let's plot bbl/ft versus EUR using the plotly library. Please note that the data frame followed by ".iplot()" syntax can be used. Pass on kind, x, y, size, and color inside the parenthesis. As illustrated below, your mouse can be hovered over any point, and the selected points will be shown on the plot.

```
df_Shale.iplot(kind='scatter',x='bbl/ft',y='EUR',
mode='markers', size=4, color='red',xTitle='Water Loading (bbl/ft)',
yTitle='EUR (BCF)', title='Water Loading Vs. EUR')
Python output=Fig. 2.33
```

Let's do a quick box plot of well spacing using the plotly library. This time, pass in "kind = 'box'".

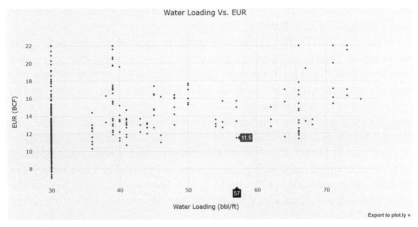

FIGURE 2.33 Scatter plot of water loading versus EUR using the plotly library.

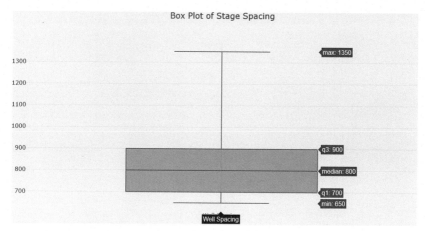

FIGURE 2.34   Box plot of well spacing using the plotly library.

```
df_Shale['Well Spacing'].iplot(kind='box',color='red',
title='Box Plot of Stage Spacing')
```
Python output=Fig. 2.34

A box plot of all parameters can also be plotted as **shown** below. Please note that due to the scale of majority of parameters with lower values, only some of the parameters can be easily visualized. To plot a few selected parameters, use "df_Shale[['Parameter 1 name','Parameter 2 name']].-iplot(kind = 'box')." This will allow to select a few parameters as opposed to all parameters.

```
df_Shale.iplot(kind='box', title='Box Plot of all Parameters')
```
Python output=Fig. 2.35

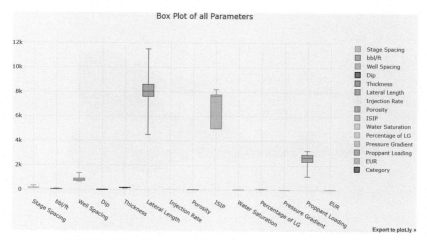

FIGURE 2.35   Box plot of all parameters using the plotly library.

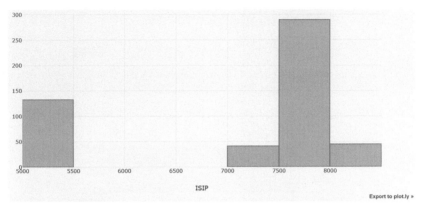

ISIP

FIGURE 2.36 ISIP distribution plot using the plotly library.

To plot a histogram plot of ISIP, the following lines of code can be written:

```
df_Shale['ISIP'].iplot(kind='hist', xTitle='ISIP')
Python output=Fig. 2.36
```

Let's also plot a bubble plot of water loading versus sand loading and use EUR as the criteria for bubble size as illustrated below. Please note that "kind = 'bubble'" can be used to define a bubble plot.

```
df_Shale.iplot(kind='bubble',x='Proppant Loading',y='bbl/ft',size=
'EUR',title='Bubble Plot Based on EUR',xTitle='Proppant Loading
(#/ft)',yTitle='Water Loading (bbl/ft)',zTitle='EUR')
Python output=Fig. 2.37
```

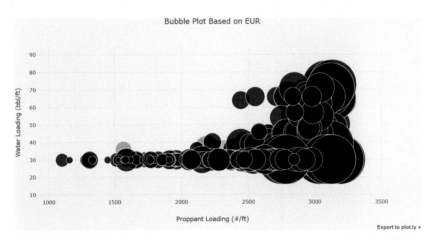

FIGURE 2.37 Bubble plot of water loading versus sand loading (based on bubble size of EUR) using the plotly library.

In addition to using iplot, other forms of plotly coding can be used to plot various plots. First, let's import the following two lines of code. Please note that "plotly.graph_objs" has several useful functions for creating graph options as will be demonstrated shortly.

```
import plotly.graph_objs as go
import plotly.offline as pyo
```

To create a plotly graph, "data" and "layout" variables must be first defined. The "data" is defined below as EUR (y-axis) versus stage spacing (x-axis) using "mode = 'markers'," marker size, color, and symbol of 9, rgb(23, 190, 207), and square. x and y axes and title are defined under the "layout" variable. Finally, pass in "data" and "layout" under "go.Figure()" and use pyo.plot() to create a .html plot as shown below. Please note that "mode = lines" can be used to convert this scatter plot into a line plot.

```
data=[go.Scatter(x=df_Shale['Stage Spacing'], y=df_Shale['EUR'],
mode='markers',marker=dict(size=9,  color='rgb(23,  190,  207)',
symbol='square', line={'width':2}))]
layout=go.Layout(title='Stage Spacing Vs. EUR', xaxis=dict(title=
'Stage Spacing'), yaxis=dict(title='EUR'), hovermode='closest')
fig=go.Figure(data=data, layout=layout)
pyo.plot(fig, filename='scatter.html')
```
Python output=Fig. 2.38

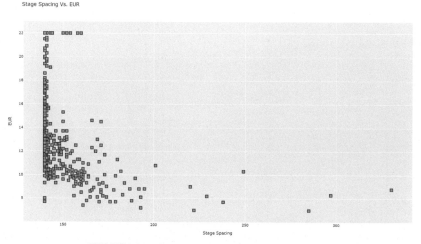

FIGURE 2.38   Scatter plot of EUR versus stage spacing.

Some of the rgb default colors in plotly are listed below:

rgb(23, 190, 207) = blue teal, rgb(188, 189, 34) = curry yellow-green, rgb(127, 127, 127) = middle gray, rgb(227, 119, 194) = raspberry yogurt pink, rgb(140, 86, 75) = chestnut brown, rgb(148, 103, 189) = muted purple, rgb(214, 39, 40) = brick red, rgb(44, 160, 44) = cooked asparagus green, rgb(255, 127, 14) = safety orange, rgb(31, 119, 180) = muted blue (refer to https://plotly.com/python/)

Now, let's create a scatter plot with both stage spacing and water/ft on the same x-axis using Trace 1 and 2 as follows:

```
trace1=go.Scatter(x=df_Shale['Stage Spacing'], y=df_Shale['EUR'],
mode='markers', name='Stage Spacing')
trace2=go.Scatter(x=df_Shale['bbl/ft'], y=df_Shale['EUR'],
mode='markers', name='Water per ft') data=[trace1,trace2]
layout=go.Layout(title='Scatter Charts of Stage Spacing and Water
per ft')
fig=go.Figure(data=data,layout=layout)
pyo.plot(fig)
```
Python output=Fig. 2.39

Next, let's create a bubble plot that shows stage spacing on the x-axis, EUR on the y-axis, bubble size by lateral length, and bubble color by water per ft. The reason for dividing "size = df_Shale['Lateral Length']" by 300 is to reduce the size of the bubble. If the size was categorized by a small number, it might be necessary to multiply by a number to increase the size of the bubble. Therefore, feel free to change this number until a desired bubble size is obtained.

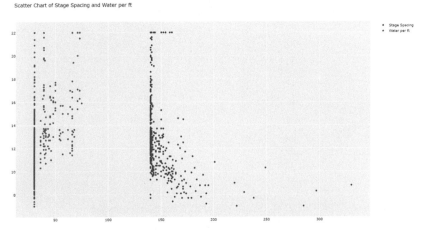

FIGURE 2.39 Scatter plot of stage spacing and water per ft.

```
data=[go.Scatter(x=df_Shale['Stage Spacing'], y=df_Shale['EUR'],
mode='markers', marker=dict(size=df_Shale['Lateral Length']/300,
color=df_Shale['bbl/ft'], showscale=True))]
layout=go.Layout(title='Stage Spacing Vs. EUR, Sized by Lateral
Length, and Colored by Water per ft', xaxis=dict(title='Stage
spacing'), yaxis=dict(title='EUR'), hovermode='closest')
fig=go.Figure(data=data, layout=layout)
pyo.plot(fig)
```
Python output=Fig. 2.40

Next, a box plot of proppant loading can be coded below. Please note that "boxpoints = 'all'" displays all the original proppant loading data points which is used to provide more insight into the distribution of that feature. "boxpoints = 'outliers'" will only show the outlier jitter points. "jitter" ranges between 0 and 1. The smaller the number, the less spread out the jitter points would be left and right and vice versa. "pointpos" dictates where the jitter points will be located within the plot. A value of 0 means the points will be on top of the box plot, +1 or +2 indicates the jitter points will be on the right, and −1 or −2 indicates the jitter points will be on the left (a range of −2 to +2 can be specified).

```
data=[go.Box(y=df_Shale['Proppant  Loading'],  boxpoints='all',
jitter=0.5,pointpos=-2)]
layout=go.Layout(title='Proppant Loading', hovermode='closest')
fig=go.Figure(data=data, layout=layout)
pyo.plot(fig)
```
Python output=Fig. 2.41

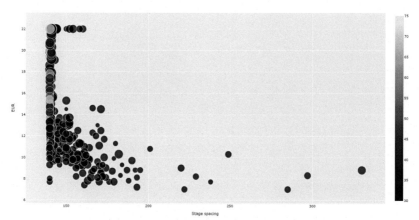

FIGURE 2.40   Bubble chart.

Proppant Loading

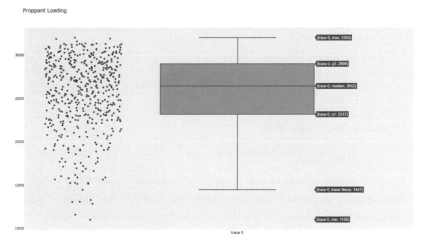

FIGURE 2.41 Box plot of proppant loading (including jitter points).

In addition, let's code the box plots of porosity and water saturation as follows:

```
data=[go.Box(y=df_Shale['Porosity'],name='Porosity'),
go.Box(y=df_Shale['Water Saturation'],name='Water Saturation')]
layout=go.Layout(title='Porosity and Water Saturation Box Plots',
hovermode='closest')
fig=go.Figure(data=data, layout=layout)
pyo.plot(fig)
```
Python output=Fig. 2.42

Porosity and Water Saturation Box Plots

FIGURE 2.42 Porosity and water saturation box plots.

Please also spend some time exploring "go.Histogram" for distribution plots, "go.Bar" for bar plots, and "go.Heatmap" for heat maps using the plotly library. Please also note that plotly express (import plotly.express as px) can also be used as an abbreviated version of the plotly library for these discussed plots. These are some of the main plots that will be used throughout the book. Other customization plots can be found through matplotlib, seaborn, and plotly library documentations.

References

Freedman, D., & Diaconis, P. (1981). *On the histogram as a density estimator: L 2 theory.* https://doi.org/10.1007/BF01025868

Kokoska, S., & Zwillinger, D. (2000). *CRC standard probability and statistics tables and formulae.* Student Edition. CRC Press, 9780849300264.

seaborn.lmplot. (n.d.). Retrieved May 10, 2020, from https://seaborn.pydata.org/generated/seaborn.lmplot.html.

Pearson's Correlation Coefficient. (2008). *Encyclopedia of public health.* https://doi.org/10.1007/978-1-4020-5614-7

Visualizing the distribution of a dataset. (n.d.). Retrieved May 15, 2020, from https://seaborn.pydata.org/tutorial/distributions.html.

Chapter 3

Machine learning workflows and types

Introduction

Before diving into various machine learning (ML) algorithms and types, it is crucial to understand typical ML workflows and types. A common ML workflow includes but is not limited to: (i) data gathering and integration, (ii) data cleaning which includes (a) data visualization, (b) outlier detection and, (c) data imputation, (iii) feature ranking/selection, (iv) normalization/standardization, (v) cross-validation, (vi) model development, (vii) grid search for parameter fine tuning and optimization, and (viii) finally, implementing the trained model. In this chapter, each ML step will be discussed in detail to provide a practical workflow that can be applied to any ML project. In addition, various types of ML such as supervised, unsupervised, semi-supervised, and reinforcement learning will be thoroughly examined. The combination of data science knowledge and domain expertise is the key to a successful ML project implementation. Unfortunately, the O&G industry has seen many ML projects fail due to data scientists and domain experts not properly working together and in silos.

Machine learning workflows

Data gathering and integration

Companies across the globe have different workflows when applying AI and ML projects to their problems; however, the goal of this chapter is to provide a practical workflow that can be used to tackle any problem. Data gathering and integration is one of the most important steps in any ML project. Many ML projects simply fail due to lack of data. Therefore, before diving into the project, it is important to understand data availability within an organization. If the data are simply not available or very challenging to obtain, it is recommended to focus on projects with available data and projects that are highly likely to provide tangible results that can demonstrate the potential of ML within an organization. If your organization is new to the application of AI and ML, it is recommended to find some practical use cases, prove the concept

Machine Learning Guide for Oil and Gas Using Python. https://doi.org/10.1016/B978-0-12-821929-4.00001-9

with a few projects, and show tangible value associated with practical applications of ML and AI. Since data come from many different sources, it is imperative for organizations to have a central data warehouse that can be relied on. A centralized data warehouse allows everyone within an organization to use the same data for different projects. The O&G and other industries must move away from using the excel spreadsheets to using a centralized data warehouse such as SQL to store all the data. Unfortunately, because of inefficient data storage systems, in many companies data scientists have to spend 80% of their time collecting the data and only 20% analyzing the data. Wouldn't it be so powerful and rewarding to spend less than 5% of the time obtaining the necessary information while spending most of the time analyzing the data and finding out of the box problems to apply ML to solve various complex problems? When the data are housed in a centralized location, there is no need to visit multiple sources to obtain the necessary data to perform ML analysis. If your company is small and privately held, it is worthwhile to invest into a data warehouse where data from different sources are compiled into one source. The setup might be a lengthy project, but will pay off in the long run. Even if your company's intention is to sell the asset and cash out, it is still important to have the data warehouse setup because this would be a selling point to any major operator. Data are the new gold and whoever has the best system in terms of storing and analyzing data would be the most powerful in terms of adding significant financial benefit.

Before proceeding to data cleaning, it is important to understand the term "data center." A data center is simply a building, multiple buildings, or a dedicated space within a building that houses hardware, data storage, servers, computer systems, etc., for an organization. There are different varieties of data centers as follows:

- **On-premises data center**, also referred to as on-prem data center, is a group of servers owned and controlled privately by an organization. Therefore, there is complete control over the hardware. In an on-prem data center, there is an upfront capital investment which is one of the biggest disadvantages of this kind of data center. The upfront capital investment is used to buy server and networking hardware required to set up the data center. In addition, there are ongoing maintenance costs associated with maintaining the hardware since they will age and need to be maintained/replaced. On-prem data centers are typically found to have limited scalability due to the time needed to bring additional servers online. As observed, an on-prem data center uses space and power regardless of its usage amount. The security of the data center depends entirely on the local IT team. In addition, updates are not automatic and rely on the IT team. The main advantage of an on-prem data center is that the cost of use is easier to understand (when compared to cloud data centers), privacy can be maintained, and there is complete control over hardware. An on-prem data center is like buying a house. If anything in the purchased house breaks down, it is solely your responsibility to fix and maintain. On-prem data centers are typically suited for smaller companies that already have a heavy capital investment in IT infrastructure.

- **Cloud data center**: As opposed to an on-prem data center, in a cloud data center, the actual hardware is managed and run by the cloud company in question. There are various cloud data center providers such as Amazon Web Services (AWS), Microsoft Azure (also known as Azure), and Google Cloud Platform (GCP). As opposed to on-prem data center where a heavy amount of capital is invested upfront, there is little upfront investment with cloud data centers, and it is a scalable model. Cloud data centers require little knowledge and less manpower to get started, have a rapid implementation period, have independent platforms, have automatic updates, have easy remote access, and the customer pays for what is used. Some of the disadvantages are less control over the hardware and operating system, complicated cost structure, and closely monitoring privacy and access. Some companies are simply not comfortable with giving up too much control with a cloud data center, and they like to have full internal control to hardware, data, etc. Please note that major cloud providers have extensive security levels since billions of dollars are spent every year to ensure customer data privacy. Cloud data center can be thought of as renting an apartment where there is no upfront down payment (investment) and can be rented immediately. When equipment breaks down, the property management company will fix those broken items. Cloud data centers are suitable for most organizations of any size.
- **Hybrid data center**: This is simply a combination of on-prem data center, private clouds, or public clouds. Please note that public clouds are offered to multiple clients (companies) by a cloud provider while a private cloud is a cloud service that is not shared with any other organizations. Note that in public clouds, each customer's data and applications remain hidden from other cloud customers, so no data is shared. On a private cloud, the services and infrastructure are maintained on a private network. In addition, both hardware and software are dedicated to your specific organization.

Cloud vs. edge computing

After generating the data on an end device, the data can then be traveled to the cloud for processing through public internet *or* the data can be processed at the site through an edge device. In the case of the second scenario where the data are sent to a device that is locally present, the device is referred to as an edge device. When the data are processed at the cloud, it is referred to as cloud computing. However, when the data are processed at the edge for near real-time or real-time data processing, it is referred to as edge computing.

Some advantages of cloud computing are that it is someone else's headache, and it is a scalable model as previously discussed. The main advantage of edge computing is that all data processing can be done at the edge and the control command can be sent back to the edge device for real-time or near real-time feedback control loop. There is no latency of sending the data to the cloud where processing occurs and sending the control command back down to the edge. The main advantage is therefore near real-time control close to the command. Another advantage is to process an extensive amount of data at the edge and send the remaining data back to the cloud. Near real-time feedback loop and saving bandwidth by performing most of the processing at the edge are two of the main advantages of edge computing. The main disadvantage of edge computing is that, since edge computing by definition is localized, if you are trying to scale at a larger level, it is better to send all those data to the cloud where algorithms can be run at the field level. So, to get a higher level of visibility as well as a broader level of optimization at the field level, cloud computing is better. Deploying machine learning models on the edge will have an enormous impact on real-time value proposition in the O&G industry.

Data cleaning

After data gathering and integration, the next crucial step is data cleaning. Many ML projects fail due to spending inadequate time interpreting and cleaning the data prior to applying ML algorithms. There is no such a thing as quick ML analysis because the proper application of ML to any project requires careful data interpretation and cleaning. Within data cleaning, there are a few main steps which are as follows:

- **Data visualization**: The first step in data cleaning is data visualization. Many ML models fail simply because proper data cleaning was not performed prior to starting the analysis. Quality checking and controlling the data is perhaps the most important piece of any ML analysis. Having bad data can easily break a model and yield unexpected erroneous results. As illustrated in Chapter 2, Visualization library packages such as Matplotlib, Seaborn, and Plotly in Python are ideal for visualization which would lead to finding outlier points. Understanding the distribution of each feature is also crucial. The recommended plots to evaluate are distribution, box, and scatter.
- **Outlier detection**: As previously mentioned, data visualization leads to finding outlier points. In addition, looking at parameters using distribution, scatter, cross plots, heat maps, and box plots provides easy visualizations of the parameters that have reasonable ranges versus the ones that are completely off the chart. For example, if most of the proppant per foot data lies between 1000 and 4000 #/ft, and there are two data points at 10,000 #/ft, it is important to investigate those anomalous points. Either investigate

the validity of these outlier points or simply remove them before proceeding to the next step. In addition, basic statistics of the data including frequency, minimum, maximum, average, median, and standard deviation of each parameter must be obtained before and after data cleaning and outlier detection to make sure the final range of parameters are satisfactory.

- **Data imputation**: The various approaches to missing data are as follows:
  1) Simply remove any row of data with N/A values: This is the simplest method of treating missing data, and the disadvantage of this approach is that the number of samples could be drastically reduced due to the removal of many of the samples with missing info.
  2) Replacing missing values with the mean and median of the existing values. Although this approach is very simplistic and easy to perform, we do not recommend such approach to filling missing data. Unless there is proven data behind this approach, it might be better to simply remove the missing data rows as opposed to introducing erroneous data.
  3) Using the Fancyimpute package in Python and algorithms such as k-nearest neighbor (KNN) and Multivariate Imputation by Chained Equations (MICE) to filling the missing data. These algorithms will be discussed at the end of the supervised learning chapter (Chapter 5). These algorithms do a great job only when the following condition exists: There is an inherent relationship between the parameters that are missing and the existing parameters. For example, if sand per ft for one well is missing, but water per foot for the well is available, these algorithms might produce meaningful answers. The challenge associated with using these algorithms that could produce low accuracy is that when a well has missing info, most of the info could potentially be absent. The fancy impute package in Python is a powerful package when one or two parameters in a data set are sporadically missing while there are other parameters that have inherent relationship between those parameters. When most of the parameters are simply missing from a well's data, it is not recommended to use these packages because they might very well produce erroneous results. When using these packages, be sure to normalize or standardize the data before applying KNN or MICE. These algorithms will be discussed in detail under the supervised learning chapter.

Feature ranking and selection

The next important step in any ML analysis is feature ranking followed by feature selection. This is when domain expertise plays a significant role in a successful ML project. If a data scientist has no practical experience in an industry, it is extremely difficult to understand relationship between variables. This is an example showing the need for combined domain expertise and AI knowledge when problem-solving. The first step is to identify the type of problem that a data scientist is trying to solve. Is the problem land, drilling, completions, reservoir, production, facilities, or combinations of many different disciplines' problem? Let's assume the problem is a production-related ML problem. The first step is to find an experienced production engineer and pair him/her with the data scientist. This is an almost sure way of being on the right track to solving the problem. It is simply because the production engineers know every detail about a subject matter (in this case production engineering and operations). Let's just say the subject matter that is trying to be solved is a plunger lift optimization. The production engineer knows the exact detail about the parameters affecting plunger cycles and details of practical plunger functioning. On the other hand, the data scientist understands the algorithms, and underlying statistics behind them. Therefore, feature selection is tied directly with domain expertise.

In some occasions, both engineer and data scientist do not necessarily know the impact of each feature and are unsure about some of the features that must be selected in a ML project. Therefore, it is important to perform a feature ranking to observe the impact of each feature on the desired output. There are many algorithms such as random forest (RF), extreme random forest or extra trees (XRF), gradient boost (GB), as well as many other algorithms available in the literature to perform feature ranking. These algorithms and Python application of each algorithm will be discussed in detail under the supervised learning chapter.

Feature collinearity is also very important when performing feature selection. As was discussed in Chapter 2, please ensure to plot a Pearson correlation coefficient heat map of input variables to study the features collinearity. If the features are collinear, providing the model with the same information could potentially result in model confusion. Simply drop one of the collinear inputs. If both inputs are important to understand, it is advised to train two separate models with each collinear feature.

Scaling, normalization, or standardization

To make sure the learning algorithm is not biased to the magnitude of the data, the data (input and output features) must be scaled. This can also speed up the optimization algorithms such as gradient descent that will be used in model development by having each of the input values in roughly the same range. Please note that not all ML algorithms will require feature scaling. The general rule of thumb is that algorithms that exploit distances or similarities between data samples, such as artificial neural network (ANN), KNN, support vector machine (SVM), and k-means clustering, are sensitive to feature transformations. However, some tree-based algorithms such as decision tree (DT), RF, and GB are not affected by feature scaling. This is because tree-based models are not distance-based models and can easily handle varying range of features. This is why tree-based models are referred to as "scale-invariant" due to not having to perform feature scaling. Each of the mentioned algorithms will be discussed in detail in the upcoming chapters. The three main types of scaling are referred to as:

- **Feature normalization**: Feature normalization guarantees that each feature will be scaled to [0,1] interval. This is a critical step simply due to the fact that a model will be biased when features have different magnitudes. For instance, if two of the features in a ML model are surface frac treating rate and surface frac treating pressure, surface frac treating rate can range between 60 and 110 bpm while surface frac treating pressure can range between 8000 and 12,500 psi. Leaving these features without scaling between 0 and 1 would result in a model bias towards frac treating pressure which has larger values. One of the primary disadvantages of feature normalization is that smaller standard deviation is obtained when the bound is between 0 and 1 which could result in suppressing the effects of outliers (if any). The equation for feature normalization is illustrated in Box 3.1.
- **Feature standardization (z-Score normalization)**: Standardization transforms each feature with Gaussian distribution to Gaussian distribution with a mean of 0 and a standard deviation of 1. Note that standardization does not transform the underlying distribution structure of the data. Some learning algorithms, such as SVM, assume that data are distributed around 0 with the same order of variance. If this condition is not met, then the algorithm will be biased towards features with a larger variance. Feature standardization can be performed using Box 3.2.

The next question that is often asked is whether to use normalization or standardization. The answer of this question depends on the application. In clustering algorithms such as k-means clustering, hierarchical clustering, density-based spatial clustering of applications with noise (DBSCAN), etc., standardization is important due to comparing feature similarities based on distance measures. On the other hand, normalization becomes more important with certain algorithms such as ANN and image processing (Raschka, n.d.).

BOX 3.1 Feature normalization

$$Feature\ normalization(X') = \frac{X - min(X)}{max(X) - min(X)} \tag{3.1}$$

where X' is the normalized data point, $min(X)$ is the minimum of the data set, and $max(X)$ is the maximum of the data set.

BOX 3.2 Feature standardization

$$Feature\ standardization(X') = \frac{X - \mu}{\sigma} \qquad (3.2)$$

where X' is the standardized data point, μ is the mean of the data set, and σ is the standard deviation of the data set.

Cross-validation

Before applying various ML models, cross-validation must be applied when a supervised ML algorithm is used. Various types of ML algorithms will be discussed shortly, therefore, for now just remember that cross-validation is needed when supervised ML models are used. Unsupervised ML models do not require cross-validation because the data are only used for clustering. There are different types of cross-validations and they are as follows:

- **Holdout method**: This is one of the simplest type of cross-validation where a portion of the data is held out from the training data set. For example, in a data set with 100K rows of data, 70% or 70K is used for training the model, and 30K or 30% is used for testing the model. The training and testing splits are chosen at random. Note that a seed number can be defined in Python to replicate any process that relies on drawing random numbers, e.g., the training and testing split. As can be guessed, one of the main disadvantages of the holdout method is that since the training is done randomly, it cannot be defined which data rows to be used for training versus testing. If the data set being used for ML analysis is small, this could lead to entirely different ML model outcome.
- **K-fold cross-validation**: This technique is an extension of the holdout method, where the data set is divided into K subsets and the holdout method is repeated K times. In k-fold cross-validation, the data set will be shuffled randomly first and then, the slices will be used for train/test split. At each round, one of the K slices will be held as test data. For instance, for a data set with 100K rows of data, the k-fold cross-validation is performed as follows when 10-fold is selected:
  - First 10K rows of data is used as the testing set, while the remaining 90K is used as the training set.
  - Next, rows 10K−20K is used as the testing set, and the remaining data set (rows 1K−10K and 20K−100K) is used as the training set.
  - Afterward, rows 20K−30K is used as the testing set, and the remaining data set (rows 1K−20K and 30K−100K) is used as the training set.

- This process is continued 10 times or until the defined k-fold is reached. In other words, each time one of the K subsets is used as test and the remaining K-1 is used as training set.
- Finally, the evaluation metric is averaged over 10 trials (for 10-fold cross-validation).
- This simply indicates that every data point is in the test set one time and in the training set nine times (for 10-fold cross-validation).

  One of the main advantages of k-fold cross-validation is that it reduces bias and variance (will be discussed) since most of the data is being used for both training and testing. Please note that as the number of K increases, the computation time also increases. Therefore, please make sure to perform a quick sensitivity analysis to see whether the problem can be optimized with less number of K's such as fivefold or threefold without deteriorating the model performance. Fig. 3.1 illustrates a fivefold cross-validation.

- **Stratified k-fold cross-validation**: When there is a large imbalance in the output variable (also called the response variable) of a ML model, a stratified k-fold cross-validation could be used. For example, let's assume that 70% of the output of a classification model is classified as "Continue" and 30% is classified as "Evaluate." This is referred to as a **model imbalance** as the number of **output** classification is 70%−30%. Therefore stratified k-fold cross-validation ensures that each fold is representative of all data strata. This process will aim to make sure output distribution is similar across each test fold. In other words, k-fold cross-validation is varied in a manner that each fold contains approximately the same sample percentage of each target class. For instance, in a stratified k-fold cross-validation with 100K rows of data, if 30K is classified as "Evaluate" and 70K is classified as "Continue" and 10-fold cross-validation is used, each fold will have 3K (30K divided by 10-fold) of "Evaluate" class and 7K (70K divided by 10) of "Continue" class. The stratified k-fold cross-validation is not necessary when there is no model imbalance.
- **Leave-P-out cross-validation (LPOCV)**: This is another type of cross-validation where every possible combination of P test data points is evaluated. In this technique, P data points are left out of the training data. If there are N data points, N−P data points are used to train a model and P

FIGURE 3.1   Fivefold cross-validation.

P=2

1st Iteration	Test	Test	Train
2nd iteration	Test	Train	Test
3rd iteration	Train	Test	Test

FIGURE 3.2 LPOCV with a P = 2.

points are used to validate or test a model. This is repeated until all combinations are divided and tested. Afterward, the evaluation metric for all the trials are averaged. For instance, if 100 rows of data is used and $P = 5$, 95 rows (N−P) will be used as training and 5 rows will be used as testing. This process is repeated until all combinations are divided and tested.

Fig. 3.2 can be used to perform LPOCV. This is a data set with only 3 rows of data and $P = 2$ for illustration purpose and is not simply practical in the real-world. Now imagine the computation time that it would take with a data set with 100K rows. Therefore, this technique can provide a very accurate estimate of the evaluation metric but it is very exhaustive for large data sets. Therefore, leave-one-out cross-validation (LOOCV) $(P = 1)$ is more commonly used because it requires less computation time than LPOCV with P greater than 1. In LOOCV, **number of folds** will be equal to the **number of data points**. Therefore, it is recommended to use such cross-validation technique where a small data set is being used. Practically speaking, LPOCV can be very time-consuming and would not be a feasible technique where a project timing as well as the computation power are limited. Fig. 3.3 illustrates LOOCV with 9 iterations.

Blind set validation

In addition to train and test splits, the blind set validation must also be applied to evaluate a model's prediction accuracy. A blind set is a percentage of the data that is completely set aside and is not used in training nor in testing. This allows for a completely unbiased evaluation of a model's accuracy. For example, let's assume that 50 sonic logs are available to build a ML model

Iteration										Performance
1st iteration	Test	Train	Train	Train	Train	Train	Train	Train	Train	1st Iteration Performance
2nd iteration	Train	Test	Train	Train	Train	Train	Train	Train	Train	2nd Iteration Performance
3rd iteration	Train	Train	Test	Train	Train	Train	Train	Train	Train	3rd Iteration Performance
4th iteration	Train	Train	Train	Test	Train	Train	Train	Train	Train	4th Iteration Performance
5th iteration	Train	Train	Train	Train	Test	Train	Train	Train	Train	5th Iteration Performance
6th iteration	Train	Train	Train	Train	Train	Test	Train	Train	Train	6th Iteration Performance
7th iteration	Train	Train	Train	Train	Train	Train	Test	Train	Train	7th Iteration Performance
8th iteration	Train	Train	Train	Train	Train	Train	Train	Test	Train	8th Iteration Performance
9th iteration	Train	Train	Train	Train	Train	Train	Train	Train	Test	9th Iteration Performance

Avg performance

FIGURE 3.3 Leave-one-out cross-validation (LOOCV) illustration.

where shear and compression wave travel times are desired to be predicted. Before performing any analysis, it is recommended to take out 10%−20% of your data and set aside as a **blind set**. Afterward, feed in the remaining 80% −90% of the data and split that into **training** and **testing**. In this example, if 20% is used for the blind set, 20% of 50 logs or 10 logs will be set aside from the beginning and is used as a blind set. The remaining 40 logs are used to feed into the training and testing sets. After training a ML model, apply the trained model to the blind set (10 logs) to see the model's performance. Consequently, in this problem, there will be training set accuracy, testing set accuracy, and blind set accuracy. Please note that in some literature, testing and blind sets are used interchangeably.

Bias−variance trade-off

A very instrumental concept in ML is referred to as bias−variance trade-off. Before digging deep into the concept, let's define bias and variance first as follows:

Bias is referred to as model assumption simplification. In other words, the inability of a ML model to capture the true relationship is called "bias." Models with high bias are unable to capture the true relationship between input and output features, and it usually leads to oversimplification of the model. An example of a model with high bias is when linear regression is used for a nonlinear data set.

Variance refers to the variability of model prediction for a given data point. A model with high variance means it fits the training data very well but does a poor job predicting the testing data. It other words, it memorizes the training data very well and is not able to predict the test data due to low generalization.

If a model is simple (which means it does not capture most of the complexity in the data), it is referred to as an underfit model. An underfit model is another name for a model that has high bias and low variance as illustrated in Fig. 3.4. On the other hand, if a model is super complex and captures all the underlying complexity in the data in a training set, it is called an overfit model. An overfit model means it has high variance and low bias. The ideal model would have **low bias** and **low variance** as shown in Fig. 3.5.

Model development and integration

After completions of steps (i)−(v), the next step is to select the appropriate algorithm to model the problem. Note that in this phase of the ML analysis, various algorithms should be evaluated to be able to pick the most effective algorithm. There is no one solution fits all algorithm and the underlying data will dictate the type of algorithm to be used for the problem. There are

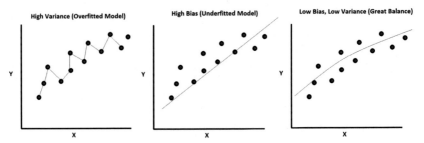

FIGURE 3.4 Overfit, underfit, and good balance model.

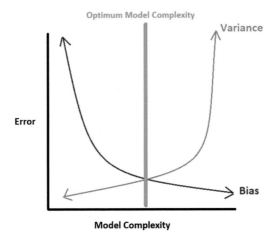

FIGURE 3.5 Optimum model complexity.

different supervised, unsupervised, and reinforcement learning algorithms that are available and discussed below. After developing a successful ML model, save and apply the model. The trained model is now ready to be deployed or integrated into a real-time operations center.

Machine learning types

Supervised learning

In this type of ML, both set of M number of inputs (x_i) and output(s) (y_i) are available. x is an M by N matrix where M is the number of data points and N is the number of input features or attributes. y_i is a vector of response features. A response feature (y_i) in the form of a number indicates a regression problem. Otherwise, it is a classification or pattern recognition problem. Let's assume that x_i is a matrix of 10 features (sand to water ratio (SWR), gas in place (GIP), cluster spacing, proppant loading/cluster, average rate/cluster, geologic complexity, BTU content, well spacing, pressure drawdown strategy, and number of clusters per stage) for 2000 wells drilled and completed in the Haynesville Shale reservoir and the y_i is the estimated ultimate recovery (EUR) per 1000 ft of lateral from these 2000 wells. If the question is to find the pattern between these features and EUR/1000 ft, it can be treated as a regression problem. However, if instead of EUR per 1000 ft of lateral, the y_i indicates having well interference or not having well interference during stimulation of these 2000 wells, then the problem becomes a classification problem. In both cases, a supervised ML algorithm can be used since pairs of input X_i and output y_i are available. Note that a supervised ML algorithm could also have multiple outputs. For the same problem indicated above, the outputs of the model could be defined as 30-day cumulative gas production per ft (CUM 30/ft), CUM 60/ft, ..., CUM 720/ft. Therefore, instead of having a supervised single output ML model with EUR/1000 ft as the output, a multi-output ML model can be trained to predict CUM 30/ft, CUM 60/ft, ..., CUM 720/ft. Examples of supervised ML algorithms are included but not limited to artificial neural network (ANN), support vector machine (SVM), decision tree (DT), random forest (RF), extreme random forest (ERF), gradient boosting (GB), adaptive gradient boost (AGB), multi-linear regression (MLR), logistic regression (LR), and KNN.

Unsupervised learning

In this type of ML, only sets of input features (x_i) are available and there is no associated label. Therefore, existing pattern is sought in the input data. The major difference between unsupervised and supervised technique is that in unsupervised, there are no output(s) to compare the predictions, while in supervised learning, the model predictions could always be compared with actual available output(s). Unsupervised learning can potentially reduce labor within an organization because unsupervised learning algorithms can be used in place of human efforts to filter through large sets of data for clustering purposes. This could lead to improving efficiency. Examples of unsupervised ML algorithms are k-means clustering, hierarchical clustering, DBSCAN, principal component analysis (PCA), and apriori algorithm. An example of using an unsupervised ML algorithm is determining the type curve boundaries within a company's acreage position. E&P companies primarily use production performance, geologic features, land, infrastructure, and capacity to determine type curve boundaries. All of these important features can simply be used as input features into an unsupervised ML model. A clustering algorithm can then be applied to cluster the data. After clustering the data, plot the resulting clusters for each well based on their latitude and longitude to determine proper type curve boundaries for each area of interest. This is an unbiased powerful unsupervised technique to provide a realistic view of type curve boundaries and regions without using human interference.

Semi-supervised learning

Semi-supervised ML is a combination of using **unsupervised** and **supervised** learning. If a large amount of unlabeled data is available, the first step is to use some type of unsupervised ML algorithm to cluster the data (label the data). Afterward, the labeled data obtained from the unsupervised technique can be used as the output for the supervised ML algorithm. Therefore, a combination of unsupervised and supervised algorithms is used to solve a problem. Please note that clustering the data using an unsupervised technique cannot necessarily lead to the best result in term of obtaining a desired labeled class. Therefore, massaging the data prior to using an unsupervised technique is extremely crucial to obtain the desired labeled class. In some cases, it might take less time to use domain expertise to label the data prior to using it in a supervised ML algorithm.

Reinforcement learning

Reinforcement learning is a learning technique that directs the action to maximize the reward of an immediate action and those following. In this type of ML algorithm, the machine trains itself continuously by using a computational approach to learning from action. Reinforcement learning is different from supervised learning because labeled input/output pairs are not needed to be present. Imagine that you are a 9-year-old basketball player watching your mentors dunking to score. Then you manage to lower the basketball hoop and use it to dunk for practicing. You have learned that dunking can be used for the positive action of scoring. Now, you try to raise the hoop and dunk again and manage to twist your ankle while landing. You then realize that it can be used for the negative action of hurting yourself when used inappropriately. This learning procedure helps the kid to learn proper use of a basketball hoop for positive and not negative action. Humans learn by interaction and reinforcement learning. Learning through trial and error with prize and punishment is the most important feature of reinforcement learning. Reinforcement learning can also be used in the O&G industry for various purposes. An example of using reinforcement learning is for producing plunger or intermittent wells. For example, reinforcement learning can be used to adjust production set points as to when to open or close a well. This can be achieved by continuous model learning from set points that lead to poor production performance (bad behavior) or set points that lead to excellent production performance (good behavior). After learning the good and bad behaviors, the set points can be automatically optimized without human intervention. Coupling reinforcement learning and edge computing devices can add a tremendous value across various O&G companies and sectors. Two of the most commonly used reinforcement learning algorithms are Markov decision process (MDP) and Q-learning.

Dimensionality reduction

Dimensionality reduction is the process of reducing the number of variables by obtaining a set of important variables. Often times when features are heavily related, the number of features can simply be reduced using some type of dimensionality reduction algorithms as opposed to using all collinear features. Dimensionality reduction is an effective approach when many correlated attributes exist within a data set. For instance, many geologic features that are collinear can be reduced to only a few important features. Examples of dimensionality reduction algorithms are PCA, nonnegative matrix factorization (NMF), and stochastic neighbor embedding (SNE).

Principal component analysis (PCA)

The desired goal in PCA is to find patterns in a data set and to reduce the dimensionality of collinear features in an attempt to increase computational efficiency while retaining all important information. The following steps are needed to perform PCA:

- Standardize the data set as discussed in the workflow section.
- Calculate the covariance matrix (also known as variance–covariance matrix).
- Calculate eigenvectors and eigenvalues from the covariance matrix.
- Sort the eigenvalues in a descending order and choose **n eigenvectors.** n is the number of dimensions after applying PCA.
- Calculate the principal components and reduce the dimensions of the data set.

PCA concept is easier to understand with illustrating a step-by-step example. Therefore, let's examine Table 3.1. As illustrated in this table, permeability values for matrix, natural fracture, and effective **permeability** values are shown for five different wells.

TABLE 3.1 Permeability properties (nd).

Matrix permeability	Natural fracture permeability	Effective permeability
30	75	120
45	90	145
20	65	120
100	140	185
50	90	130

1) The first step in PCA is standardizing the data set; however, for the sake of working with easier numbers for this illustration purpose, this step was skipped. Please make sure this is done prior to performing any real-life example or the PCA will be inaccurate.

2) The data from the table can be rewritten as matrix X:

$$X = \begin{bmatrix} 30 & 75 & 120 \\ 45 & 90 & 145 \\ 20 & 65 & 120 \\ 100 & 140 & 185 \\ 50 & 90 & 130 \end{bmatrix} \qquad (3.3)$$

3) Calculate the covariance matrix of X using the equation below:

$$cox(X, Y) = \frac{1}{n-1} \sum_{i=1}^{n} (X_i - \overline{X})(Y_i - \overline{Y}) \qquad (3.4)$$

The covariance matrix of the following scenarios must be calculated:
- Covariance(matrix perm, matrix perm), Covariance(natural fracture perm, matrix perm), Covariance(effective perm, matrix perm), Covariance(matrix perm, natural fracture perm), Covariance(natural fracture perm, natural fracture perm), Covariance(effective perm, natural fracture perm), Covariance(matrix perm, effective perm), Covariance(natural fracture perm, effective perm), Covariance(effective perm, effective perm)

In other words, Table 3.2 illustrates the covariance matrix of matrix X:

Therefore, the covariance of matrix X is as follows:

$$Covariance\ of\ matrix\ X = \begin{bmatrix} 764 & 712 & 645 \\ 712 & 666 & 610 \\ 645 & 610 & 590 \end{bmatrix} \qquad (3.5)$$

TABLE 3.2 Covariance matrix.

Perm covariance matrix	Matrix perm	Natural fracture perm	Effective perm
Matrix perm	764	712	645
Natural fracture perm	712	666	610
Effective perm	645	610	590

As illustrated in the diagonal in the covariance matrix X above, 764 is the matrix perm covariance which is the highest number as compared to natural fracture perm (666) and effective perm (590). This essentially indicates higher variability in matrix perm as compared to natural fracture perm and effective perm. There is positive covariance between all perm values. This indicates as one perm increases, the other will also increase.

4) Calculate eigenvalues and vectors of the covariance matrix shown above. Eigenvalues of a square matrix X can be calculated as follows:

$$Determinant(X - \lambda I) \tag{3.6}$$

where λ is eigenvalue, and I is an identity matrix. Let's plug in matrix X in the above equation as follows:

$$det\left(\begin{pmatrix} 764 & 712 & 645 \\ 712 & 666 & 610 \\ 645 & 610 & 590 \end{pmatrix} - \lambda \begin{pmatrix} 1 & 0 & 0 \\ 0 & 1 & 0 \\ 0 & 0 & 1 \end{pmatrix}\right) \tag{3.7}$$

Simplify the matrix excluding determinant as follows:

$$\begin{pmatrix} 764 & 712 & 645 \\ 712 & 666 & 610 \\ 645 & 610 & 590 \end{pmatrix} - \begin{pmatrix} \lambda & 0 & 0 \\ 0 & \lambda & 0 \\ 0 & 0 & \lambda \end{pmatrix} \xrightarrow{Simplify} \begin{pmatrix} 764 - \lambda & 712 & 645 \\ 712 & 666 - \lambda & 610 \\ 645 & 610 & 590 - \lambda \end{pmatrix} \tag{3.8}$$

Add determinant back to the simplified matrix:

$$det\begin{pmatrix} 764 - \lambda & 712 & 645 \\ 712 & 666 - \lambda & 610 \\ 645 & 610 & 590 - \lambda \end{pmatrix} \xrightarrow{\textit{Calculate determinant and place the equation equal to 0}}$$

$$-\lambda^3 + 2020\lambda^2 - 57455\lambda + 24950 = 0$$

$lambda_1 \approx 0.441$

$lambda_2 \approx 28.408$

$lambda_3 \approx 1991.151$

$$\tag{3.9}$$

Lambda 1, 2, and 3 represent the eigenvalues. Now it is time to calculate eigenvectors of the eigenvalues shown above. Eigenvectors are denoted as V_1, V_2, and V_3 below and were calculated using the Cramer's rule.

$$v_1 \approx \begin{pmatrix} 4.029 \\ -5.227 \\ 1 \end{pmatrix}$$

$$v_2 \approx \begin{pmatrix} -0.608 \\ -0.278 \\ 1 \end{pmatrix} \tag{3.10}$$

$$v_3 \approx \begin{pmatrix} 1.152 \\ 1.079 \\ 1 \end{pmatrix}$$

5) Sort the eigenvectors by decreasing eigenvalues calculated in step 4.

$$\begin{pmatrix} 1991.151 \\ 28.408 \\ 0.441 \end{pmatrix} \tag{3.11}$$

For this example, let's reduce the original three-dimensional feature space to two-dimensional feature subspace. Therefore, the eigenvectors associated with the maximum two eigenvalues of 1991.151 and 28.408 are as follows:

$$K = \begin{matrix} & e1 & e2 \\ \begin{bmatrix} 1.152 & -0.608 \\ 1.079 & -0.278 \\ 1 & 1 \end{bmatrix} \end{matrix} \tag{3.12}$$

$e1$ and $e2$ represent the top two eigenvectors. Finally, the last step is to take the 3×2 dimensional K matrix and transform it using the following equation:

$$Transformed\ data = Feature\ matrix \times K \tag{3.13}$$

Feature matrix represents the original standardized matrix (for this example the perm matrix was not standardized) and K represents the top K eigenvectors.

PCA using scikit-learn library

Now that the concept of PCA component is clear, let's go over the Python code and apply PCA to a completions data set in an attempt to reduce the dimensionality from 4 to 2 or 3 parameters. In a new Jupyter Notebook, let's import

	Stage Spacing	Cluster Spacing	Sand per ft (# per ft)	Water per ft (gal per ft)
count	144.000000	144.000000	144.000000	144.000000
mean	197.908333	39.713194	2949.275000	57.986111
std	28.411963	5.702597	1414.559452	37.618435
min	146.200000	26.000000	798.000000	5.000000
25%	173.400000	36.400000	1197.000000	15.000000
50%	193.800000	39.000000	3351.600000	65.000000
75%	217.600000	42.900000	4069.800000	90.000000
max	268.600000	57.200000	5506.200000	125.000000

FIGURE 3.6 df properties description.

the important libraries that will be used for PCA. Afterward, import "Chapter3_Completions_DataSet" located at the following link:

https://www.elsevier.com/books-and-journals/book-companion/
9780128219294

```
import numpy as np
import pandas as pd
import matplotlib.pyplot as plt
import seaborn as sns
from scipy import stats
%matplotlib inline
df=pd.read_excel('Chapter3_Completions_DataSet.xlsx')
df.describe()
Python output=Fig. 3.6
```

The next step is to standardize the data set using "standard scaler" library. Let's import the standard scaler library and apply it to the data frame "df." Please note that "scaler.transform(df)" applies the standardization to the df data set. When this process is done, the data are transferred into a numpy array. Therefore, the next step is to change the scaled features (standardized data) from a numpy array to a data frame for ease of using in the PCA.

```
from sklearn.preprocessing import StandardScaler
scaler=StandardScaler()
scaler.fit(df)
scaled_features=scaler.transform(df)
scaled_features=pd.DataFrame(scaled_features, columns=['Stage
Spacing', 'Cluster Spacing', 'Sand per ft (# per ft)', 'Water per ft
(gal per ft)'])
scaled_features.head()
Python output=Fig. 3.7
```

	Stage Spacing	Cluster Spacing	Sand per ft (# per ft)	Water per ft (gal per ft)
0	-0.865617	1.018309	-1.299677	-1.280053
1	-1.105788	-0.125501	-1.299677	-1.280053
2	-1.345959	0.332023	-1.356287	-1.280053
3	-1.466045	0.103261	-1.243066	-1.280053
4	-0.985703	1.247071	-1.299677	-1.280053

FIGURE 3.7 Standardized features with a mean of 0 and a standard deviation of 1.

Next, let's import PCA from "sklearn.decomposition" and apply two principal components to the scaled features as follows. Please note that n_components show the number of principal components that would like to be achieved. For this example, two principal components have been achieved which essentially indicate transforming the data set from four features to two features and reducing the data sets' dimensionality. Please note that this library's default uses singular value decomposition (SVD) to improve computational efficiency which is the LAPACK implementation of the full SVD or a randomized truncated SVD (Halko et al., 2009).

```
from sklearn.decomposition import PCA
PCA= PCA(n_components=2)
PCA.fit(scaled_features)
Transformed_PCA=PCA.transform(scaled_features)
Transformed_PCA
```
Python output=Fig. 3.8

```
Transformed_PCA

array([[-2.20401503,  0.49230415]
       [-2.01847823, -0.65257072]
       [-2.3001967 , -0.3263588 ]
       [-2.23481144, -0.5804819 ]
       [-2.32844843,  0.65610591]
       [-2.01570154,  1.4935046 ]
       [-2.37840516,  0.05716489]
       [-2.17148144,  0.23678942]
       [-2.26829929, -1.09568622]
       [-2.12311515, -0.44880128]
       [-2.10815144,  1.05406075]
       [-2.26338126,  0.14507647]
       [-2.15628491, -0.70709925]
       [-2.56671864, -0.94403556]
       [-2.14337646,  1.8671572 ]
       [-2.20456595,  2.68214851]
       [-2.14710823,  1.48793811]
       [-2.12858409,  0.50028039]
       [-1.84194759,  1.41483116]
```

FIGURE 3.8 Transformed two principal components.

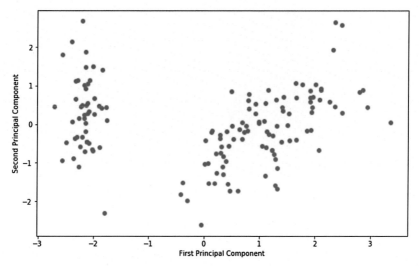

FIGURE 3.9 Plot of two principal components.

The "Transformed_PCA" show the two principal components in a numpy array. Next, let's plot these generated principal components as follows:

```
plt.figure(figsize=(10,6))
plt.scatter(Transformed_PCA[:,0],Transformed_PCA[:,1])
plt.xlabel('First Principal Component')
plt.ylabel('Second Principal Component')
Python output=Fig. 3.9
```

To get the eigenvectors of each principal component which are the co-efficients of each variable for the first component versus the coefficients for the second component, "**PCA.components_**" can be used in Python. These scores provide an insight on the variables that have the largest impact on each of the generated two principal components. These scores can range between −1 and 1 with scores of −1 and 1 indicating the largest influence on the components.

```
df_components=pd.DataFrame(PCA.components_,columns=['Stage
Spacing', 'Cluster Spacing', 'Sand per ft (# per ft)', 'Water per ft
(gal per ft)'])
df_components
Python output=
```

	Stage Spacing	Cluster Spacing	Sand per ft (# per ft)	Water per ft (gal per ft)
0	0.519427	-0.271276	0.580313	0.565545
1	0.387659	0.919532	0.024583	0.059802

PCA components.

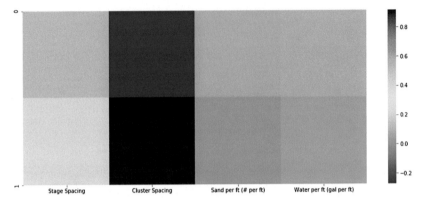

FIGURE 3.10 Heat map of principal components.

As illustrated, cluster spacing has the largest impact on the second principal component with a **score of 0.919532**. Let's also visualize this using a heat map via the seaborn library.

```
plt.figure(figsize=(15,6))
sns.heatmap(df_components,cmap='coolwarm')
Python output=Fig. 3.10
```

The heat map illustrates the relationship between correlations of various features and the principal components. Each principal component is shown as a row and the principal components with the higher scores are more correlated to a specific feature in the columns. This helps identifying the features that are specifically important for each principal component. The reduced versions of these principal components can now be used for feeding into a ML model. Just remember that if the reduced features are used, it is hard to draw conclusion on the significance of those features after building a ML model. For example, if eight features were reduced to two features and most of the data is maintained using the reduced dimensionality, it is important to note that understanding the impact of those features on the target variable would be difficult. This is one of the main drawbacks with PCA. Therefore, if the goal is to understand the sensitivity of a particular variable (feature) on the output of a ML model in a regression or a classification scenario, it is important to keep those features that are going to be used for sensitivity analysis. Finally, to export the two generated principal components into a CSV file, let's first transform the numpy array to a data frame followed by writing it into a csv file in the same directory that the Jupyter Notebook is located.

```
transformed_PCA=pd.DataFrame(Transformed_PCA, columns=['First
Principal Component', 'Second Principal Component'])
transformed_PCA.to_csv('Two Principal Components.csv')
Python output=A CSV file will get generated in the same folder that
Jupyter Notebook resides.
```

Nonnegative matrix factorization (NMF)

The main goal in NMF is to decompose a matrix into two matrices. NMF is a matrix factorization technique. As was previously discussed, PCA creates factors that can be both positive and negative; however, NMF only creates positive factors. PCA is desirable when a high-dimensional data set is transferred into low dimensions as long as losing some of the original features are acceptable. Some of the applications of NMF is in image processing, transcription processes, text mining, cryptic encoding/decoding, and decomposition of videos and music. Let's assume matrix A is factorized into two matrices called X and Y as follows:

$$A \approx X \times Y \tag{3.14}$$

Therefore, for matrix A of dimension **mbyn** ($m \times n$) (each element of the matrix must be greater than 0 meaning $a_{ij} >= 0$), NMF decomposes matrix A into two matrices X and Y of dimensions **mbyr** ($m \times r$) and **rbyn** ($r \times n$) subsequently where each element $x_{ij} >= 0$ and $y_{ij} >= 0$.

$$\begin{bmatrix} \\ \\ \\ \end{bmatrix} \times \begin{bmatrix} \\ \\ \end{bmatrix} = \begin{bmatrix} \\ \\ \\ \end{bmatrix} \tag{3.15}$$
$$\qquad X \qquad\qquad Y \qquad\qquad\qquad A$$

For example, let's assume that A is composed of m rows of $a_1, a_2, \dots, a_m$, X is composed to n rows of $x_1, x_2, \dots, x_n$, and Y is composed of K rows of $y_1, y_2, \dots,$ and y_k. Each row in A can be considered as a data point.

$$A = \begin{bmatrix} a1 \\ a2 \\ a3 \\ \dots \\ an \end{bmatrix} \quad X = \begin{bmatrix} \bar{x}1 \\ \bar{x}2 \\ \bar{x}3 \\ \dots \\ \bar{x}n \end{bmatrix} \quad Y = \begin{bmatrix} y1 \\ y2 \\ y3 \\ \dots \\ yn \end{bmatrix} \tag{3.16}$$

where $\overline{xi} = [xi1\, xi2 \cdots xin]$. Let's summarize as follows:

$$ai = \begin{bmatrix} xi1 & xi2 & \dots & xin \end{bmatrix} \times \begin{bmatrix} y1 \\ y2 \\ \dots \\ yn \end{bmatrix} = \sum_{j=1}^{n} xij \times yj \tag{3.17}$$

The objective in NMF is to minimize the squared distance of *A-XY* with respect to *X* and *Y*, noting that *X* and *Y* must be greater or equal to 0. In other words, the objective is to minimize the following function:

$$\|A - XY\|^2 \tag{3.18}$$

Note that Euclidean or Frobenius (will be discussed in later chapters) can be used to minimize such function. There are two popular approaches for solving this problem numerically. The first approach is called coordinate descend where *X* is fixed and *Y* is optimized with gradient descend. Then *Y* is fixed and *X* is optimized with gradient descend. This process is repeated until a tolerance is met. The other approach is called multiplicative technique. Please note that in both methodologies, a local minimum is found which are often useful enough (Cichocki & Phan, 2009).

Nonnegative matrix factorization using scikit-learn

Let's import the same data set used for PCA but this time apply NMF. If a new Jupyter Notebook is being created for this exercise, make sure to import pandas and numpy libraries prior to proceeding to the next step. Afterward, let's normalize the data by importing the preprocessing library from sklearn package. This library is used for data normalization. Please note that "feature_range=(0,1)" will essentially normalize the data to be between 0 and 1 using the normalization formula discussed earlier in this chapter.

```
df=pd.read_excel('Chapter3_Completions_DataSet.xlsx')
from sklearn import preprocessing
scaler=preprocessing.MinMaxScaler(feature_range=(0,1))
scaler.fit(df)
df_scaled=scaler.transform(df)
# After applying data normalization, df_scaled will be in an array
format. Therefore, use pandas library to convert into a data frame as
shown below.
df_scaled=pd.DataFrame(df_scaled, columns=['Stage Spacing',
'Cluster Spacing', 'Sand per ft (# per ft)', 'Water per ft (gal
per ft)'])
df_scaled
Python output=Fig. 3.11
```

Next, let's import the NMF library from "sklearn.decomposition" and apply two components to the df_scaled data set. Afterward, apply NMF using two components, cd solver (coordinate descent solver), and frobenius loss function. Please note that "mu" which is a multiplicative update solver can also be used under solver.

	Stage Spacing	Cluster Spacing	Sand per ft (# per ft)	Water per ft (gal per ft)
0	0.222222	0.625000	0.067797	0.041667
1	0.166667	0.416667	0.067797	0.041667
2	0.111111	0.500000	0.050847	0.041667
3	0.083333	0.458333	0.084746	0.041667
4	0.194444	0.666667	0.067797	0.041667
...	...	...	...	...
139	0.722222	0.458333	0.745763	0.833333
140	0.666667	0.458333	0.779661	0.958333
141	0.722222	0.458333	0.694915	0.916667
142	0.416667	0.291667	0.694915	0.750000
143	0.694444	0.500000	0.830508	0.916667

FIGURE 3.11   Normalized features.

	Stage Spacing	Cluster Spacing	Sand per ft (# per ft)	Water per ft (gal per ft)
0	1.4408	0.1877	1.9898	1.9848
1	0.6368	2.3017	0.0526	0.0000

FIGURE 3.12   NMF components.

```
from sklearn.decomposition import NMF
nmf=NMF(n_components=2, init=None, solver='cd',beta_loss=
'frobenius',random_state=100)
nmf_transformed=nmf.fit_transform(df_scaled)
df_scaled_components=pd.DataFrame(np.round(nmf.components_,4),
columns=df_scaled.columns)
df_scaled_components
Python output=Fig. 3.12
```

As observed, the created components are all positive. Let's plot these components in a heat map as follows (make sure the visualization libraries are imported if a new Jupyter Notebook is used):

```
plt.figure(figsize=(15,6))
sns.heatmap(df_scaled_components,cmap='coolwarm')
Python output=Fig. 3.13
```

Finally, the final reduced two-dimensional components can be obtained as follows:

```
nmf_transformed.
Python output=Fig. 3.14.
```

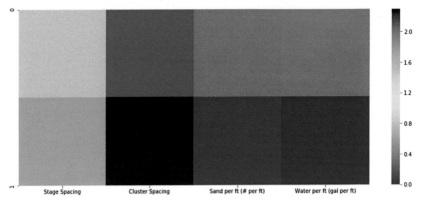

FIGURE 3.13 Heat map of NMF components.

	0	1
0	0.026080	0.270967
1	0.027337	0.180380
2	0.013042	0.211297
3	0.018034	0.190421
4	0.020670	0.286060
...	...	...
139	0.401350	0.169978
140	0.426674	0.157624
141	0.408085	0.167790
142	0.340134	0.083775
143	0.430479	0.177035

FIGURE 3.14 Final reduced two-dimensional components using NMF.

References

Cichocki, A., & Phan, A.-H. (2009). *Fast local algorithms for large scale nonnegative matrix and tensor factorizations* (Vol.E92-A (3), pp. 708−721). The Institute of Electronics, Information, and Communication Engineers. https://doi.org/10.1587/transfun.E92.A.708.

Halko, N., Martinsson, P.-G., & Tropp, J. (2009). Society of industrial and applied mathematics. In *Finding structure with randomness: Probabilistic algorithms for constructing approximate matrix decompositions*. https://arxiv.org/pdf/0909.4061.pdf.

Raschka, S. (n.d.). *About feature scaling and normalization*. Retrieved February 15, 2020, from https://sebastianraschka.com/Articles/2014_about_feature_scaling.html#z-score-standardization-or-min-max-scaling. (Original work published 2014).

Chapter 4

Unsupervised machine learning: clustering algorithms

Introduction to unsupervised machine learning

As discussed in Chapter 3, unsupervised **Machine Learning (ML)** algorithms can be powerfully applied in clustering analysis. There are different types of clustering algorithms that are most commonly used. Some of the applications of clustering algorithms in the O&G industry are liquid-loading detection using k-means clustering proposed by Ansari et al., area of interest (type curve) clustering, and lithology classification clustering. The idea behind using a clustering algorithm is to cluster or partition the data. For instance, if dividing a producing gas well into its loaded versus unloaded condition is the desirable outcome, a clustering algorithm can be used to make this division. One might ask Turner and Coleman rates can be used for such segmentation using empirical equations, but those methodologies were developed for vertical conventional wells that are not applicable in deviated and horizontal wells with multistage hydraulic fracturing. The use of a clustering technique allows one to perform such analysis independent of empirical equations developed many decades ago and use the power of data to segment each well to loaded versus unloaded condition. Another example of using an unsupervised clustering technique in the O&G industry is type curve clustering. E&P operators use their knowledge of the area such as geologic features, production performance, BTU content, proximity to pipeline, etc., to define each formation's type curve area and boundaries. As can be imagined, this process can be very difficult to comprehend, considering all the features that would affect the outcome. Therefore, the power of data and unsupervised ML algorithms can be used to cluster like-to-like or similar areas together. All of the aforementioned features can be used as the input features of the unsupervised ML model, and the clustering algorithm will indicate to which cluster each row of data will belong. Another application of unsupervised algorithms in the O&G is lithology classification. Both supervised and unsupervised ML models can be used for lithology classification. If unsupervised ML models are used for this purpose, the idea is to automate the process of identifying geologic formations

Machine Learning Guide for Oil and Gas Using Python. https://doi.org/10.1016/B978-0-12-821929-4.00002-0

such as sandstone, limestone, shale, etc., using an automated technique as opposed to manually having a geologist going through the process. If a model has been clustered properly, these clustering algorithms should provide similar classification results when compared to human interpretation. Clustering algorithms can also be used for clustering one second frac stage data to help screen out detection. In all of these applications, it is important to note that some of the algorithms will not yield the desirable outcome on the first trial. It is crucial to change/massage the data until the algorithm understands the best way to cluster the data. For example, when clustering the data for liquid-loading detection, feeding the model the basic properties such as gas rate, casing pressure, tubing pressure, and line pressure might not suffice. Other parameters such as slope and standard deviation of the gas rate might be necessary for the model to correctly cluster the data based on domain expertise. ML analysis heavily relies on domain expertise to understand and deeply comprehend the problem. Without domain expertise, most ML projects could potentially lead to unsatisfactory results. In this chapter, k-means clustering, hierarchical clustering, and density-based spatial clustering of applications with noise (DBSCAN) will be discussed. In addition, other unsupervised anomaly detection algorithms will be discussed.

K-means clustering

K-means is a powerful yet simple unsupervised ML algorithm used to cluster the data into various groups. K-means clustering can also be used for outlier detection. One of the biggest challenges of k-means clustering is identifying the optimum number of clusters that must be used. Domain expertise plays a key role in determining the number of clusters. For example, if the intent is to cluster the data into loaded versus unloaded conditions for dry gas wells, the domain expertise instructs one to use two clusters. On the other hand, if clustering is desired to be used to cluster type curve regions, various number of clusters should be applied and the results visualized before determining the number of clusters needed to define the type curve boundaries. If a problem is supercomplex and determining the number of clusters is simply not feasible, the "elbow" method can be used to get some understanding of the potential number of clusters that will be needed. In this approach, k-means algorithm is run multiple times at various clusters (2 clusters, 3 clusters, etc.). Afterward, plot "number of clusters" on the x-axis versus "within cluster sum of squared errors" on the y-axis until an elbow point is observed. Increasing the number of clusters beyond the elbow point will not tangibly improve the result of the k-means algorithm. Fig. 4.1 illustrates the elbow point or the optimum number of clusters that occurred with 4 clusters. This method provides some insight into the number of clusters to choose, but it is very crucial to test higher

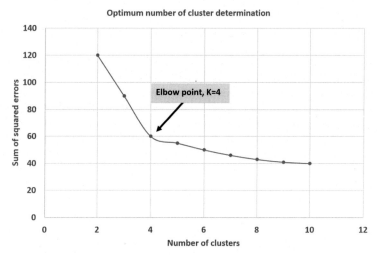

FIGURE 4.1 Cluster determination using elbow point.

number of clusters in the event that more granularity is needed. On the other hand, if your particular domain expertise indicates that less number of clusters is needed, the domain expertise will supersede the elbow point methodology. The notion of the elbow point technique is to provide guidance when unsure of the number of clusters that will be needed, and it is not to provide an exact solution to a problem.

Please note that k-means algorithm's goal is to select centroids that will lead to minimizing "inertia." Another term used in lieu of "inertia" is "within cluster sum of squared errors" and it is defined **for cluster** j as follows:

$$\sum_{i=0}^{n} \min\left(\left\|x_i - \mu_j\right\|^2\right) \tag{4.1}$$

x_i is referred to as **the** ith **instance in cluster** j and μ_j is referred to as the mean of the samples **or "centroid" of cluster** j. Inertia essentially measures how internally consistent clusters are. Please note that a lower inertia number is desired. 0 is optimal; however, in a high-dimensional problem, inertia could be high. In those problems, applying PCA prior to k-means clustering could alleviate this problem which is often referred to as "curse of dimensionality" (Clustering, n.d.).

After applying k-means clustering, visualization is the key to making sure the desired clustering outcome is achieved. Another useful approach in using k-means clustering is clustering (labeling) or partitioning the data prior to feeding the labeled data as the output of a supervised ML algorithm. This is the semi-supervised ML approach that was discussed.

How does K-means clustering work?

Understanding k-means clustering is actually very simple when broken down into the step-by-step procedure. Therefore, let's go over the step-by-step procedure when applying k-means clustering to any data:

1) Standardize the data since similarities between features based on distance measures are the key in k-means clustering. Hence, having different scales can skew the result in favor of features with larger values.

2) Determine the number of clusters, using (i) elbow method, (ii) silhouette, or (iii) hierarchical clustering. If unsure, write a "for loop" to calculate the sum of squared errors versus number of clusters. Afterward, determine the number of clusters that will be used.

3) The initialization of centroids within a data set can be either **initialized** randomly or selected purposefully. The default initialization method for most open-source ML software including Python's scikit learn library is random initialization. If random initialization does not work, carefully selecting the initial centroids could potentially help the model.

4) Next, find the distance between each data point (instance) and the randomly selected (or carefully selected) cluster centroids. Afterward, assign each data point to each cluster centroid based on the distance calculations (euclidean is commonly used) presented below. For example, if two centroids have been randomly selected within a data set, the model will calculate the distance from each data point to centroid #1 and #2. In this case, each data point will be clustered under either centroid #1 or #2 based on the distance from each data point to randomly initialized cluster centroids. There are various ways to calculate the distance. This step can be summarized as assigning each data point to each cluster centroid based on their distance. The ones closest to centroid #1 will be assigned to centroid #1 and the ones closest to centroid #2 will be assigned to centroid #2. The following distance functions are commonly used, the most common being the euclidean distance function:

$$\text{Euclidean distance function} = \sqrt{\sum_{k=1}^{n} \left(x_{ik} - \mu_{jk}\right)^2}$$

$$\text{Manhattan distance function} = \sum_{k=1}^{n} \left|x_{ik} - \mu_{jk}\right|$$

$$\text{Minkowsky distance function} = \left(\sum_{k=1}^{n} \left(\left|x_{ik} - \mu_{jk}\right|\right)^q\right)^{1/q}$$

(4.2)

where assuming each instance has n features, the distance of the ith instance (xi) to centroid of cluster j (μj) can be calculated. q represents the order of the norm. The scenario where q is equal to 1 represents Manhattan distance and the case where q is equal to 2 represents Euclidean distance.

To illustrate the euclidean distance calculation for a data set with 3 features, let's apply the euclidean distance to the following two vectors: $(3,9,-5)$ and $(8,-1,12)$

$$\sqrt{\sum_{i=1}^{k} (x_i - y_i)^2} = \sqrt{(8-3)^2 + (-1-9)^2 + (12-(-5))^2} \qquad (4.3)$$

$$= \sqrt{25 + 100 + 289} = \sqrt{414} = 20.35$$

5) Afterward, find the average value of instances assigned to each cluster centroid (that was assigned in step 4) and recalculate a new centroid for each cluster by moving the cluster centroid to the average of instances' average for each cluster.

6) Since new centroids have been created in step 5, in this step, reassign each data point to the newly generated centroids based on one of the distance functions.

7) Steps 5 and 6 are repeated until the model converges. This indicates that additional iteration will not lead to significant modification in the final centroid selection. In other words, cluster centroids will not move any further.

Fig. 4.2 illustrates a step-by-step workflow on applying k-means clustering on a two-dimensional space data set.

K-means like many other algorithms is not flawless and some of the main disadvantages of k-means are as follows:

- K-means is very sensitive to outlier points. Therefore, before applying k-means clustering, make sure to investigate the outliers. This goes back to one of the first steps in applying any ML algorithm which was data visualization. If the outliers are invalid, make sure to remove them prior to using k-means algorithm. Otherwise, use an algorithm such as DBSCAN (will be discussed) which is more robust to noise.
- K-means requires the number of clusters to be defined. This could also be classified as a disadvantage.

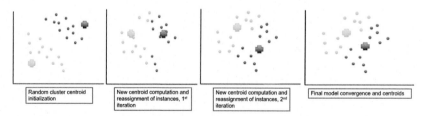

| Random cluster centroid initialization | New centroid computation and reassignment of instances, 1st iteration | New centroid computation and reassignment of instances, 2nd iteration | Final model convergence and centroids |

FIGURE 4.2   K-means clustering illustration.

K-means clustering application using the scikit-learn library

One of the discussed applications of k-means clustering is type curve clustering. Type curve clustering can be determined based on various factors such as geologic features, production performance, BTU, etc. The current industry practice is using production performance knowledge and geologic domain expertise to define the type curve boundaries. While this might be an acceptable solution, using a clustering algorithm such as k-means could provide much deeper insights into type curve boundaries without human cognition bias. Therefore, below is a data set that contains 438 wells with their respective geological features. Although other features such as production performance could be added to this analysis, the goal of this clustering is solely using geologic features to cluster the data. This synthetically generated data set can be obtained using the link below:

https://www.elsevier.com/books-and-journals/book-companion/9780128219294

This data set includes seven parameters including gamma ray, bulk density, resistivity, water saturation, Phih (porosity*thickness), TOC (total organic content), and TVD. The excel file is called "Chapter4_Geologic_DataSet." Let's start by importing the necessary libraries as follows:

```
import pandas as pd
import numpy as np
import matplotlib.pyplot as plt
import seaborn as sns
%matplotlib inline
```

Next, let's import the excel data set called "Chapter4_Geologic_DataSet" and start visualizing the distribution of each parameter to understand the underlying data that will be used for clustering analysis.

```
df=pd.read_excel('Chapter4_Geologic_DataSet.xlsx')
df.describe()
```
Python output=Fig. 4.3

To visualize the distribution of each parameter, use the code below and change the column name to plot each feature.

	GR_API	Bulk Density, gcc	Resistivity, ohm-m	Water Saturation, fraction	PhiH, ft	TOC, fraction	TVD, ft
count	438.000000	438.000000	438.000000	438.000000	438.000000	438.000000	438.000000
mean	157.972603	2.242265	22.438356	0.159863	20.283105	0.063221	9935.125571
std	30.396528	0.019978	7.971895	0.037465	3.187825	0.008410	827.981530
min	66.000000	2.209100	5.000000	0.100000	10.000000	0.032000	8046.000000
25%	139.000000	2.226425	17.000000	0.130000	19.000000	0.057000	9372.250000
50%	155.000000	2.239300	22.000000	0.150000	20.000000	0.065000	9844.500000
75%	178.000000	2.255925	26.000000	0.190000	22.000000	0.070000	10440.000000
max	259.000000	2.319600	49.000000	0.310000	33.000000	0.077000	12474.000000

FIGURE 4.3 Geologic data description.

```
# Below is for showing the distribution of gamma ray. Simply use the
# other column names to plot other parameters
sns.distplot(df['GR_API'],label='Clustering Data',norm_hist=True,
    color='g')
```
**Python output=please plot the distribution of all parameters and
Fig. 4.4 will be the outcome**

Next, let's plot a heat map of all parameters versus one another to find potential collinear features.

```
fig=plt.figure(figsize=(15,8))
sns.heatmap(df.corr(), cmap='coolwarm', annot=True, linewidths=4,
    linecolor='black')
```
Python output=Fig. 4.5

As shown in Fig. 4.5, TOC and bulk density have a negative Pearson correlation coefficient of -0.99. This makes sense as TOC is a calculated feature derived from bulk density. Therefore, let's use the lines of code below to remove TOC from the analysis because TOC and bulk density would provide the same information when clustering. The lines of code below will permanently drop the TOC from the "df" data frame. Please remember to use "inplace = True" when the desired column removal outcome is intended to be permanent.

```
df.drop(['TOC, fraction'], axis=1, inplace=True)
```

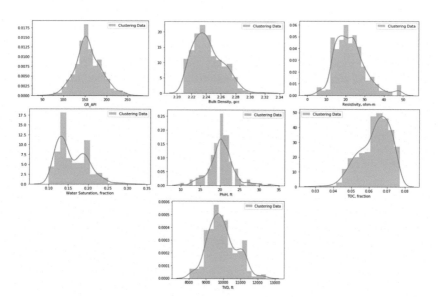

FIGURE 4.4 The distribution of all geologic features.

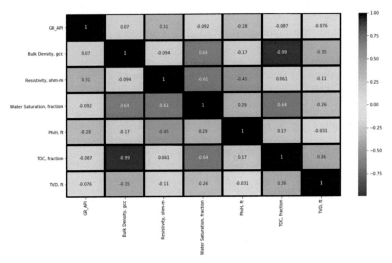

FIGURE 4.5 Heat map of geologic features.

Next, let's import the StandardScaler library and standardize the data prior to feeding the data into the k-means algorithm:

```
from sklearn.preprocessing import StandardScaler
scaler=StandardScaler()
df_scaled=scaler.fit(df)
df_scaled=scaler.transform(df)
df_scaled
```
Python output=Fig. 4.6

Next, let's import the k-means library and write a for loop for calculating within cluster sum of squared errors. Afterward, let's use the matplotlib library to plot number of clusters on the x-axis versus within cluster sum of squared errors:

```
from sklearn.cluster import KMeans
distortions=[]
for i in range (1,21):
    km=KMeans(n_clusters=i,random_state=1000,
    init='k-means++', n_init=1000, max_iter=500)
    km.fit(df_scaled)
    distortions.append(km.inertia_)
plt.plot(range(1,21),distortions, marker='o')
plt.xlabel('Number of Clusters')
plt.ylabel('Distortion (Within Cluster Sum of Squared Errors)')
plt.title('The Elbow Method showing the optimal k')
```
Python output=Fig. 4.7

```
array([[-1.31654221, -1.63691997,  0.07053355, -0.79799783,  0.85324677,
          0.88856422],
       [ 0.39613573, -1.19092098, -0.18063471,  0.27088   , -0.08890975,
         -0.00377924],
       [ 0.26439127, -0.81507913, -1.05972362,  1.07253837,  0.22514243,
         -0.22746968],
       ...,
       [-0.03203375,  0.96891686, -0.30621884,  0.27088   , -0.7170141 ,
         -1.12102229],
       [-0.52607547,  0.50287296, -1.05972362,  1.33975783,  0.5391946 ,
          0.55846969],
       [-0.32845878, -0.45928217, -0.43180297,  0.27088   ,  1.16729895,
         -0.65913177]]])
```

FIGURE 4.6 Standardized geologic data.

n_clusters represents the number of clusters that is desirable to be chosen when applying k-means clustering. In this example, since the goal is to plot various number of clusters from 1 to 20, n_clusters was set to "i" and "i" is defined as a range between 1 and 21. The term "init" refers to a method for initialization that can be set to "random" that will randomly initialize the centroids. A more desirable approach that was used in this example is called "k-means++" which, according to the scikit library definition, refers to selecting the initial cluster centers in an intelligent way to speed up convergence. Using "k-means++" initializes the centroids to be far from one another which could potentially lead to better results than random initialization (Clustering, n.d.). "max_iter" refers to the maximum number of iterations for a single run. "random_state" was set to 1000 to have a repeatable approach in determining the random number generation for centroid initialization. In other words, the outcome of k-means will be the same if run again, since the same seed number is being used to generate the random number. The default value

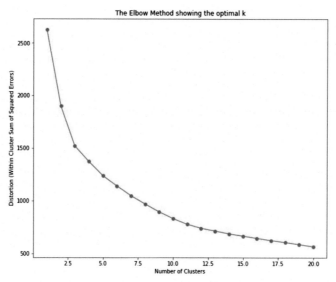

FIGURE 4.7 Elbow point.

for "n_jobs" is 1 which means 1 processor will be used when k-means clustering is performed and run. If −1 is chosen, all available processors will be used. Please note that choosing "n_jobs = −1" could result in CPU hogging by Python and other tasks will be less responsive as a result; therefore, determine the number of processors that your computer can handle and choose the n_jobs accordingly. In the for loop above, an instance of k-means clustering class is created with the defined arguments and assigned to variable "km". Afterward, method "km.fit" was called with argument "df_Scaled" (which is the standardized data). Next, the inertia results have been appended to the empty list called "distortions" that was initially defined. The rest of the code is simply plotting x and y in a line plot. To get a list of inertia numbers, simply type "print(distortions)".

```
print(distortions)
Python output=Fig. 4.8
```

The purpose of the elbow technique is to find the number of desired clusters to choose. In this synthetically generated geologic data set, there is no prior knowledge on the number of clusters to choose. Therefore, from the elbow point shown in Fig. 4.7, 10 clusters were chosen. As illustrated from the distortions, as the number of clusters increases, the difference in inertia value between the current and prior cluster point decreases. Please note that 10 clusters is not an exact solution and more or less clusters could be selected to see the cluster distribution across these wells as a function of number of clusters.

The next step is to assume 10 clusters and proceed with the next phase of labeling the data set and obtaining the centroids for each cluster. In the code below, the same assumptions (random_state = 1000, init = 'k-means++', n_init = 1000, max_iter = 500) were used with 10 clusters. Afterward, "kmeans.cluster_centers_" is used to obtain the cluster centroids for each of the 10 clusters and 6 geologic features. Note that these centroids are the standardized version and it must be converted back to its original form using an inverse transform to make sense.

```
n=10
kmeans=KMeans(n_clusters=n,random_state=1000,init='k-means++',
    n_init=1000, max_iter=500)
kmeans=kmeans.fit(df_scaled)
print(kmeans.cluster_centers_)
Python output=Fig. 4.9
```

[2628.0, 1900.0636934990364, 1522.4518114963212, 1374.229107843318, 1239.1811054000568, 1143.0882975129625, 1048.46512569521
59, 970.4133087183873, 896.0488099115373, 832.60979063326, 779.3373555936982, 741.3035042464755, 715.5112230818482, 689.2615
091696939, 668.1997271560688, 647.2713751646522, 626.2988895440585, 608.0596346197101, 588.1268896977692, 566.9805505987356]

FIGURE 4.8 Inertia numbers.

```
[[-0.81376234 -0.84668052  0.35502005 -0.76527708  0.34691776 -0.3705674 ]
 [-0.63019609  1.182783    -0.5776426   0.96478859 -0.16489011 -0.19384786]
 [-0.11678936 -0.20043572 -0.82697437  0.63429846  0.19583089 -0.31751012]
 [-0.93899675 -0.870945    -0.27133436 -0.74851275 -0.01330459  1.5069037 ]
 [ 1.49714297  1.42135886   2.61809734 -0.89343335 -2.06295198 -0.64341298]
 [ 0.93027358  1.88836765  -0.92867931  2.10656149  0.29341464 -0.76343302]
 [ 0.93004747 -0.96638294   0.37457934 -0.77924559  0.21963274  0.61264753]
 [ 1.82650412 -0.05790694   0.36356319 -0.61985153 -1.55448656  0.7942515 ]
 [ 0.29438666  0.14662858   1.25909764 -0.59758324 -0.67775757 -0.76365738]
 [-0.43702597  0.55465562  -0.72948387  0.94387715  2.20250797 -0.35747435]]
```

FIGURE 4.9   Standardized cluster centroids.

The next step is to obtain the labels for each well. Simply call "kmeans.labels_" as follows:

```
labels=kmeans.labels_
labels
Python output=Fig. 4.10
```

Next, let's convert "df_scaled" from an array to a data frame and add the labeled clusters per well to that data frame as follows:

```
df_scaled=pd.DataFrame(df_scaled,columns=df.columns[0:6])
df_scaled['clusters']=labels
df_scaled
Python output=Fig. 4.11
```

```
array([3, 2, 2, 2, 0, 9, 4, 0, 6, 6, 1, 9, 1, 1, 6, 8, 3, 1, 3, 8, 8, 0,
       3, 7, 8, 2, 9, 2, 3, 2, 5, 1, 6, 1, 2, 5, 8, 6, 0, 9, 9, 1, 1, 2,
       8, 9, 7, 8, 3, 1, 8, 8, 0, 0, 0, 6, 3, 0, 6, 1, 2, 0, 3, 9, 2, 6,
       3, 4, 4, 6, 2, 2, 3, 0, 7, 4, 3, 1, 5, 8, 9, 2, 6, 6, 6, 3, 9, 1,
       8, 0, 8, 4, 3, 6, 6, 1, 1, 1, 1, 1, 9, 6, 6, 0, 2, 9, 0, 0, 5, 9,
       9, 7, 0, 0, 0, 8, 3, 7, 6, 8, 9, 1, 2, 6, 2, 8, 2, 8, 8, 6, 0, 8,
       9, 2, 8, 0, 1, 8, 3, 7, 3, 7, 5, 0, 2, 2, 3, 8, 9, 0, 1, 3, 1, 9,
       7, 0, 8, 3, 2, 8, 5, 6, 3, 8, 4, 2, 5, 9, 3, 8, 0, 2, 8, 1, 1, 2,
       3, 0, 3, 6, 3, 0, 5, 8, 1, 9, 9, 1, 2, 2, 1, 2, 3, 0, 2, 5, 5, 7,
       3, 4, 3, 4, 0, 6, 8, 8, 3, 3, 6, 3, 8, 8, 5, 0, 1, 2, 2, 1, 2, 7,
       1, 1, 1, 1, 9, 8, 8, 4, 5, 5, 5, 0, 6, 6, 3, 6, 2, 3, 2, 2, 8, 6,
       8, 4, 8, 5, 2, 2, 1, 2, 3, 7, 4, 8, 6, 6, 1, 2, 2, 1, 2, 2, 2, 2,
       1, 1, 6, 0, 6, 0, 8, 2, 2, 1, 2, 9, 3, 7, 0, 3, 2, 3, 2, 5, 8, 1,
       1, 9, 5, 9, 3, 3, 8, 7, 0, 6, 2, 2, 8, 8, 8, 2, 5, 6, 6, 6, 3, 0,
       3, 1, 2, 2, 4, 0, 6, 7, 2, 3, 1, 2, 1, 6, 3, 3, 3, 6, 6, 7, 6, 2,
       2, 9, 9, 8, 1, 0, 6, 3, 8, 7, 6, 8, 0, 3, 6, 0, 2, 2, 1, 1, 2, 1,
       1, 6, 3, 6, 3, 6, 6, 2, 2, 2, 0, 3, 6, 6, 6, 8, 8, 6, 6, 3, 7, 2,
       2, 1, 1, 2, 2, 2, 4, 1, 1, 2, 9, 2, 0, 3, 6, 5, 1, 5, 0, 8, 8, 0,
       6, 1, 5, 5, 0, 1, 1, 1, 2, 7, 6, 7, 6, 6, 3, 8, 1, 4, 8, 8, 8, 0,
       8, 0, 7, 2, 5, 0, 0, 3, 1, 8, 0, 7, 3, 2, 2, 0, 1, 1, 1, 2])
```

FIGURE 4.10   K-means labels.

	GR_API	Bulk Density, gcc	Resistivity, ohm-m	Water Saturation, fraction	PhiH, ft	TVD, ft	clusters
0	-1.316542	-1.636920	0.070534	-0.797998	0.853247	0.888564	3
1	0.396136	-1.190921	-0.180635	0.270880	-0.088910	-0.003779	2
2	0.264391	-0.815079	-1.059724	1.072538	0.225142	-0.227470	2
3	0.264391	-0.815079	-1.059724	1.072538	0.225142	-0.227470	2
4	-0.756628	-0.599596	-0.055051	0.003661	1.167299	-0.862267	0
...	...	...	...	...	...	...	...
433	-1.151862	-0.579552	0.196118	-0.530778	0.539195	-0.399167	0
434	-0.559012	0.878715	-0.306219	1.072538	-0.402962	0.182428	1
435	-0.032034	0.968917	-0.306219	0.270880	-0.717014	-1.121022	1
436	-0.526075	0.502873	-1.059724	1.339758	0.539195	0.558470	1
437	-0.328459	-0.459282	-0.431803	0.270880	1.167299	-0.659132	2

438 rows × 7 columns

FIGURE 4.11 Standardized labeled data.

Next, let's return the data to its original (unstandardized form) by multiplying each variable by the standard deviation of that variable and adding the mean of that variable as illustrated below. Note that "scaler.inverse_transform()" in scikit-learn could have also been used to transform the data back to its original form. Please ensure the codes listed below are continuous when you replicate them in Jupyter Notebook. For example, "df_scaled['Water Saturation, fraction']" is split into two lines in the code shown below due to space limitation. Therefore, ensure to have continuous code lines to avoid getting an error.

```
df_scaled['GR_API']=(df_scaled['GR_API']*(df['GR_API'].std())+
   df['GR_API'].mean())
df_scaled['Bulk Density, gcc']=(df_scaled['Bulk Density, gcc']*
   (df['Bulk Density, gcc'].std())+df['Bulk Density, gcc'].mean())
df_scaled['Resistivity, ohm-m']=(df_scaled['Resistivity, ohm-m']*
   (df['Resistivity, ohm-m'].std())+df['Resistivity, ohm-m'].mean())
df_scaled['Water Saturation, fraction']=(df_scaled['Water
   Saturation, fraction']*(df['Water Saturation, fraction'].std())+
   df['Water Saturation, fraction'].mean())
df_scaled['PhiH, ft']=(df_scaled['PhiH, ft']*(df['PhiH, ft'].std())
   +df['PhiH, ft'].mean())
df_scaled['TVD, ft']=(df_scaled['TVD, ft']*(df['TVD, ft'].std())+
   df['TVD, ft'].mean())
```

To obtain a comprehensible version of the cluster centroids, let's groupby "clusters" and take the mean of each cluster as follows:

```
Group_by=df_scaled.groupby(by='clusters').mean()
Group_by
Python output=Fig. 4.12
```

Clusters	GR_API	Bulk Density, gcc	Resistivity, ohm-m	Water Saturation, fraction	PhiH, ft	TVD, ft
0	133.237053	2.225350	25.268539	0.131192	21.389018	9628.302607
1	138.816830	2.265895	17.833450	0.196009	19.757464	9774.623125
2	154.422612	2.238261	15.845803	0.183627	20.907380	9672.233056
3	129.430362	2.224865	20.275307	0.131820	20.240692	11182.814005
4	203.480551	2.270661	43.309553	0.126390	13.706774	9402.391509
5	186.249690	2.279991	15.035022	0.238786	21.218460	9303.017133
6	186.242817	2.222959	25.424463	0.130668	20.983256	10442.386413
7	213.491986	2.241108	25.336644	0.136640	15.327673	10592.751143
8	166.920935	2.245194	32.475750	0.137474	18.122532	9302.831362
9	144.688531	2.253346	16.622987	0.195226	27.304316	9639.143409

FIGURE 4.12   Comprehensible cluster centroids.

As illustrated in Fig. 4.12, each cluster centroid represents the average of each feature's average. For example, cluster 3 (since indexing starts with 0) has an average GR of 154.422 API, a bulk density of 2.238 g/cc, a resistivity of 15.845 Ω-m, a water saturation of 18.3627%, a Phi*H of 20.907 ft, and a TVD of 9672.233 ft.

The next step is to understand the number of counts per each cluster. Let's use the following lines of code to obtain it.

```
df_scaled.groupby(by='clusters').count()
Python output=Fig. 4.13
```

The last step in type curve clustering is to plot these wells based on their latitude and longitude on a map to evaluate the clustering outcome. In addition, the domain expertise plays a key role in determining the optimum number of clusters to successfully define the type curve regions/boundaries. For example, if there are currently 10 type curve regions within your company's acreage position, 10 clusters can be used as a starting point to evaluate k-means clustering's outcome. For this synthetic data set, the last step of plotting and evaluating the clustering's outcome is ignored. However, please make

Clusters	GR_API	Bulk Density, gcc	Resistivity, ohm-m	Water Saturation, fraction	PhiH, ft	TVD, ft
0	49	49	49	49	49	49
1	62	62	62	62	62	62
2	75	75	75	75	75	75
3	54	54	54	54	54	54
4	14	14	14	14	14	14
5	23	23	23	23	23	23
6	57	57	57	57	57	57
7	21	21	21	21	21	21
8	56	56	56	56	56	56
9	27	27	27	27	27	27

FIGURE 4.13   Number of wells in each cluster centroid.

sure to always visualize the clustering outcome and adjust the selected number of clusters accordingly.

K-means clustering application: manual calculation example

The production data such as gas rate, casing pressure, tubing pressure, and line pressure from 500 wells were gathered into a single data source. After applying k-means clustering using k-means++ initialization and **euclidean distance** function using **two clusters**, the data were converged into the two cluster centroids as shown in Table 4.1. Please indicate which cluster (1 or 2) the following numbers will fall under:

Solution: Step 1) Calculate the euclidean distance function from the provided points to **cluster 1** as follows: Gas rate= 4200 Mscf/D, CSG pressure= 800 psi, TBG pressure= 700 psi, line pressure= 500 psi

Cluster 1 distance to provided points =

$$\sqrt{(4200 - 4300)^2 + (800 - 640)^2 + (700 - 520)^2 + (500 - 360)^2} = 360$$

(4.4)

Step 2) Calculate the euclidean distance function from the provided points to **cluster 2** as follows:

Cluster 2 distance to provided points =

$$\sqrt{(4200 - 3900)^2 + (800 - 2050)^2 + (700 - 590)^2 + (500 - 500)^2} = 1194$$

(4.5)

Step 3) If the calculated distance is the smallest, assign to cluster 1. In this example, the calculated distance for cluster 1 (step 1) is smaller than the calculated distance for cluster 2 (step 2). Therefore, these proposed numbers (provided in this example) will be clustered as cluster 1. If there are multiple cluster centroids, simply calculate the distance from each interested new row

TABLE 4.1 K-means clusters.

Attribute	Cluster 1	Cluster 2
Gas rate (MSCF/D)	4300	3900
CSG pressure (casing pressure, psi)	640	2050
TBG pressure (tubing pressure, psi)	520	590
Line pressure (psi)	360	500

Gas rate = 4200 MSCF/D, CSG pressure = 800 psi, TBG pressure = 700 psi, line pressure = 500 psi.

of data to each cluster centroid, and use if statements to assign each row of data to each cluster. The smallest distance will be assigned to the first cluster and the largest distance will be assigned to the last cluster and the ones in the middle can be assigned in the order of their distance magnitudes. Please note that the resultant k-means clustering centroids can be easily programmed real-time to detect liquid loading.

Silhouette coefficient

Another metric to evaluate the quality of clustering is referred to as silhouette analysis. Silhouette analysis can be applied to other clustering algorithms as well. Silhouette coefficient ranges between -1 and 1, where a higher silhouette coefficient refers to a model with more **coherent** clusters. In other words, silhouette coefficients close to $+1$ means the sample is far away from the neighboring clusters. A value of 0 means that the sample is on or very close to the decision boundary between two neighboring clusters. Finally, negative values indicate that the samples could have potentially been assigned to the wrong cluster (Rousseeuw, 1987).

To calculate silhouette coefficient, cluster cohesion (a) and cluster separation (b) must be calculated. Cluster cohesion refers to the average distance between an instance (sample) and all other data points within the same cluster while cluster separation refers to the average distance between an instance (sample) and all other data points in the nearest cluster. Silhouette coefficient can be calculated as shown below. Silhouette coefficient is essentially the difference between cluster separation and cohesion divided by the maximum of the two (Rousseeuw, 1987).

$$ s = \frac{b - a}{\max(a, b)} \tag{4.6} $$

Silhouette coefficient in the scikit-learn library

Let's apply silhouette coefficient and use the graphical tool to plot a measure of how tightly grouped the samples in the clusters are. **Please make sure to place this code before unstandardizing the data**. The "df_scaled" used in "silhouette_vals = silhouette_samples(df_scaled,labels,metric = 'euclidean')" refers to the "df_scaled" prior to unstandardizing the data.

```
from matplotlib import cm
from sklearn.metrics import silhouette_samples
cluster_labels=np.unique(labels)
n_clusters=cluster_labels.shape[0]
silhouette_vals=silhouette_samples(df_scaled,labels,
    metric='euclidean')
y_ax_lower, y_ax_upper=0,0
```

```
yticks=[]
for i, c in enumerate (cluster_labels):
    c_silhouette_vals=silhouette_vals[labels==c]
    c_silhouette_vals.sort()
    y_ax_upper +=len(c_silhouette_vals)
    color=cm.jet(float(i)/n_clusters)
    plt.barh(range(y_ax_lower,y_ax_upper),c_silhouette_vals,
        height=1,edgecolor='none',color=color)
    yticks.append((y_ax_lower+y_ax_upper)/2.)
    y_ax_lower +=len(c_silhouette_vals)
silhouette_avg=np.mean(silhouette_vals)
plt.axvline(silhouette_avg,color="red",linestyle="--")
plt.yticks(yticks, cluster_labels +1)
plt.ylabel('Cluster')
plt.xlabel('silhouette coefficient')
Python output=Fig. 4.14
```

Reference: Raschka and Mirjalili, 2019, Python Machine Learning: Machine Learning and Deep Learning with Python, scikit-learn, and TensorFlow 2, Packt.

As illustrated in Fig. 4.14, the average silhouette coefficient across all clusters is ~0.41. The same process can be repeated with various numbers of clusters to find the average silhouette coefficient when different number of clusters are used. The number of clusters with maximum average silhouette value is desirable.

Hierarchical clustering

Another powerful unsupervised ML algorithm is referred to as hierarchical clustering. Hierarchical clustering is an algorithm that groups similar instances

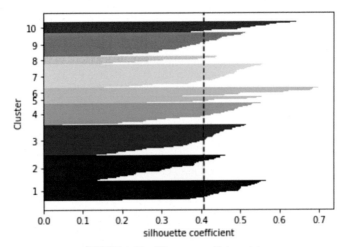

FIGURE 4.14 Silhouette coefficient plot.

into clusters. Hierarchical clustering just like k-means clustering uses a **distance-based algorithm** to measure the distance between clusters. There are two main types of hierarchical clustering as follows:

1) **Agglomerative hierarchical clustering (additive hierarchical clustering):** In this type, each point is assigned to a cluster. For instance, if there are 10 points in a data set, there will be 10 clusters at the beginning of applying hierarchical clustering. Afterward, based on a distance function such as euclidean, the closest pair of clusters are merged. This iteration is repeated until a single cluster is left.

2) **Divisive hierarchical clustering:** This type of hierarchical clustering works the opposite way of agglomerative hierarchical clustering. Hence, if there are 10 data points, all data points will initially belong to one single cluster. Afterward, the farthest point is split in the cluster and this process continues until each cluster has a single point.

To further explain the concept of hierarchical clustering, let's go through a step-by-step example of applying agglomerative hierarchical clustering to a small data set of 4 wells with their respective EURs as shown below. Note that since this is a one-dimensional data, the data was not standardized prior to calculating the distances. In other words, it is OK to not standardize the data for this particular example.

4 well data set

Well number	EUR/1000 ft (BCF/1000 ft)
1	1.4
2	1.2
3	2.5
4	2.0

Step 1) The first step in solving this problem is creating **proximity matrix**. Proximity matrix simply stores the distances between each two points. To create a proximity matrix for this example, a square matrix of n by n is created. n represents the number of observations. Therefore, a proximity matrix of 4*4 can be created as shown in Table 4.2.

The diagonal elements of the matrix will be 0 because the distance of each element from itself is 0. To calculate the distance between point 1 and 2, let's use the euclidean distance function as follows:

$$\sqrt{(1.2 - 1.4)^2} = 0.2 \qquad (4.7)$$

Similarly, that's how the rest of the distances were calculated in Table 4.2. Next, the smallest distance in the proximity matrix is identified, and the points with the smallest distance are merged. As can be seen from this table, the smallest distance is 0.2 between points 1 and 2. Therefore, these two points

TABLE 4.2 Proximity matrix, first iteration.

Well number	1	2	3	4
1	0	**0.2**	1.1	0.6
2	**0.2**	0	1.3	0.8
3	1.1	1.3	0	0.5
4	0.6	0.8	0.5	0

can be merged. Let's update the clusters followed by updating the proximity matrix. To merge points 1 and 2 together, average, maximum, or minimum can be chosen. For this example, maximum was chosen. Therefore, the maximum EUR/1000 ft between well numbers 1 and 2 is 1.4.

Updated clusters (wells 1 and 2 were merged into one cluster)

Well number	EUR/1000 ft (BCF/1000FT)
(1,2)	1.4
3	2.5
4	2.0

Let's recreate the proximity matrix with the new merged clusters as illustrated in Table 4.3.

Clusters 3 and 4 can now (as shown in bold in Table 4.3) be merged into one cluster with the smallest distance of 0.5. The maximum EUR/1000 ft between well numbers 3 and 4 is 2.5. Let's update the table as follows:

Updated clusters (wells 3 and 4 were merged into one cluster)

Well number	EUR/1000 ft (BCF/1000 ft)
1,2	1.4
3,4	2.5

Finally, let's recreate the proximity matrix as shown in Table 4.4. Now, all clusters 1, 2, 3, and 4 can be combined into one cluster. This is essentially how

TABLE 4.3 Proximity matrix, second iteration.

Well number	1,2	3	4
1,2	0	1.1	0.6
3	1.1	0	**0.5**
4	0.6	**0.5**	0

TABLE 4.4 Proximity matrix, third iteration.

Well number	1,2	3,4
1,2	0	1.1
3,4	1.1	0

agglomerative hierarchical clustering functions. The example problem started with four clusters and ended with one cluster.

Dendrogram

A dendrogram is used to show the hierarchical relationship between objects and is the output of the hierarchical clustering. A dendrogram could potentially help with identifying the number of clusters to choose when applying hierarchical clustering. Dendrogram is also helpful in obtaining the overall structure of the data. To illustrate the concept of using a dendrogram, let's create a dendrogram for the hierarchical clustering example above. As illustrated in Fig. 4.15, the distance between well numbers 1 and 2 is 0.2 as shown on the y-axis (distance) and the distance between well numbers 3 and 4 is 0.5. Finally, merged clusters 1,2 and 3,4 are connected and have a distance of 1.1.

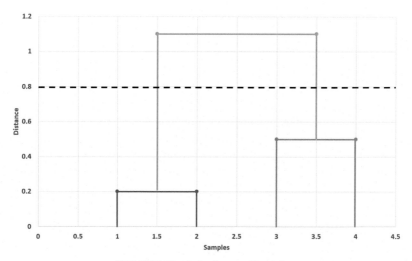

FIGURE 4.15   A dendrogram illustration.

Longer vertical lines in the dendrogram diagram indicate larger distance between clusters. As a general rule of thumb, identify clusters with the longest distance or branches (vertical lines). Shorter branches are more similar to one another. For instance, in Fig. 4.15, one cluster combines two smaller branches (clusters 1 and 2) and another cluster combines the other two smaller branches (clusters 3 and 4). Therefore, two clusters can be chosen in this example. Please note that the optimum number of clusters is subjective and could be influenced by the problem, domain knowledge of the problem, and application.

Implementing dendrogram and hierarchical clustering in scikit-learn library

Let's use the scikit-learn library to apply dendrogram and hierarchical clustering. Please create a new Jupyter Notebook and start importing the main libraries and use the link below to access the hierarchical clustering data set which includes 200 wells with their respective Gas in Place (GIP) and EUR/1000 ft.

https://www.elsevier.com/books-and-journals/book-companion/9780128219294

```
import pandas as pd
import numpy as np
import matplotlib.pyplot as plt
import seaborn as sns
%matplotlib inline
df=pd.read_excel('Chapter4_ GIP_EUR_DataSet.xlsx')
df.describe()
```
Python output=Fig. 4.16

	GIP (BCFperSection)	EURper1000ft
count	200.00000	200.000000
mean	199.84800	2.201754
std	86.67358	1.132611
min	49.50000	0.043860
25%	136.95000	1.524123
50%	202.95000	2.192982
75%	257.40000	3.201754
max	452.10000	4.342105

FIGURE 4.16 df descriptions.

Next, let's standardize the data prior to applying hierarchical clustering as follows:

```
from sklearn.preprocessing import StandardScaler
scaler=StandardScaler()
df_scaled=scaler.fit(df)
df_scaled=scaler.transform(df)
```

Next, let's create the dendrogram. First, "import scipy.cluster.hierarchy as shc" library. Please make sure to pass along "df_scaled" which is the standardized version of the data set in "dend = shc.dendrogram(shc.linkage(df_scaled, method = 'ward'))."

```
import scipy.cluster.hierarchy as shc
plt.figure(figsize=(15, 10))
dend=shc.dendrogram(shc.linkage(df_scaled, method='ward'))
plt.title("Dendrogram")
plt.xlabel('Samples')
plt.ylabel('Distance')
Python output=Fig. 4.17
```

As illustrated in Fig. 4.17, the dashed black line intersects 5 vertical lines. This dashed black horizontal line is drawn based on longest distance or branches observed and is subjective. Therefore, feel free to alter "n_clusters"

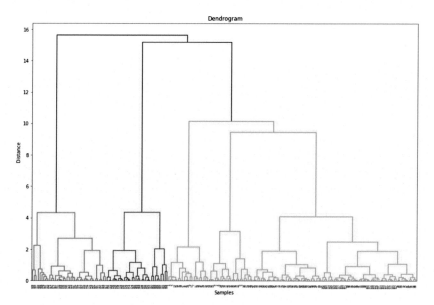

FIGURE 4.17 Resulting dendrogram.

and visualize the clustering outcome. Let's now import agglomerative clustering and apply agglomerative clustering to "df_scaled" data frame. Under "AgglomerativeClustering," number of desired clusters can be accessed with attribute "n_clusters," "affinity" returns the metric used to compute the linkage. In this example, euclidean distance was selected. The linkage determines which distance to use between sets of observation. The "linkage" parameter can be set as (i) ward, (ii) average, (iii) complete or maximum, and (iv) single.

According to scikit-learn library, "ward" minimizes the variance of the clusters being merged. "average" uses the average of the distances of each observation of the two sets. "complete" or "maximum" linkage uses the maximum distances between all observations of the two sets. "single" uses the minimum of the distances between all observations of the two sets. The default linkage of "ward" was used in this example. After defining the hierarchical clustering criteria under "HC," apply "fit_predict()" to the standardized data set (df_scaled).

```
from sklearn.cluster import AgglomerativeClustering
HC=AgglomerativeClustering(n_clusters=5, affinity='euclidean',
  linkage='ward')
HC=HC.fit_predict(df_scaled)
HC
Python output=Fig. 4.18.
```

Let's convert the "df_scaled" to a data frame using panda's "pd.DataFrame." Afterward, apply the silhouette coefficient to this data set as illustrated below.

```
df_scaled=pd.DataFrame(df_scaled,columns=df.columns[0:2])
df_scaled['Clusters']=HC
df_scaled
Python output=Fig. 4.19
```

```
array([4, 3, 4, 3, 4, 3, 4, 3, 4, 3, 4, 3, 4, 3, 4, 3, 4, 3, 4, 3, 4, 3,
       4, 3, 4, 3, 4, 3, 4, 3, 4, 3, 4, 3, 4, 3, 4, 3, 4, 3, 4, 3, 4, 2,
       4, 2, 2, 2, 2, 2, 2, 2, 2, 2, 2, 2, 2, 2, 2, 2, 2, 2, 2, 2, 2, 2,
       2, 2, 2, 2, 2, 2, 2, 2, 2, 2, 2, 2, 2, 2, 2, 2, 2, 2, 2, 2, 2, 2,
       2, 2, 2, 2, 2, 2, 2, 2, 2, 2, 2, 2, 2, 2, 2, 2, 2, 2, 2, 2, 2, 2,
       2, 2, 2, 2, 2, 2, 2, 2, 2, 2, 2, 2, 2, 1, 2, 1, 2, 1, 0, 1, 0, 1,
       2, 1, 0, 1, 0, 1, 0, 1, 0, 1, 2, 1, 0, 1, 2, 1, 0, 1, 0, 1, 0, 1,
       0, 1, 0, 1, 0, 1, 2, 1, 0, 1, 0, 1, 0, 1, 0, 1, 0, 1, 0, 1, 0, 1,
       0, 1, 0, 1, 0, 1, 0, 1, 0, 1, 0, 1, 0, 1, 0, 1, 0, 1, 0, 1, 0, 1,
       0, 1], dtype=int64)
```

FIGURE 4.18 Resulting hierarchical clustering array.

	GIP (BCFperSection)	EURper1000ft	Clusters
0	-1.738999	-0.434801	4
1	-1.738999	1.195704	3
2	-1.700830	-1.715913	4
3	-1.700830	1.040418	3
4	-1.662660	-0.395980	4
...	...	...	...
195	2.268791	1.118061	1
196	2.497807	-0.861839	0
197	2.497807	0.923953	1
198	2.917671	-1.250054	0
199	2.917671	1.273347	1

FIGURE 4.19 Hierarchical clustering data frame.

```
from matplotlib import cm
from sklearn.metrics import silhouette_samples
cluster_labels=np.unique(HC)
n_clusters=cluster_labels.shape[0]
silhouette_vals=silhouette_samples\
(df_scaled,HC,metric='euclidean')
y_ax_lower, y_ax_upper=0,0
yticks=[]
for i, c in enumerate (cluster_labels):
    c_silhouette_vals=silhouette_vals[HC==c]
    c_silhouette_vals.sort()
    y_ax_upper +=len(c_silhouette_vals)
    color=cm.jet(float(i)/n_clusters)
    plt.barh(range(y_ax_lower,y_ax_upper),
      c_silhouette_vals,height=1,edgecolor='none',color=color)
    yticks.append((y_ax_lower+y_ax_upper)/2.)
    y_ax_lower +=len(c_silhouette_vals)
silhouette_avg=np.mean(silhouette_vals)
plt.axvline(silhouette_avg,color="red",linestyle="--")
plt.yticks(yticks, cluster_labels +1)
plt.ylabel('Cluster')
plt.xlabel('silhouette coefficient')
Python output=Fig. 4.20
```

As illustrated in Fig. 4.20, the silhouette coefficient on this data set is high (close to 0.7). It is recommended to use silhouette coefficient to provide insight on the clustering effectiveness of the data set.

Next, let's unstandardize the data and show the mean of each cluster:

```
df_scaled['GIP (BCFperSection)']=(df_scaled['GIP (BCFperSection)']\
  *(df['GIP(BCFperSection)'].std())+df['GIP(BCFperSection)'].mean())
df_scaled['EURper1000ft']=(df_scaled['EURper1000ft']*\
  (df['EURper1000ft'].std())+df['EURper1000ft'].mean())
Group_by=df_scaled.groupby(by='Clusters').mean()
Group_by
Python output=Fig. 4.21
```

As illustrated above, the first cluster (cluster 0) represents low EUR/high GIP, the second cluster (cluster 1) represents high EUR/high GIP, the third cluster (cluster 2) represents medium EUR/medium GIP, the forth cluster (cluster 3) represents high EUR/low GIP, and finally the last cluster (cluster 4) represents low EUR and low GIP.

Please note that dendrogram in hierarchical clustering can be used to get a sense of the number of clusters to choose prior to applying to k-means clustering. In other words, if unsure of selecting the number of clusters prior to applying k-means clustering, a dendrogram can be used to find out the number of clusters. This is another approach in addition to the elbow method and silhouette analysis that were previously discussed under k-means clustering section of this book.

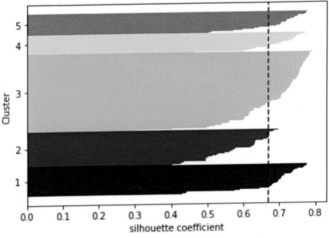

FIGURE 4.20 Silhouette coefficient after applying hierarchical clustering.

Clusters	GIP (BCFperSection)	EURper1000ft
0	295.279503	0.680128
1	285.792052	3.605628
2	184.139503	2.154681
3	82.520600	3.514146
4	86.520674	0.914015

FIGURE 4.21   Mean of the resulting clusters using hierarchical clustering.

Density-based spatial clustering of applications with noise (DBSCAN)

DBSCAN is a density-based clustering algorithm in which the data points are partitioned into high-density regions separated by low-density regions. This algorithm is proposed by Martin Ester, Hans-Peter Kriegel, Jörg Sander, and Xiaowei Xu in 1996 (Ester et al., 1996, pp. 226–231). The main reasons for using this algorithm are as follows:

- Clusters with arbitrary shapes can be detected
- DBSCAN is very efficient for very large data sets

In DBSCAN, density of each instance is measured based on the number of samples within a specific radius (ε) of that instance. Each circle with radius epsilon (ε) has minimum number of data points. DBSCAN is defined by **two parameters**: (i) epsilon (ε) and (ii) MinPts. Parameter ε specifies the radius of the circle and how close observations (points) must be from one another to be considered as part of the same cluster. MinPts is defined as the **minimum** number of observations (neighbors) that should be within the radius of a data point to be considered as a core point. As illustrated in Fig. 4.22, there are three main classifications of a point as follows:

- **Core point:** a core point specifies a dense area. A core point must have at least MinPts number of neighboring samples within radius ε from it (Al-Fuqaha, n.d.).
- **Border point**: a border point has fewer than a specified number of points (MinPts) within epsilon but it is in the neighborhood of a core point.
- **Noise point**: a noise point is a point that is neither a core point nor a border point.

One of the biggest advantages of DBSCAN is that noise points can be removed and each point is not assigned to a cluster (unlike k-means clustering).

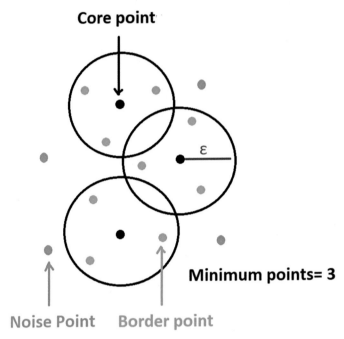

FIGURE 4.22 DBSCAN illustration.

In addition, the number of clusters are not required to be defined since epsilon (ε) and MinPts will be used to determine the **desired density of clusters resulting in specific** number of clusters. However, choosing these hyperparameters requires domain expertise. Visualization and interpretation of the resulting clustered data is crucial after applying DBSCAN. If epsilon parameter is set to be too high, DBSCAN might interpret only one cluster. On the other hand, if epsilon parameter is set to be too low, DBSCAN could yield many unnecessary clusters. In addition, if MinPts (min sample) is set too high, DBSCAN could interpret good neighboring points as noise. Therefore, one of the biggest challenges in DBSCAN application is setting epsilon and MinPts parameters when using the algorithm. It is highly recommended to perform several iterations to fine tune these hyperparameters followed by visualizing until satisfactory clustering is achieved. In addition, the following guidelines can be used to estimate the hyperparameters (Schubert et al., 2017):

For two-dimensional data, use default value of MinPts = 4
For larger dimensional data sets, MinPts = 2*dimensions

After determining MinPts, epsilon can be determined as follows:

Plot samples sorted by distance on the x-axis versus k-nearest neighbor (KNN) distance on the y-axis where K = MinPts (Rahmah & Sitanggang, n.d.)

Finally, select the "elbow" in the graph where the k-distance value is the epsilon value as illustrated in Fig. 4.23.

How does DBSCAN work?

Below are the steps when applying DBSCAN algorithm:

1) DBSCAN starts with an arbitrary point. The neighboring observations are then retrieved from the epsilon (ε) parameter.
2) If this point contains **MinPts** based on the defined epsilon radius, DBSCAN starts forming clusters. Otherwise, this point is labeled as "noise." Please note that this point can later be found within the radius of investigation of another point. Therefore, it could potentially become part of a cluster.
3) If a point is found to be a core point, the points within the radius of investigation (epsilon) is part of the cluster.
4) This process is continued until the **density-connected cluster** is found.
5) A new point can restart the workflow which could either be classified as part of a new cluster or labeled as noise.

A density-connected cluster as defined by Ester et al. is as follows:

*A point **a** is density-connected to a point **b** with respect to **epsilon** and **MinPts**, if there is a point **c** such that, both **a** and **b** are density reachable from **c** w.r.t. **epsilon** and **MinPts**.*

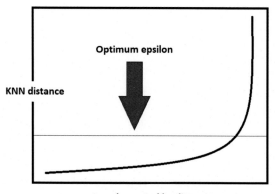

FIGURE 4.23   Optimum epsilon determination.

DBSCAN implementation and example in scikit-learn library

Let's apply DBSCAN using the scikit-learn to a data set with 1007 wells including Young's modulus, Poisson's ratio, and closure pressure (minimum horizontal stress). The exercise is to use DBSCAN to cluster the data. First, import the DBSCAN clustering csv file using the following link:

https://www.elsevier.com/books-and-journals/book-companion/9780128219294

Let's start a new Jupyter Notebook and import all the necessary libraries as well as **"Chapter4_Geomechanics_DataSet.csv"** file as follows:

```
import numpy as np
import pandas as pd
import matplotlib.pyplot as plt
import seaborn as sns
%matplotlib inline
df=pd.read_csv('Chapter4_Geomechanics_DataSet.csv')
df.describe()
Python output=Fig. 4.24
```

Next, use the seaborn library to visualize the distribution of each parameter.

```
sns.distplot(df['Closure Pressure (psi)'],label='Clustering Data',
  norm_hist=True, color='r')
plt.legend()
# Repeat the same code for the other parameters but change to parameter
  names "'YM (MMpsi)" and "PR"
Python output after plotting all the features=Fig. 4.25
```

	Closure Pressure (psi)	YM (MMpsi)	PR
count	1007.000000	1007.000000	1007.000000
mean	9500.000000	5.000000	0.250000
std	2212.788225	0.395503	0.066015
min	1479.148671	4.474815	0.034032
25%	7924.313997	4.686613	0.206557
50%	9737.579335	4.776741	0.249805
75%	11321.568010	5.451288	0.291797
max	13264.890570	5.858267	0.458470

FIGURE 4.24 DBSCAN description.

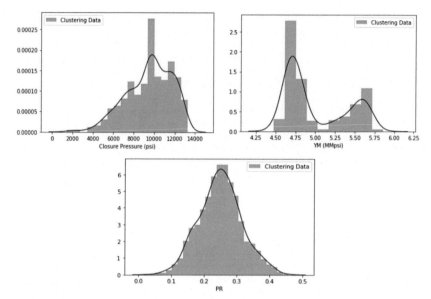

FIGURE 4.25   Feature distributions.

Next, let's standardize the data as follows:

```
from sklearn.preprocessing import StandardScaler
scaler=StandardScaler()
df_scaled=scaler.fit(df)
df_scaled=scaler.transform(df)
```

Before applying DBSCAN, it is important to estimate the value of epsilon using the KNN library. Therefore, first import the library followed by applying KNN with 6 neighbors. The reason for picking 6 is because this is a three-dimensional database where 2*number of dimensions is recommended to be used for MinPts. Since MinPts is 6, the number of neighbors chosen to be used in the KNN algorithm is also 6. Therefore, after creating a KNN model with K = 6, that created model can be used to fit "df_scaled" (standardized data). In unsupervised KNN algorithm, the distance from each point to its closest neighbors is calculated. In this example, since K = 6, the distance from each point to the 6 nearest neighbors is calculated. KNN method yields two arrays. The first array contains the distance to the closest nearest neighbor points, and the second array is simply the index for each of those points. Please note that "NearestNeighbors" implements unsupervised nearest neighbors learning. KNN can also be applied for supervised regression and classification problems. Supervised KNN algorithm will be discussed in detail in the next chapter.

```
from sklearn.neighbors import NearestNeighbors
Neighbors=NearestNeighbors(n_neighbors=6)
nbrs=Neighbors.fit(df_scaled)
distances, indices=nbrs.kneighbors(df_scaled)
```

Next, sort and plot the results as follows:

```
fig=plt.figure(figsize=(10,8))
distances=np.sort(distances, axis=0)
distances=distances[:,1]
plt.plot(distances)
plt.title('Finding the optimum epsilon')
plt.xlabel('Samples sorted by distance')
plt.ylabel('6th NN distance')
Python output=Fig. 4.26
```

As illustrated in Fig. 4.26, optimum epsilon is approximately 0.3. Next, let's import the DBSCAN library and apply DBSCAN to "df_scaled." Note that "eps" is the term used to represent epsilon in the DBSCAN algorithm within the scikit-learn library. According to the scikit-learn library, "min_-sample" refers to "the number of samples in a neighborhood for a point to be considered as a core point which includes the point itself." In other words, "min_sample" is referred to as **MinPts** that was discussed earlier. For this example, "euclidean distance" was used for distance measurements.

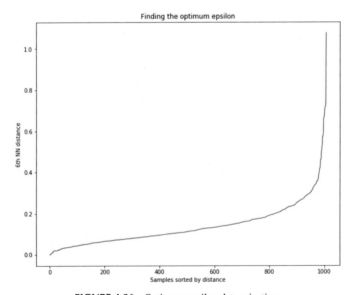

FIGURE 4.26 Optimum epsilon determination.

```
from sklearn.cluster import DBSCAN
Clustering=DBSCAN(eps=0.3,min_samples=6, metric='euclidean')
DB=Clustering.fit_predict(df_scaled)
```

To get the label for each data row and convert to a data frame, the following lines can be executed:

```
labels=pd.DataFrame(DB,columns=['clusters'])
labels
```
Python output=Fig. 4.27

Next, let's add the clusters to the standardized data (df_scaled). Once the data are in a data frame, convert the input features (YM, PR, and closure pressure) to its original state (prior to standardization).

```
df_scaled=pd.DataFrame(df_scaled,columns=df.columns[0:3])
df_scaled['clusters']=DB
df_scaled['Closure Pressure (psi)']=(df_scaled['Closure Pressure
   (psi)']*(df['Closure Pressure (psi)'].std())+df['Closure
   Pressure (psi)'].mean())
df_scaled['YM (MMpsi)']=(df_scaled['YM (MMpsi)']*(df['YM (MMpsi)'].
   std())+df['YM (MMpsi)'].mean())
df_scaled['PR']=(df_scaled['PR']*(df['PR'].std())+df
   ['PR'].mean())
```

	clusters
0	0
1	0
2	0
3	0
4	0
...	...
1002	-1
1003	0
1004	0
1005	0
1006	0

FIGURE 4.27 Labeled clusters using DBSCAN.

Next, let's take the average and groupby the cluster column. As illustrated, there are 9 resulting clusters. As was discussed earlier, the number of clusters is determined and calculated based on epsilon and MinPts defined. Therefore, after applying any type of clustering algorithm, it is extremely important to visualize the data to make sure the resulting clusters make sense. Since this is a synthetically generated database, the resulting clusters are not plotted based on their latitude and longitude. However, please make sure to plot the resulting clusters on any database to obtain more knowledge from the resulting clusters. The "-1" cluster below indicates the outlier points. DBSCAN is one of the many methodologies that can be used for **outlier detection**.

```
Group_by_mean=df_scaled.groupby(by='clusters').mean()
Group_by_mean
Python output=Fig. 4.28
```

Please note that all three clustering techniques that have been discussed are primarily suitable for continuous variables. Distance calculations between categorical features (converted to binary values) can be done but some more preprocessing such as OneHotEncoder is needed to convert the categorical values to numeric values.

clusters	Closure Pressure (psi)	YM (MMpsi)	PR
-1	8559.325853	5.093442	0.251312
0	9103.763751	4.721670	0.269255
1	9923.806912	4.698112	0.379960
2	8606.155092	5.569215	0.193501
3	11175.672378	5.502061	0.244701
4	6054.320378	4.748625	0.088967
5	4656.636445	4.632924	0.140017
6	9868.815507	4.483571	0.268379
7	12062.134910	5.625208	0.159870

FIGURE 4.28 DBSCAN clustering (average per cluster).

Important notes about clustering

Although unsupervised clustering techniques are very powerful in deciphering information from the data, it is important to note that there are occasions where clustering would not be the most suitable option and supervised ML which will be discussed in the next chapter could potentially make more sense. The first item to consider is whether the data have any labeled class. If the data already have a label associated with each row and those labels were either automatically or manually created, a supervised ML could potentially lead to a better result. On the other hand, some problems are simply too time-consuming to label. In those cases, unsupervised clustering techniques could facilitate the process. One of the problems that have been successfully tackled in the O&G industry has been liquid-loading clustering. As opposed to using a labeled class using Turner and Coleman techniques, k-means clustering was successfully applied to cluster the data into two separate groups of loaded versus unloaded. This is very powerful because clustering techniques independently divide the data into two separate clusters of loaded versus unloaded as opposed to using Turner and Coleman calculations which were developed many decades ago for vertical conventional wells.

Cluster validation is another important step in clustering. As was discussed, silhouette coefficient is one way of analyzing the resulting clusters. If the average silhouette coefficient across all the resulting clusters turns out to be 0.2 (for example), clustering might not provide too much insight into the data. In those cases, a supervised ML model could probably shed more light.

Finally, as discussed throughout this chapter, it is extremely important to apply basic statistical measures such as mean of each resultant cluster after applying any clustering technique. For example, let's assume that a database is divided into two separate clusters of loaded versus unloaded using k-means clustering. Let's assume gas rate, casing-tubing pressure, line pressure, and water rate were the features fed into the unsupervised model. The resulting average values for clusters 1 and 2 are shown in Table 4.5. As illustrated in this

TABLE 4.5 Resulting cluster example.

Clusters	Gas rate (MSCF/D)	Casing-tubing pressure (psi)	Line pressure (psi)	Water rate (BBL/D)
Cluster 0 (loaded)	1000	200	500	20
Cluster 1 (unloaded)	1002	198	499	19

table, the resulting numerical values for each cluster appear to be very similar. Did clustering really help correctly classify the data into two separate groups in this example? The answer is obviously no. This is why it is crucial to understand the mean of each cluster and visualize the data to make sure there is a complete separation between the resulting clusters.

Outlier detection

Outlier detection algorithms are incredibly powerful and useful tools for detecting outliers within a data set. First, let's define some terminologies prior to diving in deeper into various outlier detection algorithms:

- **Outlier:** an outlier is simply a point that is vastly different from other points in a data set. For instance, if the average GIP in a particular area ranges between 100 and 200 BCF/section, a point with a GIP of 20 or 350 BCF/section could be considered as outlier points.
- **Anomaly detection:** is the process of finding outliers in a data set.

Anomaly detection has many applications in different industries including banking, finance, manufacturing, insurance, oil and gas industry, etc. One of the main applications of anomaly detection in the banking industry is detecting large deposits or strange transactions for customers. If you have owned a credit card, you might have received a call from the bank inquiring about a fishy transaction that you most likely had nothing to do with. Therefore, banking industry saves billions of dollars a year detecting frauds using anomaly detection algorithms to find outliers. Anomaly detection can also be used in the insurance industry for detecting fraudulent insurance claims and payments. In the manufacturing and oil and gas industry, anomaly detection can be used for detecting abnormal behaviors for machines and equipment in terms of preventative maintenance and predicting failures. For example, finding abnormal behaviors in frac pumps can lead to preventative maintenance which could potentially lead to a longer life span for frac equipment. Anomalous activity detection can also be used in the IT industry to detect intruders into networks. Now that some of the use cases for anomaly detection have been identified in different industries, let's review some of the algorithms that can be used for anomaly detection. Please note that DBSCAN is one of the algorithms that can be used for anomaly detection, and since it has already been addressed, it won't be revisited under this section.

Isolation forest

Isolation forest is a type of unsupervised ML algorithm that can be used for anomaly detection which works based on the principle of isolating anomalies (Tony Liu et al., 2008). Isolation forest is based on a decision tree algorithm which will be discussed in the next chapter. As the name indicates, this algorithm isolates features by randomly selecting a feature from a given data set and randomly selecting a **split** value between min and max values of that feature. In other words, it selects a random **split** value within the range of that feature. Next, if the chosen value keeps the point above, change the minimum value of the feature range to that value, otherwise, if the chosen value keeps the point below, change the maximum value of the feature range to that value. Repeat these last two steps until the points are isolated. The number of times that these steps are repeated is called "isolation number." The lower the isolation number, the more anomalous a point is. This is because the random feature partitioning will result in a shorter tree paths for anomalous points. In other words, the anomalous points require less number of random partitions in order to be detected as anomaly. As will be illustrated shortly within Python, isolation forest uses ensemble of isolation trees (through a parameter called "n_estimators") to isolate anomalies. To calculate the anomaly score in an isolation forest, the equation below can be used.

$$s(x, n) = 2^{-\frac{E(h(x))}{c(n)}} \tag{4.8}$$

where:

$$c(n) = 2H(n-1) - \left(\frac{2(n-1)}{n}\right)$$

n is defined as the number of data points, $c(n)$ is a reference metric that normalizes the score between 0 and 1 for ease of interpretation. $E(h(x))$ is referred to as the average of path lengths from the isolation forest. Path length indicates whether a point is normal or outlier. H is the harmonic number and can be estimated as follows:

$$H(i) = \ln(i) + \gamma$$

where $\gamma = 0.5772156649$ (Euler −Mascheroni constant)

Isolation forest using scikit-learn

Let's go over an example of using isolation forest to find the outliers in a petroleum engineer's income versus spending habit data set. The data set can be found using the link below:

https://www.elsevier.com/books-and-journals/book-companion/9780128219294

Let's import all the necessary libraries (pandas, numpy, and regular visualization libraries) to get started and import "Chapter4_PE_Income_Spending_DataSet.csv" data set. As illustrated below, this data set has 4 columns of data and one of the columns is a categorical feature of petroleum engineer's gender. Therefore, it is crucial to convert this categorical feature to a dummy/indicator variable using the "pd.get_dummies" method in pandas library.

	Petroleum_Engineer_Age	Petroleum_Engineer_Income (K$)	Spending_Habits (From 1 to 100)
count	200.000000	200.000000	200.000000
mean	38.850000	140.000000	47.690000
std	13.969007	60.717651	24.532346
min	18.000000	34.676354	0.950000
25%	28.750000	95.937913	33.012500
50%	36.000000	142.173052	47.500000
75%	49.000000	180.317041	69.350000
max	70.000000	316.710700	94.050000

FIGURE 4.29 Petroleum engineer income/spending data set.

```
import pandas as pd
import numpy as np
import seaborn as sns
import matplotlib.pyplot as plt
%matplotlib inline
df=pd.read_csv('Chapter4_PE_Income_Spending_DataSet.csv')
df.describe()
Python output=Fig. 4.29
```

"pd.get_dummies" will create a column for each category and replace each category with 0 and 1 instead in each respective column. In this example, since petroleum_engineer_gender column has two categories, after applying "pd.get_dummies" to the df data set, two new columns will be added as follows:

- Petroleum_Engineer_Gender_Male
- Petroleum_Engineer_Gender_Female

Please note that only k-1 (k minus 1) categorical features will be needed to represent a categorical feature column. For example, if a categorical feature has two categories of **male** and **female**, only k-1 or 1 additional column will be necessary to provide the adequate info for the ML model. One easy and quick way to perform this is to set "drop_first = **True**" inside the "pd.get_dummies" method (please note that the default "drop_first" in pandas is "False"). This will automatically drop one of the columns and will only add one column which is the "Petroleum_Engineer_Gender_Male".

```
df=pd.get_dummies(df,drop_first=True)
df
Python output=Fig. 4.30
```

Let's also look at the box plot of "Petroleum_Engineer_Age," "Petroleum_Engineer_Income (K$)," and "Spending_Habits (From 1 to 100)". Make sure to place each box plot line of code shown below independently, otherwise, the box plots will overlay.

	Petroleum_Engineer_Age	Petroleum_Engineer_Income (K$)	Spending_Habits (From 1 to 100)	Petroleum_Engineer_Gender_Male
0	19	34.676354	37.05	1
1	21	34.676354	76.95	1
2	20	36.988111	5.70	0
3	23	36.988111	73.15	0
4	31	39.299868	38.00	0
...	...	...	...	...
195	35	277.410832	75.05	0
196	45	291.281374	26.60	0
197	32	291.281374	70.30	1
198	32	316.710700	17.10	1
199	30	316.710700	78.85	1

FIGURE 4.30   df after applying get_dummies.

```
sns.boxplot(df['Petroleum_Engineer_Age'], color='orange')
sns.boxplot(df['Petroleum_Engineer_Income (K$)'], color='orange')
sns.boxplot(df['Spending_Habits (From 1 to 100)'], color='orange')
Python output=Fig. 4.31
```

Next, let's import the isolation forest algorithm from the scikit-learn library and apply it to the "df" data frame as shown below. "n_estimators" refers to the number of trees in the ensemble that will get built in the forest. "max_-sample" refers to the number of samples to draw to train each base estimator. If max_samples is set to be larger than the total number of samples provided, all samples will be used for all trees. In other words, no sampling will be done.

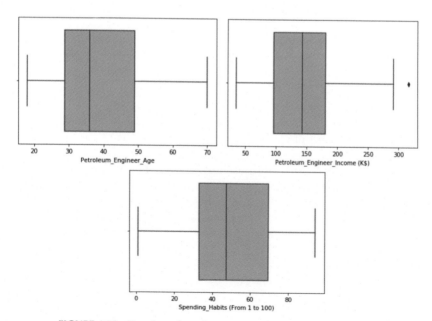

FIGURE 4.31   Box plots of age, income, and petroleum engineer's habits.

The next important parameter that has a **significant** impact on outlier detection in a data set is referred to as "contamination." "contamination" refers to the expected proportion of the outliers and it ranges between 0 and 1. The lower the number, the less number of outliers that will be identified. Higher "contamination" values will lead to a higher number of outliers in a data set. Therefore, please spend a few minutes and change the contamination value from 0.1 to 0.2, 0.3, ..., 0.9 to observe the difference in outlier detection. Once the clf model is defined, use "clf.fit(df)" to apply this defined model to the "df" data frame.

```python
from sklearn.ensemble import IsolationForest
clf=IsolationForest(n_estimators=100,
  max_samples=250,random_state=100, contamination=0.1)
clf.fit(df)
```
Python output=Fig. 4.32

As shown in Fig. 4.32, the warning message essentially indicates that since "max_sample" of 250 was entered as the input and it is greater than the total number of samples (which is 200 in this exercise), "max_sample" will be set to 200.

Next, let's obtain the isolation forest scores using "clf.decision_function(df)," place it in the "df" data frame, and use a histogram to plot the "df['Scores']" as follows. The negative score values indicate presence of anomalous points.

```python
df['Scores']=clf.decision_function(df)
plt.figure(figsize=(12,8))
plt.hist(df['Scores'])
plt.title('Histogram of Average Anomaly Scores: Lower Scores=More\
  Anomalous Samples')
```
Python output=Fig. 4.33

Please feel free to call "df" to observe the resulting "df" data frame with the added "Scores" column. The next step is to predict the anomalies by applying "clf.predict(df.iloc[:,:4])" to the first 4 columns of "PE_Age," "PE_Income," "PE_Spending habits," and "PE_gender_male." Do not apply the "predict" function to "df" since anomaly scores were added in the previous step. After applying the "predict" function to the first 4 columns, let's look at the resulting anomalous rows that were classified as "-1." "-1" means presence of anomalies and "1" represents normal data points.

```
C:\Users\Hoss\anaconda3\lib\site-packages\sklearn\ensemble\_iforest.py:281: UserWarning: max_samples (250) is greater than t
he total number of samples (200). max_samples will be set to n_samples for estimation.
  % (self.max_samples, n_samples))

IsolationForest(behaviour='deprecated', bootstrap=False, contamination=0.1,
                max_features=1.0, max_samples=250, n_estimators=100,
                n_jobs=None, random_state=100, verbose=0, warm_start=False)
```

FIGURE 4.32 Isolation forest warning and input assumptions.

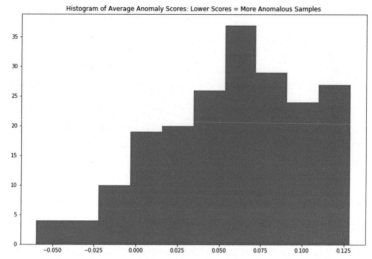

FIGURE 4.33   Histogram of average anomaly scores.

```
df['Anomaly']=clf.predict(df.iloc[:,:4])
Anomaly=df.loc[df['Anomaly']==-1]
Anomaly_index=list(Anomaly.index)
Anomaly
Python output=Fig. 4.34
```

Finally, let's visualize the outlier points on a scatter plot of PE_Income and PE_Spending_Habits and color coded by the anomaly points.

	Petroleum_Engineer_Age	Petroleum_Engineer_Income (K$)	Spending_Habits (From 1 to 100)	Petroleum_Engineer_Gender_Male	Scores	Anomaly
0	19	34.676354	37.05	1	-0.032601	-1
1	21	34.676354	76.95	1	-0.004939	-1
2	20	36.988111	5.70	0	-0.006046	-1
6	35	41.611625	5.70	0	-0.006933	-1
8	64	43.923382	2.85	1	-0.058308	-1
10	67	43.923382	13.30	1	-0.058579	-1
12	58	46.235139	14.25	0	-0.012855	-1
14	37	46.235139	12.35	1	-0.010101	-1
18	52	53.170410	27.55	1	-0.008263	-1
30	60	69.352708	3.80	1	-0.011495	-1
32	53	76.287979	3.80	1	-0.000866	-1
33	18	76.287979	87.40	1	-0.024124	-1
40	65	87.846764	33.25	0	-0.001817	-1
140	57	173.381770	4.75	0	-0.014385	-1
162	19	187.252312	4.75	1	-0.008069	-1
178	59	214.993395	13.30	1	-0.005844	-1
194	47	277.410832	15.20	0	-0.009086	-1
196	45	291.281374	26.60	0	-0.033078	-1
198	32	316.710700	17.10	1	-0.044971	-1
199	30	316.710700	78.85	1	-0.040957	-1

FIGURE 4.34   Anomalous data.

```
plt.figure(figsize=(12,8))
groups=df.groupby('Anomaly')
for name, group in groups:
    plt.plot(group['Petroleum_Engineer_Income (K$)'],
        group['Spending_Habits (From 1 to 100)'],
        marker='o', linestyle='', label=name)
plt.xlabel('Petroleum Engineer Income')
plt.ylabel('Spending Habits')
plt.title('Isolation Forests - Anomalies')
Python output=Fig. 4.35
```

Local outlier factor (LOF)

LOF is another useful unsupervised ML algorithm that identifies outliers with respect to the local neighborhoods as opposed to using the entire data distribution (Breunig et al., 2000). LOF is a density-based technique that uses the nearest neighbor search to identify the anomalous points. The advantage of using an LOF is identifying points that are outliers relative to a local cluster of points. For instance, when using the local outlier factor technique, neighbors of a certain points are identified and compared against the density of the neighboring points. The following steps can be applied when using an LOF model:

1) Calculate distance between P and all the given points using a distance function such as euclidean or Manhattan.

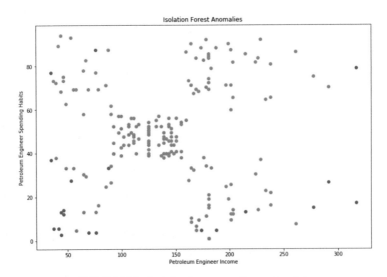

FIGURE 4.35 Isolation forest anomalies scatter plot.

2) Find the k (k-nearest neighbor) closest point. For example, if K = 3, find the third nearest neighbor's distance.
3) Find the k closest points.
4) Find local reachability density using the following equation:

$$lrd_k(O) = \frac{\|N_k(O)\|}{\sum\limits_{O' \in N_k(O)} reachdist_k(O' \leftarrow O)} \qquad (4.10)$$

where reachable distance can be calculated as follows:

$$reachdist_k(O' \leftarrow O) = \max\{dist_k(O), \ dist(O, \ O')\}$$

Please note that *Nk(O)* refers to the number of neighbors.
5) The last step is to calculate the local outlier factor as follows:

$$LOF_k(O) = \frac{\sum\limits_{O' \in N_k(O)} \dfrac{lrd_k(O')}{lrd_k(O)}}{\|N_k(O)\|} \qquad (4.11)$$

Local outlier factor using scikit-learn

Scikit-learn library can also be used to apply local outlier factor algorithm. One of the main steps that should not be forgotten before applying LOF is to make sure to standardize the data prior to applying the algorithm. LOF is based on a nearest neighbor approach where distance calculations are used. Therefore, features with higher magnitudes will dominate the distance calculation. Hence, let's use the same data set that was used to apply isolation forest algorithm and standardize the data set prior to applying LOF. To avoid confusion, please start a new Jupyter Notebook, import the necessary libraries, import the isolation forest data set (call it df), apply "pd.get_dummies" exactly as was shown previously, and standardize the data as follows:

```
from sklearn.preprocessing import StandardScaler
scaler=StandardScaler()
df_scaled=scaler.fit(df)
df_scaled=scaler.transform(df)
```

Next, let's import the LOF library, define the model with 40 neighbors, contamination of 0.1, use euclidean distance calculation, and apply "fit(df_scaled)" as shown below. Please note that a large **n_neighbor or K** could lead to points that are far away from the highest density regions to be misclassified as outliers even though those points could be part of a cluster of points. On the other hand, if K is too small, it could lead to misclassification of outliers with respect to a very small local region of points. Therefore, it is highly recommended to alternate K until a satisfactory level of outlier detection is achieved.

```
from sklearn.neighbors import LocalOutlierFactor
clf=LocalOutlierFactor(n_neighbors=40, contamination=.1,
   metric='euclidean')
clf.fit(df_scaled)
```

The abnormality scores of the training samples can be obtained using "negative_outlier_factor_" attribute. Next, let's plot "clf.negative_outlier_factor_" in a histogram. Please note that **smaller** abnormality score indicates more **anomalous points**. To find the outlier points, simply rank the abnormality score from highest to lowest. The smallest values indicate the outlier points.

```
df_scaled=pd.DataFrame(df_scaled,
columns=['Petroleum_Engineer_Age',
'Petroleum_Engineer_Income (K$)',
'Spending_Habits (From 1 to 100)',
'Petroleum_Engineer_Gender_Male'])
df_scaled['Scores']=clf.negative_outlier_factor_
plt.figure(figsize=(12,8))
plt.hist(df_scaled['Scores'])
plt.title('Histogram of Negative Outlier Factor')
Python output=Fig. 4.36
```

Next, let's use "clf.fit_predict" and apply it to the first 4 columns (excluding the abnormality score column that was just added). Afterward, locate the anomaly by identifying anomaly of "-1." Please note that the anomalies of −1 represent outlier points and the anomalies of 1 represent normal points. Also note that since the data were standardized prior to applying LOF, the new added columns were simply added to the "df_scaled"

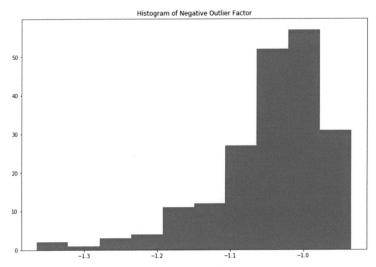

FIGURE 4.36 Histogram of negative outlier factor.

data frame as opposed to the "df" data frame. Therefore, the numbers shown in the table and graph below will be in the standardized form. To convert it back, simply return it back to its normal form (unstandardized form) by multiplying by standard deviation of each original column and adding the average.

```
df_scaled['Anomaly']=clf.fit_predict(df_scaled.iloc[:,:4])
Anomaly=df_scaled.loc[df_scaled['Anomaly']==-1]
Anomaly_index=list(Anomaly.index)
Anomaly
```
Python output=Fig. 4.37

Finally, the last step is to plot to evaluate the resulting anomalous points. Please spend some time changing "n_neighbors" and "contamination" and observing the new outlier points. In addition, spend some time comparing the isolation forest outlier results with LOF.

```
plt.figure(figsize=(12,8))
groups=df_scaled.groupby("Anomaly")
for name, group in groups:
    plt.plot(group['Petroleum_Engineer_Income (K$)'], group
        ['Spending_Habits (From 1 to 100)'], marker="o", linestyle="",
        label=name)
plt.xlabel('Petroleum Engineer Income')
plt.ylabel('Petroleum Engineer Spending Habits')
plt.title('Local Outlier Factor Anomalies')
```
Python output=Fig. 4.38

	Petroleum_Engineer_Age	Petroleum_Engineer_Income (K$)	Spending_Habits (From 1 to 100)	Petroleum_Engineer_Gender_Male	Scores	Anomaly
2	-1.352802	-1.700830	-1.715913	-0.886405	-1.364613	-1
6	-0.276302	-1.624491	-1.715913	-0.886405	-1.260564	-1
7	-1.137502	-1.624491	1.700384	-0.886405	-1.196362	-1
8	1.804932	-1.586321	-1.832378	1.128152	-1.186293	-1
10	2.020232	-1.586321	-1.405340	1.128152	-1.154745	-1
11	-0.276302	-1.586321	1.894492	-0.886405	-1.235479	-1
12	1.374332	-1.548152	-1.366519	-0.886405	-1.162219	-1
19	-0.276302	-1.433644	1.855671	-0.886405	-1.206570	-1
22	0.513132	-1.357305	-1.754735	-0.886405	-1.188463	-1
162	-1.424569	0.780183	-1.754735	1.128152	-1.172355	-1
188	0.154298	1.619911	-1.288876	-0.886405	-1.158835	-1
190	-0.348068	1.619911	-1.055946	-0.886405	-1.156661	-1
192	-0.419835	2.001605	-1.638270	1.128152	-1.165080	-1
193	-0.061002	2.001605	1.583920	-0.886405	-1.155576	-1
194	0.584899	2.268791	-1.327697	-0.886405	-1.270311	-1
195	-0.276302	2.268791	1.118061	-0.886405	-1.155690	-1
196	0.441365	2.497807	-0.861839	-0.886405	-1.285226	-1
197	-0.491602	2.497807	0.923953	1.128152	-1.197818	-1
198	-0.491602	2.917671	-1.250054	1.128152	-1.269415	-1
199	-0.635135	2.917671	1.273347	1.128152	-1.323574	-1

FIGURE 4.37 Anomalies identified using LOF.

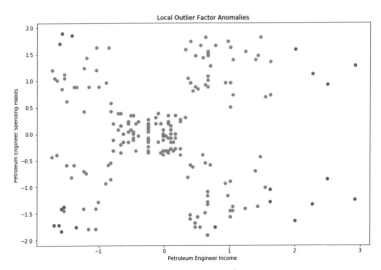

FIGURE 4.38 Local outlier factor anomaly scatter plot.

References

Al-Fuqaha, A. (n.d.). *Partitional (K-means), hierarchical, density-based (DBSCAN) clustering analysis.* Retrieved April 25, 2020, from https://cs.wmich.edu/alfuqaha/summer14/cs6530/lectures/ClusteringAnalysis.pdf.

Breunig, M. M., Kriegel, H.-P., Ng, R. T., & Sander, J. (2000). *LOF: Identifying density-based local outliers.* https://doi.org/10.1145/342009.335388

Clustering. (n.d.). Retrieved April 11, 2020, from https://scikit-learn.org/stable/modules/clustering.html (Original work published 2019).

Ester, M., Kriegel, H.-P., Sander, J., & Xu, X. (1996). *A density-based algorithm for discovering clusters in large spatial databases with noise.* AAAI Press. https://dl.acm.org/doi/10.5555/3001460.3001507.

Rahmah, N., & Sitanggang, I. S. (n.d.). Determination of optimal epsilon (Eps) value on DBSCAN algorithm to clustering data on peatland hotspots in sumatra. *IOP Conf. Series: Earth and Environmental Science.* IOP Publishing. https://doi.org/10.1088/1755-1315/31/1/012012

Rousseeuw, P. (1987). Silhouettes: a graphical aid to the interpretation and validation of cluster Analysis. *Computational and Applied Mathematics, 20*(53), 65. https://doi.org/10.1016/0377-0427(87)90125-7

Schubert, E., Sander, J., Ester, M., Kriegel, H. P., & Xu, X. (2017). *DBSCAN revisited, revisited: Why and how you should (still) use DBSCAN* (Vol. 42). Association for Computing Machinery. https://doi.org/10.1145/3068335

Tony Liu, F., Ming Ting, K., & Zhou, Z.-H. (2008). In *2008 eighth IEEE international conference on data mining. Isolation forest.* https://doi.org/10.1109/ICDM.2008.17

Chapter 5

Supervised learning

Overview

In this chapter, the following supervised machine learning (ML) algorithms will be discussed:

  (i) Multilinear regression
  (ii) Logistic regression
 (iii) K-nearest neighbor (KNN)
 (iv) Support vector machine (SVM)
  (v) Decision tree
 (vi) Random forest
 (vii) Extra trees
(viii) Gradient boosting
 (ix) Extreme gradient boosting
  (x) Adaptive gradient boosting

After discussing the concept for each algorithm and providing the mathematical background, scikit-learn library will be used to demonstrate the implementation of each algorithm with multiple, practical oil and gas problems such as production optimization, human resource employee sustainability prediction, sonic log (shear wave and compression wave travel times) prediction, economic (net present value) prediction, and stage fracability identification and prediction. Finally, at the end of this chapter, various imputation techniques such as k-nearest neighbor (KNN), multivariate imputation by chained equations (MICE), and iterative imputer will be discussed to deal with missing information.

Linear regression

There are different types of regression algorithms such as linear regression, ridge regression, polynomial regression, lasso regression, etc., each suitable for different purposes. For example, linear regression is mostly suited for drawing a linear line across a graph that shows a linear relationship between two variables. Ridge regression is used when there is a high degree of collinearity between features. Polynomial regression is suited when the relationship between features is nonlinear. Linear regression is one of the simplest and most commonly used forms of regression ML algorithms. The equation for linear regression is as follows:

$$y = mx + b \qquad (5.1)$$

where m is the slope of the line, b is the intercept, x is the independent variable (input feature), and y is the dependent variable (output feature). In linear regression, "least square method" is used to find the slope and y-intercept. The least square method minimizes the squared distance from each data point to the line being fitted. For multilinear regression models with multiple input features, the equation below can be used:

$$y = m_1 x_1 + m_2 x_2 + \ldots + m_n x_n + b \qquad (5.2)$$

Please note that when using multilinear regression model, it assumes that there is a linear relationship between input and output features. In addition, it assumes that independent variables (input features) are continuous as opposed to discrete. The majority of petroleum engineering—related problems have nonlinear relationships between the input and output features. Therefore, this algorithm must be used with care when tackling some of the nonlinear problems. We will discuss other algorithms that are best suited for nonlinear problems.

Regression evaluation metrics

Before going deeper into the implementation of a multilinear regression model in scikit-learn, let's review some of the regression evaluation metrics in addition to R and R^2 that have already been elaborated.

1) **Mean absolute error (MAE):** is the mean of the **absolute value** of the errors. It is simply the average of absolute value of difference between actual and predicted values. MAE is also referred to as a loss function since the goal is to minimize this loss function and is defined as follows:

$$\frac{1}{n} \sum_{i=1}^{n} |y_i - \widehat{y}_i| \qquad (5.3)$$

where y_i is the ith sample point output and $\widehat{y}_i$ is the estimated value obtained from the model.

2) **Mean squared error (MSE):** as the name indicates, it is referred to as the mean of the squared error as shown below. MSE is also considered a loss function that needs to be minimized. One of the reasons that MSE is heavily used in real-world ML applications is because the larger errors are penalized more when using MSE as the objective function compared to MAE.

$$\frac{1}{n} \sum_{i=1}^{n} \left(y_i - \widehat{y}_i \right)^2 \qquad (5.4)$$

3) **Root mean squared error (RMSE)**: RMSE is basically the square root of MSE as shown below. Please note that RMSE is also a very popular loss function because of interpretive capability.

$$\sqrt{\frac{1}{n}\sum_{i=1}^{n}\left(y_i - \widehat{y}_i\right)^2}$$

(5.5)

Application of multilinear regression model in scikit-learn

Multilinear regression can be easily implemented in scikit-learn. For this exercise, a data set with the following geologic information is used:

- Porosity (%)
- Matrix permeability (nd)
- Acoustic impedance (kg/m2s × 10^6)
- Brittleness ratio
- TOC (%)
- Vitrinite reflectance (%)

The goal is to train a **multilinear regression** model to understand the impact of each of these variables on production performance. A√K/lateral ft (md^1/2 × ft) is used as the output feature in this exercise. A√K is a productivity metric obtained from rate transient analysis (RTA) in unconventional reservoirs and it is equivalent to kh in conventional reservoirs. A√K is simply the cross-sectional area multiplied by the square root of permeability. It is essentially a productivity metric that shows the strength of each well.

The data set for this exercise can be found using the link below:

https://www.elsevier.com/books-and-journals/book-companion/9780128219294

Open a new Jupyter Notebook and let's start importing the important libraries and set "sns.set_" style to "darkgrid" as illustrated below. Note that they are 5 preset seaborn themes such as darkgrid, whitegrid, dark, white, and ticks that can be chosen based on personal preference.

```
import numpy as np
import pandas as pd
import matplotlib.pyplot as plt
%matplotlib inline
import seaborn as sns
sns.set_style("darkgrid")
df=pd.read_csv('Chapter5_Geologic_DataSet.csv')
df.describe().transpose()
Python output=Fig. 5.1
```

	count	mean	std	min	25%	50%	75%	max
Porosity (%)	200.0	10.493805	2.079824	4.585000	9.03875	10.549000	12.181750	16.485000
Matrix Perm (nd)	200.0	433.075000	173.101415	113.000000	312.25000	403.500000	528.750000	987.000000
Acoustic impedance (kg/m2s*10^6)	200.0	3.265735	0.623574	1.408000	2.80225	3.250500	3.679500	5.093000
Brittleness Ratio	200.0	57.794340	16.955346	13.128000	45.30600	59.412000	69.915000	101.196000
TOC (%)	200.0	3.970700	1.907119	0.100000	2.47000	4.120000	5.400000	8.720000
Vitrinite Reflectance (%)	200.0	1.571440	0.240662	0.744000	1.41600	1.568000	1.714000	2.296000
Aroot(K)	200.0	50.000000	11.505310	24.437856	41.96103	49.692285	58.986667	77.270733

FIGURE 5.1 Basic statistics of the data set.

Next, let's visualize the distribution of each parameter using subplots as follows. As illustrated, the distribution of each parameter appears to be normal which is desired.

```
f, axes=plt.subplots(4, 2, figsize=(12, 12))
sns.distplot(df['Porosity (%)'], color="red", ax=axes[0, 0])
sns.distplot(df['Matrix Perm (nd)'], color="olive", ax=axes[0, 1])
sns.distplot(df['Acoustic impedance (kg/m2s*10^6)'],
  color="blue", ax=axes[1, 0])
sns.distplot(df['Brittleness Ratio'], color="orange", ax=axes[1, 1])
sns.distplot(df['TOC (%)'], color="black", ax=axes[2, 0])
sns.distplot(df['Vitrinite Reflectance (%)'], color="green",
  ax=axes[2, 1])
sns.distplot(df['Aroot(K)'], color="cyan", ax=axes[3, 0])
plt.tight_layout()
```
Python output=Fig. 5.2

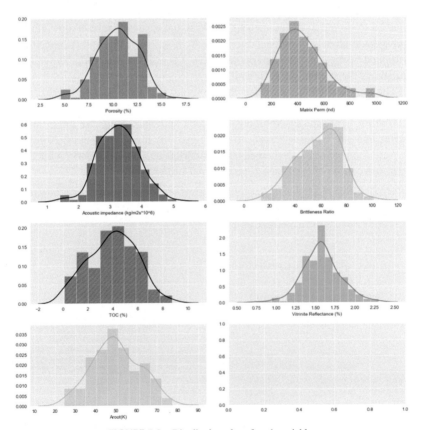

FIGURE 5.2 Distribution plot of each variable.

In addition to distribution plots, let's also visualize the box plots for each parameter as follows:

```python
f, axes=plt.subplots(4, 2, figsize=(12, 12))
sns.boxplot(df['Porosity (%)'], color="red", ax=axes[0, 0])
sns.boxplot(df['Matrix Perm (nd)'], color="olive", ax=axes[0, 1])
sns.boxplot(df['Acoustic impedance (kg/m2s*10^6)'], color="blue",
  ax=axes[1, 0])
sns.boxplot(df['Brittleness Ratio'], color="orange", ax=axes[1, 1])
sns.boxplot(df['TOC (%)'], color="black", ax=axes[2, 0])
sns.boxplot(df['Vitrinite Reflectance (%)'], color="green",
  ax=axes[2, 1])
sns.boxplot(df['Aroot(K)'], color="cyan", ax=axes[3, 0])
plt.tight_layout()
```
Python output=Fig. 5.3

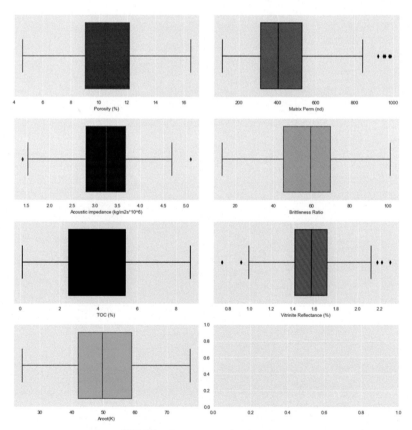

FIGURE 5.3 Box plot of each variable.

It is also important to get an understanding of the relationship between input features and Aroot(K) (output feature) in scatter plots as follows:

```
f, axes=plt.subplots(3, 2, figsize=(12, 12))
sns.scatterplot(df['Porosity (%)'],df['Aroot(K)'], color="red",
  ax=axes[0, 0])
sns.scatterplot(df['Matrix Perm (nd)'],df['Aroot(K)'],
  color="olive", ax=axes[0, 1])
sns.scatterplot(df['Acoustic impedance (kg/m2s*10^6)'],
  df['Aroot(K)'],color="blue", ax=axes[1, 0])
sns.scatterplot(df['Brittleness Ratio'], df['Aroot(K)'],
  color="orange", ax=axes[1, 1])
sns.scatterplot(df['TOC (%)'], df['Aroot(K)'],color="black",
  ax=axes[2, 0])
sns.scatterplot(df['Vitrinite Reflectance (%)'],df['Aroot(K)'],
  color="green", ax=axes[2, 1])
plt.tight_layout()
```
Python output=Fig. 5.4

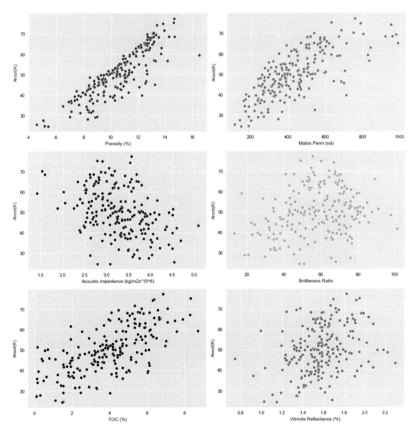

FIGURE 5.4 Scatter plot of each variable versus Aroot(K).

To find whether there are collinear input features, let's use the seaborn library to draw a heat map with Pearson correlation coefficient inside of each square. As illustrated, porosity and matrix perm are highly correlated with a Pearson correlation coefficient of 0.76. In addition, porosity and TOC have a strong Pearson correlation coefficient of 0.71. Since matrix perm and TOC provide similar information as porosity, it is recommended to drop these two features prior to proceeding with the next step.

```
plt.figure(figsize=(14,10))
sns.heatmap(df.corr(),linewidths=2, linecolor='black',cmap=
  'viridis', annot=True)
```
Python output=Fig. 5.5

To drop matrix perm and TOC, pandas' drop syntax can be used as follows. Please make sure "inplace=True" to make sure these two columns are dropped permanently.

```
df.drop(['TOC (%)', 'Matrix Perm (nd)'], axis=1, inplace=True)
```

Next, let's normalize the input features prior to applying multilinear regression. Scikit-learn library has a preprocessing portion where the interested features can be normalized to have a range between 0 and 1 as illustrated

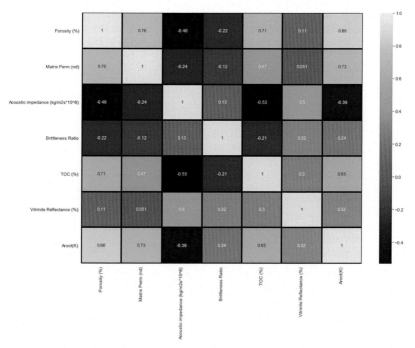

FIGURE 5.5 Heat map showing Pearson correlation coefficient.

below. First, "from sklearn import preprocessing," then define the range for MinMax scaler to be between 0 and 1 and assign it to variable "scaler." Afterward, apply the "scaler.fit" function to the "df" data frame (which includes both inputs and output features) to create the scaler model based on the data (df). Next, transform the df data frame using "scaler.transform(df)" to have a range between 0 and 1 ([0,1]) and call this "df_scaled." Finally, print "df_scaled" to see the scaled data frame which will be in an array format between 0 and 1 for each feature. Note that scaler.transform method will return a numpy array. Therefore, in the next step, "pd.DataFrame" is used to convert the data (df_scaled) back to pandas data frame.

```
from sklearn import preprocessing
scaler=preprocessing.MinMaxScaler(feature_range=(0,1))
scaler.fit(df)
df_scaled=scaler.transform(df)
print(df_scaled)
```
Python output=Fig. 5.6

```
array([[0.32529412, 0.45373134, 0.9600763 , 0.71134021, 0.45177576],
       [0.34294118, 0.57910448, 0.48003815, 0.48969072, 0.31917731],
       [0.43941176, 0.81492537, 0.84289413, 0.92268041, 0.47793158],
       [0.65411765, 0.40298507, 0.39337784, 0.48969072, 0.65669029],
       [0.64529412, 0.56716418, 0.        , 0.5       , 0.28514944],
       [0.46941176, 0.42089552, 0.5812781 , 0.3814433 , 0.50238885],
       [0.40823529, 0.49253731, 0.71903529, 0.4742268 , 0.43843026],
       [0.29588235, 0.5880597 , 0.5731026 , 0.51546392, 0.30503973],
       [0.35117647, 0.34328358, 0.74710451, 0.54123711, 0.42118734],
       [0.39411765, 0.72537313, 0.75296362, 0.88659794, 0.47808031],
       [0.49941176, 0.28059701, 0.68360812, 0.43298969, 0.56731787],
       [0.56705882, 0.30149254, 0.51996185, 0.47938144, 0.66317086],
       [0.60411765, 0.45373134, 0.75909524, 0.54123711, 0.66004008],
       [0.63764706, 0.37910448, 0.61983922, 0.64948454, 0.75497868],
       [0.42823529, 0.36716418, 0.75323614, 0.36597938, 0.48654097],
       [0.28176471, 0.64179104, 0.64164055, 0.62886598, 0.31453193],
       [0.76470588, 0.53731343, 0.33451424, 0.5257732 , 0.72852016],
       [0.53117647, 0.37313433, 0.25194168, 0.48969072, 0.37073953],
       [0.48117647, 0.52238806, 0.76958714, 0.54123711, 0.5328626 ],
       [0.34823529, 0.67462687, 0.51614661, 0.82989691, 0.37235832],
       [0.67823529, 0.32835821, 0.58073307, 0.59793814, 0.80599842],
       [0.57470588, 0.33731343, 0.48834991, 0.40721649, 0.58482678],
       [0.35705882, 0.32835821, 0.06458646, 0.28350515, 0.08318242],
       [0.20470588, 0.6       , 0.36517237, 0.39690722, 0.08962854],
       [0.71235294, 0.03283582, 0.57787164, 0.40206186, 0.86469824],
       [0.04235294, 0.6358209 , 0.5943589 , 0.38659794, 0.05359477],
       [0.54294118, 0.37910448, 0.5681973 , 0.57216495, 0.6288773 ],
       [0.54      , 0.73432836, 0.28559749, 0.62886598, 0.41542785],
       [0.34294118, 0.83880597, 0.27878458, 0.58762887, 0.14341982],
       [0.37294118, 0.48358209, 0.54912113, 0.53608247, 0.37724263],
       [0.70176471, 0.46865672, 0.48984875, 0.78350515, 0.82375655],
       [0.48176471, 0.34925373, 0.35318163, 0.34020619, 0.36633696],
```

FIGURE 5.6 Normalized features (top portion of the data).

Next, let's convert the numpy array to a data frame as follows:

```
df_scaled=pd.DataFrame(df_scaled, columns=['Porosity (%)',
  'Acoustic impedance (kg/m2s*10^6)', 'Brittleness Ratio',\
  'Vitrinite Reflectance (%)', 'Aroot(K)'])
```

Since the data have now been converted to a data frame, to train a supervised ML model, let's define x features (input features) which are porosity, acoustic impedance, brittleness ratio, and vitrinite reflectance in this example. The only y feature is Aroot(K). "x_scaled" represents the normalized input features. "y_scaled" represents the normalized Aroot(K) feature.

```
y_scaled=df_scaled['Aroot(K)']
x_scaled=df_scaled.drop(['Aroot(K)'], axis=1)
```

The next step in training a model is to split the data between training and testing. The model_selection library in scikit-learn can be used to randomly split the data between training and testing set as illustrated below. First, import "train_test_split" from "sklearn.model_selection." Next, to obtain the same result, let's fix the seed number (in this example 1000 was chosen but any number can be chosen as the seed number and the idea is to be able to replicate the process). X_train and y_train represent the input and output data points assigned as the **training** portion (which was randomly selected in this example, and it consists of 70% of the data). X_test and y_test represent the input and output data points assigned as the **testing** portion, and it consists of 30% of the data ("test_size=0.3"). To see which rows were selected as training versus testing data, simply type in "X_train" or "X_test" in a separate line of code, and the resulting output will show the input features' rows that were chosen as training versus testing based on the seed number of 1000.

```
from sklearn.model_selection import train_test_split
seed=1000
np.random.seed(seed)
X_train, X_test, y_train, y_test=train_test_split(x_scaled,
  y_scaled, test_size=0.3)
X_train
```
Python output=Fig. 5.7

In Fig. 5.7, index numbers such as 44, 101, 25, etc., were chosen as training rows. 140 (70% of 200 rows) rows were selected as training and the remaining 60 rows were selected as testing. Next, let's import linear regression model from the scikit-learn. Afterward, place the "lm=LinearRegression()" to initialize an instance of the linear model called "lm."

```
from sklearn.linear_model import LinearRegression
np.random.seed(seed)
lm=LinearRegression()
```

Next, let's fit the lm model to "(X_train, y_train)" which will fit the linear regression model to the training rows.

```
lm.fit(X_train,y_train)
```

	Porosity (%)	Acoustic impedance (kg/m2s*10^6)	Brittleness Ratio	Vitrinite Reflectance (%)
44	0.695294	0.382090	0.453468	0.726804
101	0.401176	0.525373	0.562611	0.407216
25	0.042353	0.635821	0.594359	0.386598
3	0.654118	0.402985	0.393378	0.489691
68	0.487647	0.641791	0.515874	0.608247
...	...	...	...	...
94	0.541176	0.459701	0.814007	0.572165
192	0.812941	0.325373	0.365445	0.546392
71	0.448235	0.489552	0.658945	0.505155
87	0.490000	0.602985	0.713585	0.567010
179	0.039412	0.692537	0.710587	0.381443

140 rows × 4 columns

FIGURE 5.7 X_train (training input features).

Next, let's get the intercept of the model by calling "lm.intercept_"

```
print(lm.intercept_)
Python output=-0.29565316262910485
```

Now, let's get the coefficients of the linear regression model by calling "lm.coef_." As shown below, porosity has a coefficient of ~1.070, acoustic impedance has a coefficient of −0.178, etc. These coefficients, along with the intercept value obtained above, can be used to formulate the multilinear regression model equation. The equation can then be used to predict the output feature (Aroot(K)). Before using these coefficients and intercept values, please make sure to normalize the data prior to feeding the data into the formulated multilinear regression equation. This is because we normalized the data prior to feeding the data into the multilinear regression model.

```
coeff_df=pd.DataFrame(lm.coef_,x_scaled.columns,
  columns=['Coefficient'])
coeff_df
Python output=Fig. 5.8
```

	Coefficient
Porosity (%)	1.069765
Acoustic impedance (kg/m2s*10^6)	-0.178141
Brittleness Ratio	0.426330
Vitrinite Reflectance (%)	0.231669

FIGURE 5.8 Linear regression model coefficients.

Next, let's apply the trained "lm" model to "X_test" in order to obtain the predicted y values associated with the testing portion of the data. To do so, "lm.predict(X_test)" is used. Afterward, we can plot the model's predicted values (y_pred) versus "y_test" which are the actual y values of the testing set. This will allow us to gauge the model's accuracy. As shown below, the predicted versus actual values seem to lie on the 45 degree line which means the testing accuracy must be high. Next, we will obtain the R^2 value by importing the metrics library within sklearn.

```
plt.figure(figsize=(10,8))
y_pred=lm.predict(X_test)
plt.plot(y_test,y_pred, 'b.')
plt.xlabel('Testing Actuals')
plt.ylabel('Testing Predictions')
plt.title('Actual Vs. Predicted, Testing Data Set (30% of the data)')
Python output=Fig. 5.9
```

Let's import the R^2 score library from sklearn.metrics and obtain the R^2 value for the testing set as follows:

```
from sklearn.metrics import r2_score
test_set_r2=r2_score(y_test,y_pred)
print('Testing r^2:',round(test_set_r2,2))
Python output=Testing r^2: 0.94
```

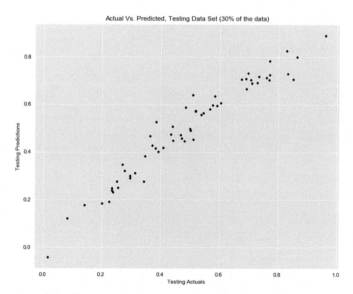

FIGURE 5.9 Actual versus predicted values, testing data set.

Let's also obtain MAE, MSE, and RMSE from the metrics library as follows:

```
from sklearn import metrics
print('MAE:', round(metrics.mean_absolute_error(y_test, y_pred),5))
print('MSE:', round(metrics.mean_squared_error(y_test, y_pred),5))
print('RMSE:', round(np.sqrt(metrics.mean_squared_error(y_test,
  y_pred)),5))
Python output=
MAE: 0.03793
MSE: 0.00266
RMSE: 0.05159
```

Now, let's create a histogram of the distribution of residuals indicating the difference between predicted y values and actual y values as follows:

```
sns.distplot((y_test-y_pred))
Python output=Fig. 5.10
```

When the residual is normally distributed, it is one sign that the model was the correct choice for the data set.

Next, let's use the statsmodel which is a Python package that provides a complement to scipy for statistical computations (Statsmodels, 2020) to get the summary stats of the multilinear regression model.

```
import statsmodels.api as sm
X=sm.add_constant(X_train)
model=sm.OLS(y_train,X).fit()
predictions=model.predict(X)
model_stats=model.summary()
print(model_stats)
Python output=Fig. 5.11
```

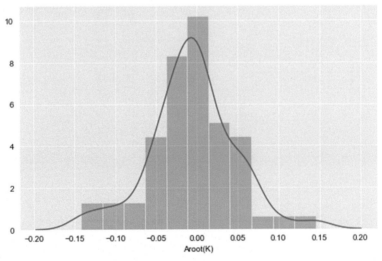

FIGURE 5.10 Residual distribution.

```
                       OLS Regression Results
==============================================================================
Dep. Variable:                Aroot(K)   R-squared:                       0.946
Model:                             OLS   Adj. R-squared:                  0.944
Method:                  Least Squares   F-statistic:                     589.0
Date:                Sat, 20 Jun 2020   Prob (F-statistic):           2.26e-84
Time:                        11:15:44   Log-Likelihood:                 218.23
No. Observations:                 140   AIC:                            -426.5
Df Residuals:                     135   BIC:                            -411.7
Df Model:                           4
Covariance Type:            nonrobust
==============================================================================
                                  coef    std err          t      P>|t|      [0.025      0.975]
------------------------------------------------------------------------------
const                          -0.2957      0.030     -9.761      0.000      -0.356      -0.236
Porosity (%)                    1.0698      0.035     30.654      0.000       1.001       1.139
Acoustic impedance (kg/m2s*10^6)  -0.1781   0.040     -4.400      0.000      -0.258      -0.098
Brittleness Ratio               0.4263      0.027     16.083      0.000       0.374       0.479
Vitrinite Reflectance (%)       0.2317      0.039      5.899      0.000       0.154       0.309
==============================================================================
Omnibus:                        6.746   Durbin-Watson:                   2.148
Prob(Omnibus):                  0.034   Jarque-Bera (JB):                9.002
Skew:                          -0.252   Prob(JB):                       0.0111
Kurtosis:                       4.136   Cond. No.                        20.9
==============================================================================

Warnings:
[1] Standard Errors assume that the covariance matrix of the errors is correctly specified.
```

FIGURE 5.11   Statsmodel summary.

As illustrated in Fig. 5.11, the intercept and coefficients for each parameter match with what was already obtained. This summary table provides a more detailed display of the model's results that is very useful to obtain.

One-variable-at-a-time sensitivity analysis

One-variable-at-a-time (OVAT) is a commonly used sensitivity analysis tool in which one variable at a time is altered to observe the resulting impact of each variable change **independently** to the model output. This means that all the remaining input variables are held constant while the input feature of interest is being sensitized on. For example, the average or the median of each parameter within the training data set can be used for each parameter while the parameter of interest is changed. In this example, an excel sheet was created and called "Sensitivity Analysis." Each input feature was sensitized as follows:

-   Porosity is increased from 4.5% to 16.5% with increments of 1% while keeping the other parameters as approximately the average of the training set to obtain the predicted values of Aroot(K) from the trained model.
-   Acoustic impedance is increased from 1.4 to 5 with increments of 0.1 while keeping the other parameters as approximately the average of the training set to obtain the predicted values of Aroot(K) from the trained model.

- Brittleness ratio is increased from 13 to 103 with increments of 10 while keeping the other parameters as approximately the average of the training set to obtain the predicted values of Aroot(K) from the trained model.
- Vitrinite reflectance is increased from 0.74 to 2.34 with increments of 0.2 while keeping the other parameters as approximately the average of the training set to obtain the predicted values of Aroot(K) from the trained model.
- Since the output feature cannot be left as blank (or Python will give an error), an arbitrary Aroot(K) value of 50 was selected for the whole column. Please note that Aroot(K) is the column that will be predicted and as long as the output cells are filled with a value, the code will be run, otherwise, Python will give an error.

NOTE: It is important to note that the sensitivity range that was selected for each input feature is between the minimum and maximum range of each feature. This is very important to understand as it is highly recommended to avoid sensitizing on ranges that are beyond the range of the training data set. In other words, do not use ML models to extrapolate points that the model has never seen before. ML models are suited for interpolation but not extrapolation.

Let's import the excel file named "Chapter5_Geologic_Sensitivity_Data-Set.xlsx" located below in the same Notebook from linear regression analysis: https://www.elsevier.com/books-and-journals/book-companion/9780128219294

```
df_sensitivity=pd.read_excel('Chapter5_Geologic_Sensitivity_
   DataSet.xlsx')
df_sensitivity.describe().transpose()
```
Python output=Fig. 5.12

Since the training data set was normalized prior to applying the multilinear regression model, let's first normalize the sensitivity data set prior to asking for prediction as follows:

```
scaled_sensitivity=scaler.transform(df_sensitivity)
```

Let's also convert the resulting numpy array (scaled sensitivity) to a data frame and define the input and output features for this sensitivity data set.

	count	mean	std	min	25%	50%	75%	max
Porosity (%)	69.0	10.500000	1.635992	4.50	10.5	10.50	10.5	16.50
Acoustic impedance (kg/m2s*10^6)	69.0	3.223188	0.787988	1.40	3.1	3.25	3.3	5.00
Brittleness Ratio	69.0	58.855072	11.020403	13.00	59.0	59.00	59.0	103.00
Vitrinite Reflectance (%)	69.0	1.505217	0.188357	0.74	1.5	1.50	1.5	2.34
Aroot(K)	69.0	50.000000	0.000000	50.00	50.0	50.00	50.0	50.00

FIGURE 5.12 df_sensitivity description.

```
scaled_sensitivity=pd.DataFrame(scaled_sensitivity, columns=
    ['Porosity (%)', 'Acoustic impedance (kg/m2s*10^6)',
    'Brittleness Ratio', 'Vitrinite Reflectance (%)', 'Aroot(K)'])
y_scaled_sensitivity=scaled_sensitivity['Aroot(K)']
x_scaled_sensitivity=scaled_sensitivity.drop(['Aroot(K)'],
    axis=1)
```

Thus far, the sensitivity data set has been imported and x and y features have been defined. Now, let's apply the "lm" model that was trained to "x_scaled_sensitivity" to predict the y values for each of the sensitivity rows in the sensitivity data frame.

```
y_pred_sensitivity=lm.predict(x_scaled_sensitivity)
```

"y_pred_sensitivity" is an array. Therefore, let's convert it to a data frame and give it a column name. Afterward, since the resulting predicted values are normalized, transform the predicted sensitivity y values back to its original form as illustrated below. This can be done by multiplying the predicted Aroot(K) values by the difference between max and min of the original data set followed by adding the resulting term to minimum of the original Aroot(K) data frame.

```
y_pred_sensitivity=pd.DataFrame(y_pred_sensitivity,columns=
    ['Predicted Aroot(K)'])
y_pred_sensitivity=y_pred_sensitivity['Predicted Aroot(K)']*(df
    ['Aroot(K)'].max()-df['Aroot(K)'].min())+df['Aroot(K)'].min()
```

Next, let's examine the impact of each variable on Aroot(K) by plotting each variable versus "y_pred_sensitivity." First, use the np.linspace to create an array between porosity values of 4.5 and 16.5 with 13 values which will result in porosity values with 1% increment as defined previously. Next, plot por versus "y_pred_sensitivity.iloc[0:13]." As discussed in Chapter 1, ".iloc [0:13]" will result in assigning the predicted Aroot(K) values from index 0 through 12 (which are the associated predicted Aroot(K) values with porosity). As illustrated below, as porosity increases, Aroot(K) significantly increases as well. Therefore, qualitative and quantitative impacts of each independent variable can be observed on the target feature which was defined as Aroot(K) for this example problem. Please note that partial dependence plots will be illustrated to perform similar sensitivity analyses in chapter 7. This methodology is useful if a set of specific input features are desired to be used to predict the output(s).

```
plt.figure(figsize=(8,5))
por=np.linspace(4.5,16.5,13)
plt.scatter(por,y_pred_sensitivity.iloc[0:13])
plt.title('Porosity Impact on Aroot(K) Sensitivity')
plt.xlabel('Porosity (%)')
plt.ylabel('Aroot(K)')
Python output=Fig. 5.13
```

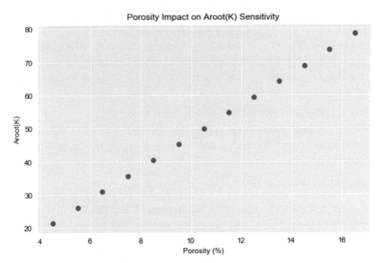

FIGURE 5.13 Porosity OVAT on Aroot(K).

Let's also understand the independent and quantitative impacts of other three variables on Aroot(K) as follows:

```
plt.figure(figsize=(8,5))
AI=np.linspace(1.4,5,37)
plt.scatter(AI,y_pred_sensitivity.iloc[13:50])
plt.title('AI Impact on Aroot(K) Sensitivity')
plt.xlabel('AI')
plt.ylabel('Aroot(K)')
Python output=Fig. 5.14
```

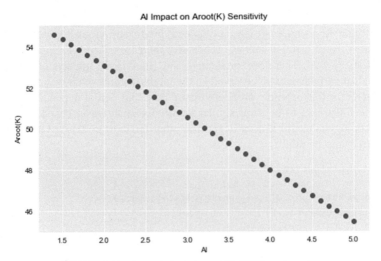

FIGURE 5.14 Acoustic impedance (AI) OVAT on Aroot(K).

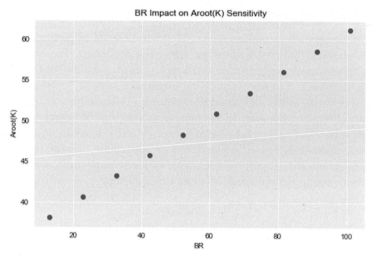

FIGURE 5.15 Brittleness ratio (BR) OVAT on Aroot(K).

```
plt.figure(figsize=(8,5))
BR=np.linspace(13,101,10)
plt.scatter(BR,y_pred_sensitivity.iloc[50:60])
plt.title('BR Impact on Aroot(K) Sensitivity')
plt.xlabel('BR')
plt.ylabel('Aroot(K)')
```
Python output=Fig. 5.15

```
plt.figure(figsize=(8,5))
VR=np.linspace(0.74,2.34,9)
plt.scatter(VR,y_pred_sensitivity.iloc[60:69])
plt.title('VR Impact on Aroot(K) Sensitivity')
plt.xlabel('VR')
plt.ylabel('Aroot(K)')
```
Python output=Fig. 5.16

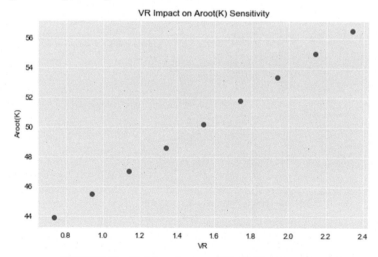

FIGURE 5.16 Vitrinite reflectance (VR) OVAT on Aroot(K).

Logistic regression

Logistic regression is another powerful supervised ML algorithm used for binary classification problems (when target is categorical). The best way to think about logistic regression is that it is a linear regression but for classification problems. Logistic regression essentially uses a logistic function defined below to model a binary output variable (Tolles & Meurer, 2016). The primary difference between linear regression and logistic regression is that logistic regression's range is bounded between 0 and 1. In addition, as opposed to linear regression, logistic regression does not require a linear relationship between inputs and output variables. This is due to applying a nonlinear log transformation to the odds ratio (will be defined shortly).

$$Logistic\ function = \frac{1}{1 + e^{-x}} \qquad (5.6)$$

In the logistic function equation, x is the input variable. Let's **feed in values** -20 to 20 into the logistic function. As illustrated in Fig. 5.17, the inputs have been transferred to between 0 and 1.

As opposed to linear regression where MSE or RMSE is used as the loss function, logistic regression uses a loss function referred to as "maximum likelihood estimation (MLE)" which is a conditional probability. If the probability is greater than 0.5, the predictions will be classified as class 0. Otherwise, class 1 will be assigned. Before going through logistic regression derivation, let's first define the **logit** function. Logit function is defined as the natural log of the odds. A probability of 0.5 corresponds to a logit of 0, probabilities smaller than 0.5 correspond to negative logit values, and probabilities greater than 0.5 correspond to positive logit values. It is important to note that as illustrated in Fig. 5.17, logistic function ranges between 0 and 1

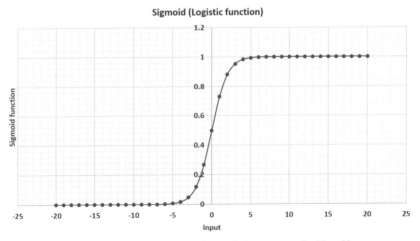

FIGURE 5.17 Logistic regression applied to a range of -20 to 20.

($P\in[0,1]$) while logit function can be any real number from minus infinity to positive infinity ($P\in[-\infty, \infty]$).

$$odds = \frac{P}{1-P} \to logit(P) = \ln\left(\frac{P}{1-P}\right) \qquad (5.7)$$

Let's set logit of P to be equal to $mx + b$, therefore:

$$logit(P) = mx + b \to mx + b = \ln\left(\frac{P}{1-P}\right)$$

$$\left(\frac{P}{1-P}\right) = e^{(mx+b)} \to P = \frac{e^{(mx+b)}}{1+e^{(mx+b)}} \to P(x) = \frac{1}{1+e^{-(mx+b)}} \qquad (5.8)$$

Before using logistic regression in scikit-learn, let's review some very important classification metrics used for evaluating a classification model such as logistic regression.

Metrics for classification model evaluation

A classification accuracy is defined as follows:

$$Accuracy = \frac{Number\ of\ correct\ predictions}{Number\ of\ all\ predictions} \qquad (5.9)$$

Although accuracy provides a general understanding of the model's accuracy in terms of predicting the correct classes, one major problem with using accuracy when evaluating a classification model is when there is unbalanced class distribution within a data set. For example, if in a data set with 1 MM rows, 100K rows are assigned to class 0 and the remaining 900K rows are assigned to class 1, this is classified as an unbalanced distribution. Therefore, confusion matrix is recommended to obtain more information in terms of evaluating a model's accuracy. Confusion matrix shows the correct and incorrect predictions on each class. For example, if there are two classes that a classification ML model predicted, a confusion matrix shows the true and false predictions on each class. In a binary classification problem when a logistic regression model is used, confusion matrix is a 2 × 2 matrix. For N output classification problems, NxN (N by N) matrix will be used. Now, let's define the terminologies used within a confusion matrix assuming a 2 × 2 confusion matrix with class A and B:

- True positive: the number of times that a ML model is correctly classifying the sample as class A
- False negative: the number of times that a ML model is incorrectly classifying the sample as class A
- False positive: the number of times that a ML model is incorrectly classifying the sample as class B
- True negative: the number of times that a ML model is correctly classifying the sample as class B

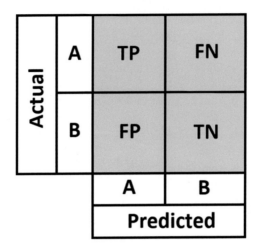

FIGURE 5.18 Confusion matrix.

The idea is to maximize true positive and true negative and minimize false positive and false negative instances. Fig. 5.18 illustrates a confusion matrix with two classes (A and B) with the terminologies described above.

Precision and recall are metrics used to evaluate a classification model. These parameters provide more insight into a model's accuracy. Precision and recall are defined below:

$$Precision = \frac{True\ positive}{Predicted\ results} = \frac{True\ positive}{True\ positive + false\ positive}$$

$$Recall = \frac{True\ positive}{Actual\ results} = \frac{True\ positive}{True\ positive + false\ negative}$$

(5.10)

As illustrated, precision measures the accuracy of a model when evaluating positive classes. In other words, out of the predicted positive classes, precision determines the number that is actually positive. Precision is preferred when the cost of false positive is very high. For instance, ML can be used for spam email detection. In an email detection, a false positive indicates that an email that is not a spam was identified as a spam. Thus, the end user might not receive important emails if precision is not high. Therefore, there is a trade-off between precision and recall. As noted, precision and recall cannot be maximized at the same time and one is preferred over the other depending on the project. These two metrics are inversely proportional.

Recall measures the number of actual positives. Recall is used when there is a high cost associated with **false negative**. In the banking industry, fraudulent detection is very important. Therefore, if a fraudulent transaction is predicted as nonfraudulent, it could lead to terrible consequences for the bank. In summary:

- High cost associated with false positive: maximize precision
- High cost associated with false negative: maximize recall

In the O&G industry, if the goal is to classify a screen-out stage ahead of time, the idea is to send a warning and make sure any instance that could potentially lead to screening-out (sanding off) a well is classified as screen-out. Otherwise, a costly screen-out could be the consequence. Therefore, there is a high cost associated with false negative and the goal is to maximize recall when training a classification ML model. This will lead to proper investigation of every alert to be on the cautious side.

When solving problems with uneven class distribution, another classification metric that can be used is referred to as F_1 score because it considers both false positive and false negative. F_1 score is also beneficial when a balance between precision and recall is desired. The equation for F_1 score is as follows:

$$F_{1,score} = 2 \frac{Precision \times recall}{Precision + recall} \qquad (5.11)$$

Logistic regression using scikit-learn

ML classification models have many different applications including human resources within the oil and gas industry. In this exercise, an HR data set which consists of 10 input and 1 output features are used to determine whether an employee will quit or not. This data set can be found in the link below:

https://www.elsevier.com/books-and-journals/book-companion/9780128219294

The input features are as follows:

Late show up percentage, Project initiative percentage, Percentage of project delivery on time, Percentage of emails exchanged, Percentage of responsiveness, Percentage of professional email response, Percentage of sharing ideas, Percentage of helping colleagues, Percentage of entrepreneurial posts on LinkedIn, Percentage of Facebook comments.

The output feature is called "Quitting" and is a binary class where 0 represents an employee staying and 1 representing an employee quitting. Please note that this data set does not require any normalization since all the input features are percentages and are already on the same scale. With that in mind, let's start a new Jupyter Notebook and start importing the usual libraries as follows and import the data set that will be used in this exercise:

```
import numpy as np
import pandas as pd
import matplotlib.pyplot as plt
import seaborn as sns
```

	count	mean	std	min	25%	50%	75%	max
Late show up percentage	1000.0	0.501025	0.187179	0.0	0.367040	0.495075	0.639074	1.0
Project initiative percentage	1000.0	0.483284	0.184640	0.0	0.359586	0.486289	0.622330	1.0
Percentage of project delivery on time	1000.0	0.427376	0.187881	0.0	0.286459	0.413932	0.552530	1.0
Percentage of emails exchanged	1000.0	0.400711	0.144444	0.0	0.295613	0.397399	0.496453	1.0
Percentage of responsiveness	1000.0	0.537204	0.182367	0.0	0.416221	0.539818	0.661523	1.0
Percentage of professional email response	1000.0	0.484969	0.182355	0.0	0.355052	0.483522	0.612790	1.0
Percentage of sharing ideas	1000.0	0.471185	0.179900	0.0	0.344950	0.458429	0.592071	1.0
Percentage of helping colleagues	1000.0	0.519861	0.194477	0.0	0.381461	0.515514	0.661990	1.0
Percentage of entrepreneurial posts on LinkedIn	1000.0	0.521589	0.193181	0.0	0.374331	0.526394	0.669513	1.0
Percentage of Facebook comments	1000.0	0.576462	0.162826	0.0	0.464761	0.586543	0.689762	1.0
Quitting	1000.0	0.500000	0.500250	0.0	0.000000	0.500000	1.000000	1.0

FIGURE 5.19 HR description.

```
%matplotlib inline
df=pd.read_excel('Chapter5_HR_DataSet.xlsx')
df.describe().transpose()
```
Python output=Fig. 5.19

Let's plot the distribution and violin plots as follows:

```
f, axes=plt.subplots(5, 2, figsize=(12, 12))
sns.distplot(df['Late show up percentage'], color="red", ax=axes
  [0, 0])
sns.distplot(df['Project initiative percentage'], color="olive",
  ax=axes[0, 1])
sns.distplot(df['Percentage of project delivery on time'],
  color="blue", ax=axes[1, 0])
sns.distplot(df['Percentage of emails exchanged'], color="orange",
  ax=axes[1, 1])
sns.distplot(df['Percentage of responsiveness'], color="black",
  ax=axes[2, 0])
sns.distplot(df['Percentage of professional email response'],
  color="green", ax=axes[2, 1])
sns.distplot(df['Percentage of sharing ideas'], color="cyan",
  ax=axes[3, 0])
sns.distplot(df['Percentage of helping colleagues'],
  color="brown", ax=axes[3, 1])
sns.distplot(df['Percentage of entrepreneurial posts on LinkedIn'],
  color="purple", ax=axes[4, 0])
sns.distplot(df['Percentage of Facebook comments'], color="pink",
  ax=axes[4, 1])
plt.tight_layout()
```
Python output=Fig. 5.20

```
f, axes=plt.subplots(5, 2, figsize=(12, 12))
sns.violinplot(df['Late show up percentage'], color="red", ax=axes
  [0, 0])
```

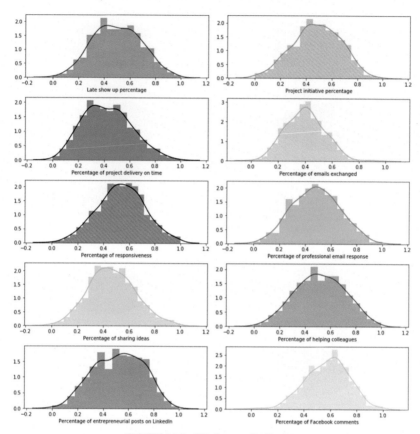

FIGURE 5.20 HR data set distribution.

```
sns.violinplot(df['Project initiative percentage'], color="olive",
  ax=axes[0, 1])
sns.violinplot(df['Percentage of project delivery on time'],
  color="blue", ax=axes[1, 0])
sns.violinplot(df['Percentage of emails exchanged'],
  color="orange", ax=axes[1, 1])
sns.violinplot(df['Percentage of responsiveness'], color="black",
  ax=axes[2, 0])
sns.violinplot(df['Percentage of professional email response'],
  color="green", ax=axes[2, 1])
sns.violinplot(df['Percentage of sharing ideas'], color="cyan",
  ax=axes[3, 0])
sns.violinplot(df['Percentage of helping colleagues'],
  color="cyan", ax=axes[3, 1])
sns.violinplot(df['Percentage of entrepreneurial posts on LinkedIn'],
  color="cyan", ax=axes[4, 0])
sns.violinplot(df['Percentage of Facebook comments'],
  color="cyan", ax=axes[4, 1])
plt.tight_layout()
```
Python output=Fig. 5.21

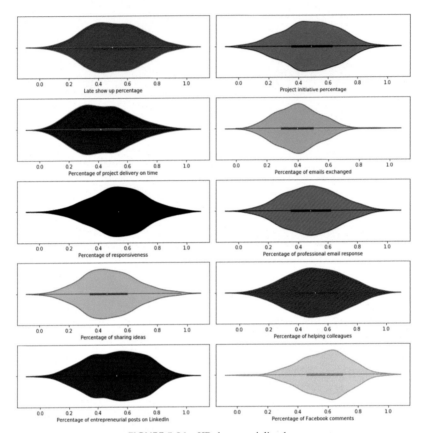

FIGURE 5.21 HR data set violin plot.

As shown in distribution plots, all the input features appear to have normal distributions and do not show any outliers. Next, let's plot the Pearson correlation coefficient heat map as follows. There appears to be no collinear input features in this data set. For this exercise, all the 10 features shown below will be used in the model. However, when discussing some other algorithms such as random forest, extra trees, gradient boosting, etc., feature ranking can be used to remove unimportant features that have a negligible impact on a model's output.

```
plt.figure(figsize=(12,10))
sns.heatmap(df.corr(), annot=True, linecolor='white',
    linewidths=2, cmap='Accent')
Python output=Fig. 5.22
```

Please note that various "cmap" or color maps can be chosen as shown in Fig. 5.23.

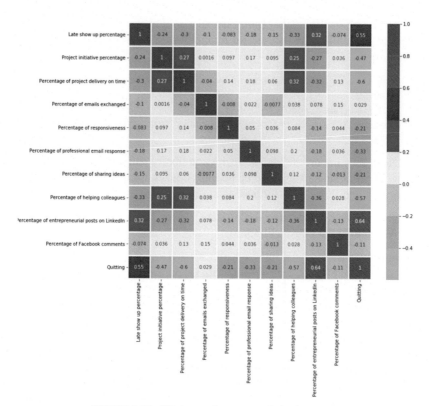

FIGURE 5.22 HR data set Pearson correlation heat map.

BuPu, BuPu_r, CMRmap, CMRmap_r, Dark2, Dark2_r, GnBu, GnBu_r, Greens, Greens_r, Greys, Greys_r, OrRd, OrRd_r, Oranges, Oranges_r, PRGn, PRGn_r, Paired, Paired_r, Pastel1, Pastel1_r, Pastel2, Pastel2_r, PiYG, PiYG_r, PuBu, PuBuGn, PuBuGn_r, PuB u_r, PuOr, PuOr_r, PuRd, PuRd_r, Purples, Purples_r, RdBu, RdBu_r, RdGy, RdGy_r, RdPu, RdPu_r, RdYlBu, RdYlBu_r, RdYlGn, RdY lGn_r, Reds, Reds_r, Set1, Set1_r, Set2, Set2_r, Set3, Set3_r, Spectral, Spectral_r, Wistia, Wistia_r, YlGn, YlGnBu, YlGnBu_ r, YlGn_r, YlOrBr, YlOrBr_r, YlOrRd, YlOrRd_r, afmhot, afmhot_r, autumn, autumn_r, binary, binary_r, bone, bone_r, brg, brg_ r, bwr, bwr_r, cividis, cividis_r, cool, cool_r, coolwarm, coolwarm_r, copper, copper_r, cubehelix, cubehelix_r, flag, flag_ r, gist_earth, gist_earth_r, gist_gray, gist_gray_r, gist_heat, gist_heat_r, gist_ncar, gist_ncar_r, gist_rainbow, gist_rain bow_r, gist_stern, gist_stern_r, gist_yarg, gist_yarg_r, gnuplot, gnuplot2, gnuplot2_r, gnuplot_r, gray, gray_r, hot, hot_r, hsv, hsv_r, icefire, icefire_r, inferno, inferno_r, jet, jet_r, magma, magma_r, mako, mako_r, nipy_spectral, nipy_spectral_ r, ocean, ocean_r, pink, pink_r, plasma, plasma_r, prism, prism_r, rainbow, rainbow_r, rocket, rocket_r, seismic, seismic_r, spring, spring_r, summer, summer_r, tab10, tab10_r, tab20, tab20_r, tab20b, tab20b_r, tab20c, tab20c_r, terrain, terrain_r, twilight, twilight_r, twilight_shifted, twilight_shifted_r, viridis, viridis_r, vlag, vlag_r, winter, winter_r

FIGURE 5.23 Color scheme options.

Now, let's define x and y variables. As indicated, all parameters are already between a scale of 0 and 1 and therefore, no normalization is necessary. As shown below, all the features except the "Quitting" column are used as features and the "Quitting" column is used as the output feature.

```
x_features=df.drop('Quitting',axis=1)
y=df['Quitting']
```

Next, import "train_test_split," define a seed number of 50, and use a 70/30 split randomly (70% is used for training and 30% will be used for testing randomly).

```
from sklearn.model_selection import train_test_split
seed=50
np.random.seed(seed)
X_train, X_test, y_train, y_test=train_test_split(x_features,y,
    test_size=0.30)
```

Next, let's import the logistic regression library and call it "lr." There are some hyperparameters that can be optimized when using a logistic regression algorithm and they are as follows:

- The solver options that scikit-learn offers in logistic regression are "newton-cg," "lbfgs," "liblinear," "sag," "saga."

newton-cg is a Newton method which uses a Hessian matrix. Hessian matrix was developed by the German mathematician named Ludwig Otto Hesse in the 19th century which uses a square matrix of second-order partial derivatives of a scalar-valued function. Please note that since second derivative is calculated in this method, this solver can be slow for large data sets.

The next frequently used algorithm which is the default algorithm in scikit-learn 0.22.0 is called **lbfgs**. lbfgs stands for limited-memory Broyden−Fletcher−Goldfarb−Shanno. lbfgs is an iterative method for solving unconstrained nonlinear optimization problems which approximates the second derivative matrix updates with gradient evaluations. It is referred to as a short-term memory because it only stores the last few updates (Avriel, 2003).

"liblinear" uses a coordinate descent algorithm which is based on minimizing a coordinate directions of a function in a loop. This solver performs well with high-dimensional data (Wright, 2015).

"sag" is stochastic average gradient descent (will be discussed in detail). "saga" is an extension of "sag" that allows for L_1 regularization (discussed below). Please note that "sag" and "saga" are faster for larger data sets.

As previously discussed, when a model is memorized and specifically tailored for a particular training set without having the capability to predict accurately when applied to a test or blind set, it is referred to as overfitting. Overfitting essentially means the model is not generalized to predict a test or bind set with high accuracy, and it is considered to be a major challenge when developing a ML model. Therefore, regularization is used to avoid overfitting. Regularization is the process of introducing additional fine-tuning parameters in an attempt to prevent overfitting. There are different techniques to avoid overfitting and some of the commonly used techniques are regularization, cross-validation, and drop out.

Regularization is simply the process of reducing complexity in a model. Reducing a model complexity can be done by penalizing the loss function.

Loss function is simply the sum of squared difference between actual versus predicted values as shown below:

$$\sum_{i=1}^{n}\left(y_i - \widehat{y}\right)^2 \tag{5.12}$$

A regularization term can be added to the above equation as follows (highlighted in red):

$$\sum_{i=1}^{n}\left(y_i - \widehat{y}\right)^2 + \lambda\sum_{i=1}^{n}\theta_i^2 \tag{5.13}$$

θ_i represents the weights of the features and λ represents the regularization strength (penalty) parameter. The regularization parameter determines the amount of penalty to apply to the weights. If no regularization is needed, simply set $\lambda=0$ which will eliminate the regularization expression in the discussed equation. A large λ could result in a model being underfit and a low λ could potentially lead to an overfit model. When grid search is discussed in the subsequent chapters, regularization parameter can be used as one of the hyper-tuning parameters used in a model optimization using grid search.

Examples of regularization parameters are L_1 and L_2 regularization. L_1 regularization is referred to as L_1 norm or Lasso regularization. In L_1 regularization, the absolute value of weights is penalized as shown below. In addition, L_1 regularization has a sparse solution that is robust to outliers and could possibly have multiple solutions. Please note that L_1 generates simple models and cannot interpret complex patterns. Therefore, for complex pattern recognition, it is recommended to use L_2 regularization.

$$\sum_{i=1}^{n}\left(y_i - \widehat{y}\right)^2 + \lambda\sum_{i=1}^{n}|\theta_i| \tag{5.14}$$

On the other hand, L_2 regularization is referred to as ridge regularization which uses the sum of square of all feature weights as highlighted below. In L_2 regularization, the weights are forced to be small. In contrast to L_1 regularization, L_2 regularization has a nonsparse solution, is not robust to outliers, and performs better with complex data patterns. L_2 regularization always has one solution.

$$\sum_{i=1}^{n}\left(y_i - \widehat{y}\right)^2 + \lambda\sum_{i=1}^{n}\theta_i^2 \tag{5.15}$$

C is the inverse of λ (regularization strength) and the default value of C is 1. C is one of the essential parameters used in SVM algorithm which will be discussed in more detail later in this chapter.

$$C = \frac{1}{\lambda} \qquad (5.16)$$

```
from sklearn.linear_model import LogisticRegression
np.random.seed(seed)
lr=LogisticRegression(penalty='l2', C=1.0,solver='lbfgs')
```

The model parameters for logistic regression were all defined under "lr." Let's apply the lr model to fit (X_train,y_train) as follows:

```
lr.fit(X_train,y_train)
```

Next, let's apply the "lr" trained model to predict the "X_test" and call the resultant array as the "y_pred" as shown below:

```
y_pred=lr.predict(X_test)
```

To generate the confusion matrix and classification metrics discussed earlier, import "classification_report" and "confusion_matrix" from "sklearn.metrics" and print the resultant confusion matrix of y_test (actual testing output values) and y_pred (predicted testing output values). In addition, generate a heat map of the resulting confusion matrix:

```
from sklearn.metrics import classification_report,confusion_matrix
print(confusion_matrix(y_test,y_pred))
sns.heatmap(confusion_matrix(y_test,y_pred), annot=True,
  cmap='viridis')
```
Python output=Fig. 5.24

```
[[143    9]
 [  4 144]]
```

`<matplotlib.axes._subplots.AxesSubplot at 0x19e1f604f48>`

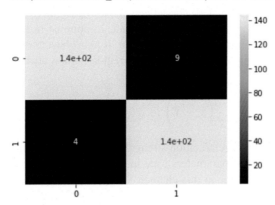

FIGURE 5.24 Confusion matrix of testing set and its heat map.

```
           precision    recall  f1-score   support

        0       0.97      0.94      0.96       152
        1       0.94      0.97      0.96       148

 accuracy                          0.96       300
macro avg       0.96      0.96      0.96       300
weighted avg    0.96      0.96      0.96       300
```

FIGURE 5.25 Classification report.

As illustrated in Fig. 5.24, 143 instances (out of 152) were correctly classified as class 0 (not quitting) and 9 were incorrectly classified under this class. On the other hand, 144 instances (out of 148) were correctly classified as class 1 (quitting) and only 4 instances were incorrectly classified.

Next, let's also get the classification report as follows:

```
print(classification_report(y_test,y_pred))
Python output=Fig. 5.25
```

The overall accuracy of the model is 96%. In addition, the precision and recall percentage for each class are high and range from 94% to 97% with a balanced class instances in each group. It is possible to improve the accuracy of this model by changing some of the fine-tuning parameters discussed either manually or using a grid search optimization (recommended approach) which will be discussed in chapter 7. For now, just change some of the fine-tuning parameters manually and observe the resultant impact on the model.

K-nearest neighbor

K-nearest neighbor also known as KNN is one of the simplest forms of supervised ML algorithm that is used for both classification and regression problems. KNN is assumed to be a nonparametric algorithm which means no assumptions are made about the underlying data (Cover & Hart, 1967). The KNN algorithm works based on the basis of similar proximity using distance calculations. The following steps are used when developing a KNN model:

1) Determine the number of nearest neighbors also known as K. For instance, if K=2, two of the closest points based on distance calculation will be chosen to determine where an instance will be assigned (in a classification problem). Selecting K can be challenging in a KNN algorithm. Choosing a small K means a higher influence on the result. On the other hand, choosing a higher K could lead to a smoother decision boundary with lower variance but increased bias. One approach in selecting K is to train a model with various K neighbors such as 1, 2 … etc., to see which K would result in the highest testing accuracy using a cross-validation technique discussed in the model validation chapter. Prior to proceeding to the next step, make sure to standardize the data.

TABLE 5.1 Oil and natural gas price categories (KNN example).

Oil price ($/BBL)	Natural gas price ($/MMBTU)	Category
90	4	Strong
40	2	Poor
60	3	Strong
80	3.5	Strong
35	1.5	Poor

2) Use a distance function such as euclidean to calculate the distance between each instance and all training samples.
3) Next rank the distance from lowest to highest. Afterward, choose the nearest neighbor(s) based on the number of nearest neighbors (K) that was selected in step 1. In other words, choose K neighbors with the lowest distance.
4) Obtain the category (classification) or numeric value (regression) of the nearest neighbors obtained in step 3.
5) Use average (in regression problems) or majority (in classification problems) of the nearest neighbors to predict a value or a class of an instance.

To illustrate this, let's go through a step-by-step KNN classification example. The following two features (oil price and gas price) belong to two categories of "Poor" and "Strong." Use 3 nearest neighbors (K=3) and euclidean distance calculation to classify whether point (20, 2.2) would be classified under "Poor" or "Strong" in Table 5.1. In other words, which category would an oil price of $20/BBL and a natural gas price of $2.2/MMBTU fall under.

Step 1) Calculate the euclidean distance between each row and the given points (20,2.2) as follows:

$$Euclidean\ distance\ between\ (90,4)\ and\ (20,2.2)$$

$$= \sqrt{(90-20)^2 + (4-2.2)^2} = 70.02$$

$$Euclidean\ distance\ between\ (40,2)\ and\ (20,2.2)$$

$$= \sqrt{(40-20)^2 + (2-2.2)^2} = 20.00$$

$$Euclidean\ distance\ between\ (60,3)\ and\ (20,2.2)$$

$$= \sqrt{(60-20)^2 + (3-2.2)^2} = 40.01$$

$$Euclidean\ distance\ between\ (80,3.5)\ and\ (20,2.2)$$

$$= \sqrt{(80-20)^2 + (3.5-2.2)^2} = 60.01$$

$$Euclidean\ distance\ between\ (35,1.5)\ and\ (20,2.2)$$

$$= \sqrt{(35-20)^2 + (1.5-2.2)^2} = 15.02$$

(5.17)

TABLE 5.2 Lowest to highest distance for KNN classification.

Oil price ($/BBL)	Natural gas price ($/MMBTU)	Category	Euclidean distance
35	1.5	Poor	15.02
40	2	Poor	20.00
60	3	Strong	40.01
80	3.5	Strong	60.01
90	4	Strong	70.02

Step 2) Rank from lowest to highest distance as shown in Table 5.2.

Step 3) Since K=3 and the two out of the three nearest neighbors (bold rows shown in Table 5.2) are classified under "Poor," point (20,2.2) would be classified under "Poor."

KNN has the following advantages:

- As was illustrated in the example above, KNN is easy to implement.
- It is considered a lazy algorithm which essentially indicates no prior training is required prior to making real-time decisions. In other words, it memorizes the training data set as opposed to learning a discriminative function from the training data set (Raschka, n.d.).
- As opposed to many other algorithms that will be discussed later in this chapter, the main two parameters are number of K and distance function.

Some of the disadvantages of using a KNN are as follows:

- Distance calculation does not work well in a high-dimensional data set; therefore, using KNN can be challenging in high-dimensional data sets.
- In addition, when KNN is used for making predictions, the cost of predictions in large training data sets is high. This is because KNN will need to search for the nearest neighbors in the entire data set.
- It can computationally become intensive since the distance of each instance to all training samples must be calculated.
- KNN algorithm does not work well with categorical features since dimension distance calculation using categorical feature is difficult.

KNN implementation using scikit-learn

Open a new Jupyter Notebook, import the necessary libraries, and import the same HR data set used in the previous example.

```
import numpy as np
import pandas as pd
import matplotlib.pyplot as plt
import seaborn as sns
%matplotlib inline
df=pd.read_excel('Chapter5_HR_DataSet.xlsx')
```

Since various visualization plots were illustrated when logistic regression was covered for the same database, let's proceed to standardizing the data. As previously discussed, KNN uses a distance-based calculation when determining the nearest neighbor(s); therefore, it is recommended to standardize the data prior to applying KNN. In addition, define x_standardized features as input variables into the model by dropping the last column of the data set. y is simply the last column and the output of the model.

```
from sklearn.preprocessing import StandardScaler
scaler=StandardScaler()
scaler.fit(df.drop('Quitting',axis=1))
y=df['Quitting']
x_standardized_features=scaler.transform(df.drop
    ('Quitting',axis=1))
x_standardized_features=pd.DataFrame(x_standardized_features,
    columns=df.columns[:-1])
x_standardized_features.head()
```
Python output=Fig. 5.26

Next, import the model selection library and use the train_test_split to divide the data into a 70/30 split. Make sure to pass in "x_standardized_features" and "y" as shown below. Since 70% of the data will be used as training, 700 rows will be randomly selected as the training data and the remaining 300 rows will be selected for testing the model.

```
from sklearn.model_selection import train_test_split
seed=1000
np.random.seed(seed)
X_train, X_test, y_train, y_test=train_test_split(
    x_standardized_features,y, test_size=0.30)
```

Next, let's import "KNeighborsClassifier" from "sklearn.neighbors" and define the KNN parameters to have one neighbor and to use the euclidean

	Late show up percentage	Project initiative percentage	Percentage of project delivery on time	Percentage of emails exchanged	Percentage of responsiveness	Percentage of professional email response	Percentage of sharing ideas	Percentage of helping colleagues	Percentage of entrepreneurial posts on LinkedIn	Percentage of Facebook comments
0	-0.123542	0.185907	-0.913431	0.319629	-1.033637	-2.308375	-0.798951	-1.482368	-0.949719	-0.643314
1	-1.084836	-0.430348	-1.025313	0.625388	-0.444847	-1.152706	-1.129797	-0.202240	-1.828051	0.636759
2	-0.788702	0.339318	0.301511	0.755873	2.031693	-0.870156	2.599818	0.285707	-0.682494	-0.377850
3	0.982841	1.060193	-0.621399	0.625299	0.452820	-0.267220	1.750208	1.066491	1.241325	-1.026987
4	1.139275	-0.640392	-0.709819	-0.057175	0.822886	-0.936773	0.596782	-1.472352	1.040772	0.276510

FIGURE 5.26 HR standardized data.

distance calculation. Afterward, apply the defined "KNN" model to "(X_train, y_train)" for training the model.

```
from sklearn.neighbors import KNeighborsClassifier
np.random.seed(seed)
KNN=KNeighborsClassifier(n_neighbors=1, metric='euclidean')
KNN.fit(X_train,y_train)
```

Before applying the model, let's create a for loop to test various number of neighbors from 1 to 50 and evaluate the outcome on the model's testing accuracy. First, import the metrics so accuracy can be obtained at the end of the loop. Next, create an empty list called "Score." Afterward, write a for loop for K in range of 1 through 50 (including 50), fit "(X_train,y_train)," predict "X_test" (30% testing data set), place the predicted values under "y_pred," append the accuracy score between "y_test" (actual output values) and "y_pred" (predicted output values) to "Scores" which is the empty list that was created initially. Please make sure that anything under the for loop is indented.

```
from sklearn import metrics
Scores=[]
for k in range(1, 51):
    KNN=KNeighborsClassifier(n_neighbors=k,metric='euclidean')
    KNN.fit(X_train, y_train)
    y_pred=KNN.predict(X_test)
    Scores.append(metrics.accuracy_score(y_test, y_pred))
```

Since the testing score values have been appended to "Score," it is important to visualize the resulting accuracy score at different trialed Ks as follows:

```
plt.figure(figsize=(10,8))
plt.plot(range(1, 51), Scores)
plt.xlabel('K Values')
plt.ylabel('Testing Accuracy')
plt.title('K Determination Using KNN', fontsize=20)
Python output=Fig. 5.27
```

As illustrated, when K is equal to 5, testing accuracy appears to be one of the highest. Other peaks are also observed at higher K values but 5 would be the first pick in this data set. Therefore, let's apply the KNN model using 5 neighbors this time and proceed with obtaining the confusion matrix and classification report. Please note that "print('\n')" simply add extra space in the output code.

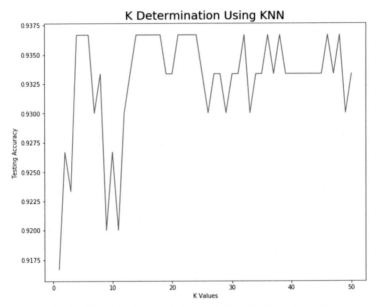

FIGURE 5.27 Number of neighbors (K) determination using KNN.

```
from sklearn.metrics import classification_report,confusion_matrix
np.random.seed(seed)
KNN=KNeighborsClassifier(n_neighbors=5,metric='euclidean')
KNN.fit(X_train,y_train)
y_pred=KNN.predict(X_test)
print('\n')
print(confusion_matrix(y_test,y_pred))
print('\n')
print(classification_report(y_test,y_pred))
```
Python output=Fig. 5.28

```
[[142    9]
 [ 10 139]]
```

	precision	recall	f1-score	support
0	0.93	0.94	0.94	151
1	0.94	0.93	0.94	149
accuracy			0.94	300
macro avg	0.94	0.94	0.94	300
weighted avg	0.94	0.94	0.94	300

FIGURE 5.28 Confusion matrix and classification report.

Support vector machine

SVM is another form of a supervised ML algorithm that can be used for both classification and regression problems. As opposed to many of the ML algorithms where the objective function is to minimize a cost function, the objective in SVM is to maximize the margin between support vectors through a separating hyperplane (Cortes & Vapnik, 1995). In Fig. 5.29, a separating hyperplane (solid black line) is drawn to separate blue instances from the red ones. This hyperplane is drawn in a manner to maximize the margin from both sides. Please note that the widest possible stretch must be drawn to separate the red and blue instances. The closest blue and red instances to the separating hyperplane are referred to as support vectors and this is why this algorithm is called support vector machine. The distance between the dashed blue and red lines to the separating hyperplane (solid black line) is called the "margin." As previously mentioned, the idea is to maximize the margin from both sides. This illustration is for a two-dimensional space and hyperplane will be n-1 for n-dimensional space. For instance, for a two-dimensional space (two features), the hyperplane will be one dimension, which is just a line. For a three-dimensional space (three input features), the hyperplane becomes a two-dimensional plane. For a four-dimensional space (four features), the hyperplane would have three dimensions. This figure illustrates the simplest form of classification problem for linearly separable cases.

The goal here is to maximize the margins between the support vectors.

Fig. 5.29 illustrates a data set with two input features for binary classification that are linearly separable. This is unfortunately not the case with most of the real-world application problems. Therefore, the following two concepts are very important to consider when the data are not linearly separable.

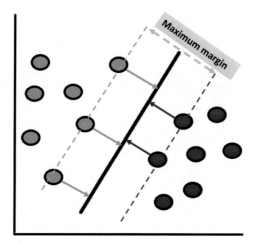

FIGURE 5.29 Support vector machine for classification illustration.

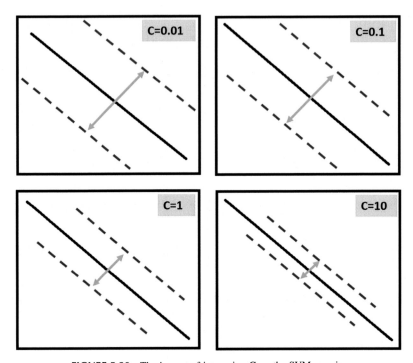

FIGURE 5.30 The impact of increasing C on the SVM margin.

1) **Soft margin:** is used to find a line for class separation but this line will tolerate one or few misclassified instances. This degree of tolerance is represented by the parameter "C" in SVM algorithm. As shown in Fig. 5.30, C essentially is a trade-off parameter used between having a **wide margin** and correctly **classifying training data**. C behaves as a regularization parameter when using SVM. C is one of the parameters that can be used for grid search optimization (which will be discussed in model validation chapter).
   - Larger C= Smaller width of the margin (smaller margin)= More incurred penalty when misclassified= Higher training accuracy and lower testing accuracy= Less general
   Lower C= Larger width of the margin (larger margin)= Less incurred penalty when misclassified= Lower training accuracy and higher testing accuracy= More general
      When C is large, the decision boundary will depend on fewer support vectors because the margin is narrower.
2) **Kernel tricks:** when the data are not linearly separable, kernel tricks utilizes existing features and applies some transformation functions to create new features. For example, linear, polynomial, or radial basis

function (rbf) transformations can be used when dealing with nonlinearly separable data.

Within the scikit-learn library, there are different Kernel functions such as "linear," "poly," "rbf," "sigmoid," etc., that can be chosen. The most popular, typically providing the highest accuracy, are poly and rbf. Therefore, let's review these two transformation functions:

- Polynomial kernel (or poly kernel):

  Polynomial kernel is basically a transformer that generates new features. SVM with a polynomial kernel can generate a nonlinear decision boundary. It might be impossible to separate a numpy array of X and Y with a line. However, after applying a polynomial transformer, a line can be drawn to separate the X and Y arrays. Polynomial kernel is defined as:

$$K(X_1, X_2) = \left(a + X_1^T X_2\right)^b \tag{5.18}$$

  In this equation, X_1 and X_2 are vectors in input space, a is constant and is greater or equal to zero, and finally b is the degree of the polynomial kernel.

- Radial basis function (rbf):

  Rbf kernel is a transformer, generating new features using distance measurement between all other dots to specific dot(s) or center(s). The most popular rbf kernel is the **Gaussian radial basis function**. For example, if X and Y arrays cannot be separated using a line, a Gaussian radial basis function transformation using two centers can be applied to generate new features. Gaussian radial basis function is defined as:

$$K(X_1, X_2) = exp\left(\gamma \|X_1 - X_2\|^2\right) \tag{5.19}$$

  In this equation, $X1$ and $X2$ are vectors in input space, $\|X_1 - X_2\|$ is the euclidean distance between X_1 and X_2. γ also known as gamma is used to control the influence of new features on the decision boundary. When gamma value is low, far away points are taken into consideration when drawing the decision boundary which would result in a more straight decision boundary. Low values of gamma tend to underfit when training a model. On the other hand, when gamma is high, only close points are taken into consideration when drawing the decision boundary (wiggly decision boundary). This is because higher gamma values will have more influence on the decision boundary. As a result, the boundaries would be more wiggly. High gamma values tend to lead to overfitting. Therefore, when using rbf kernel function, make sure to include various gamma values in the grid search optimization. Fig. 5.31 illustrates the impact of increasing gamma on the SVM boundaries. As illustrated, as gamma increases, the model tends to overfit and memorize as opposed to building a generalized model.

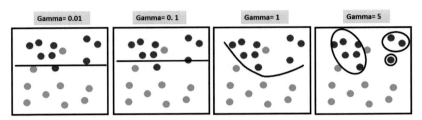

FIGURE 5.31 The impact of gamma on SVM boundary.

Support vector machine implementation in scikit-learn

In this section, building a step-by-step ML model for SVM implementation in Python will be covered. One of the most important parameters in hydraulic fracturing design is the geomechanical properties of the rock. The geomechanical properties such as Poisson's ratio, Young's modulus, and minimum horizontal stress determine fracture geometry as well as ease of fracturing operationally. Brittle rocks with high Young's modulus and low Poisson's ratio are easier to frac while ductile rocks are much harder to hydraulically fracture. The geomechanical properties of the rock can be determined using sonic logs. However, please note that running sonic log on every well is either simply not economically feasible or can become very expensive. Therefore, the existing sonic logs can be used to train a ML model to predict shear and compression wave travel times (Δt_s and Δt_c). The calculations of Young's modulus and Poisson's ratio are as follows (Belyadi et al., 2019):

$$R_v = \frac{\Delta t_s}{\Delta t_c}$$

$\Delta t_s = $ *Shear wave travel time, μsec/ft(obtained from sonic* log)

$\Delta t_c = $ *Compression wave travel time, μsec/ft(obtained from sonic* log)

(5.20)

Shear and compression wave travel times are directly obtained from sonic logs. The goal in this exercise is to train a ML model using SVM to predict shear and compression wave travel times. After training a ML model, a triple combo log with basic log properties such as gamma ray, bulk density, resistivity, total or neutron porosity, and depth can be used to predict shear and compression wave travel times in areas where there is no access to sonic logs. Therefore, as opposed to spending more money in obtaining new sonic logs, the trained ML model can be used to **predict** shear and compression wave travel times in an attempt to calculate geomechanical properties. Poisson's ratio, formation modulus, and dynamic young's modulus can be calculated as follows:

$$Poisson's\ ratio = v = \frac{0.5R_v^2 - 1}{R_v^2 - 1}$$

(5.21)

$$Formation\ modulus = G = 1.34 \times 10^{10} \times \frac{\rho_b}{\Delta t_s^2}$$

where ρ_b is bulk density in g/cc.

$$Dynamic\ Young's\ modulus = E = 2G(1 + v)$$

$$Minimum\ horizontal\ stress = P_c = \frac{v}{1 - v}(\sigma_v - \alpha P_p) + \alpha P_p + \sigma_{tectonic}$$

(5.22)

where σ_v is vertical stress in psi, α is biot's constant, P_p is pore pressure in psi, and $\sigma_{tectonic}$ is tectonic stress in psi.

Now that the concept is clear, let's go over introducing the data set that will be used in this exercise. The sonic log for MIP-3H well was obtained from the MSEEL website (http://www.mseel.org/research/research_MIP.html) and the data set used for this example can be found below:

https://www.elsevier.com/books-and-journals/book-companion/9780128219294

The input data for this exercise are as follows:

- Depth, resistivity, gamma ray, total porosity, effective porosity, and bulk density

This model will have two output features of shear (Δts) and compression wave travel times (Δt_c). Therefore, this will be a multi-output (two outputs) regression ML model. Please note that only one log is used to train such a model and the intent of this exercise is to show the step-by-step process and concept. Therefore, please make sure to use more wells for training a solid model if you have plans on applying this across your organization. Open a new Jupyter Notebook and start importing the basic libraries as follows:

```
import numpy as np
import pandas as pd
import matplotlib.pyplot as plt
%matplotlib inline
import seaborn as sns
df=pd.read_excel('Chapter5_Geomechanical_Properties_Prediction_
    DataSet.xlsx')
df.describe()
Python output=Fig. 5.32
```

	Depth	Resistivity	Gamma Ray	Total Porosity	Effective Porosity	Bulk Density	Compression Wave Travel Time	Shear Wave Travel Time
count	1904.000000	1904.000000	1904.000000	1904.000000	1904.000000	1904.000000	1904.000000	1904.000000
mean	7280.750000	146.141582	116.270437	0.059863	0.038633	2.652268	73.176217	123.929532
std	274.890887	322.464069	61.220418	0.023739	0.023542	0.072419	10.732976	19.252592
min	6805.000000	17.413170	24.463470	0.002870	0.002120	2.428950	50.805650	85.474240
25%	7042.875000	62.688825	84.264097	0.042928	0.019688	2.601138	64.791450	114.705535
50%	7280.750000	85.495010	117.570395	0.055185	0.037425	2.658085	72.718335	127.687235
75%	7518.625000	121.405745	130.947013	0.074783	0.054220	2.707355	80.846762	137.370110
max	7756.500000	7223.240720	623.150210	0.142080	0.119320	2.849760	101.455720	184.778790

FIGURE 5.32 df basic statistics.

Next, use distribution and box plots to plot each feature as follows:

```
f, axes=plt.subplots(4, 2, figsize=(12, 12))
sns.distplot(df['Depth'], color="red", ax=axes[0, 0])
sns.distplot(df['Gamma Ray'], color="olive", ax=axes[0, 1])
sns.distplot(df['Total Porosity'], color="blue", ax=axes[1, 0])
sns.distplot(df['Effective Porosity'], color="orange",
  ax=axes[1, 1])
sns.distplot(df['Resistivity'], color="black", ax=axes[2, 0])
sns.distplot(df['Bulk Density'], color="green", ax=axes[2, 1])
sns.distplot(df['Compression Wave Travel Time'], color="cyan",
  ax=axes[3, 0])
sns.distplot(df['Shear Wave Travel Time'], color="brown", ax=axes
  [3, 1])
plt.tight_layout()
```
Python output=Fig. 5.33

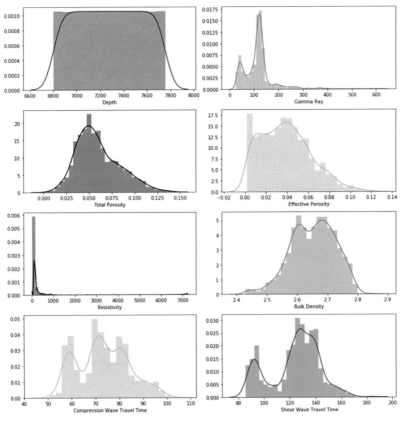

FIGURE 5.33 Distribution plots of each feature.

A quick look at the distribution plots reveals that resistivity and gamma ray curves have log normal distributions as opposed to normal distributions. Let's also visualize the box plots of each feature to find the anomalous points as follows:

```
f, axes=plt.subplots(4, 2, figsize=(12, 12))
sns.boxplot(df['Depth'], color="red", ax=axes[0, 0])
sns.boxplot(df['Gamma Ray'], color="olive", ax=axes[0, 1])
sns.boxplot(df['Total Porosity'], color="blue", ax=axes[1, 0])
sns.boxplot(df['Effective Porosity'], color="orange", ax=axes[1, 1])
sns.boxplot(df['Resistivity'], color="black", ax=axes[2, 0])
sns.boxplot(df['Bulk Density'], color="green", ax=axes[2, 1])
sns.boxplot(df['Compression Wave Travel Time'], color="cyan",
  ax=axes[3, 0])
sns.boxplot(df['Shear Wave Travel Time'], color="brown", ax=axes
  [3, 1])
plt.tight_layout()
```
Python output=Fig. 5.34

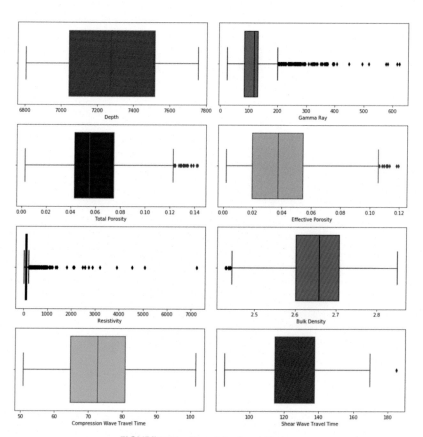

FIGURE 5.34 Box plots of each feature.

For this exercise, let's include resistivity values greater than 0 and less than
1000 Ω-m and gamma ray greater than 0 and less than 400 gAPI as follows:

```
df=df[(df['Resistivity'] > 0) & (df['Resistivity'] < 1000)]
df=df[(df['Gamma Ray'] > 0) & (df['Gamma Ray'] < 400)]
```

Next, plot a heat map of the Pearson correlation coefficient as follows:

```
plt.figure(figsize=(12,10))
sns.heatmap(df.corr(), annot=True, linecolor='white',
    linewidths=2, cmap='summer')
Python output=Fig. 5.35
```

As illustrated, Pearson correlation coefficient between effective porosity
and total porosity is 0.9. Since these parameters provide the same info, let's
drop **effective porosity** simply because the relationship between total porosity
and output features is stronger (higher Pearson correlation coefficients) than
the relationship between effective porosity and output features.

```
df.drop(['Effective Porosity'],axis=1, inplace=True)
```

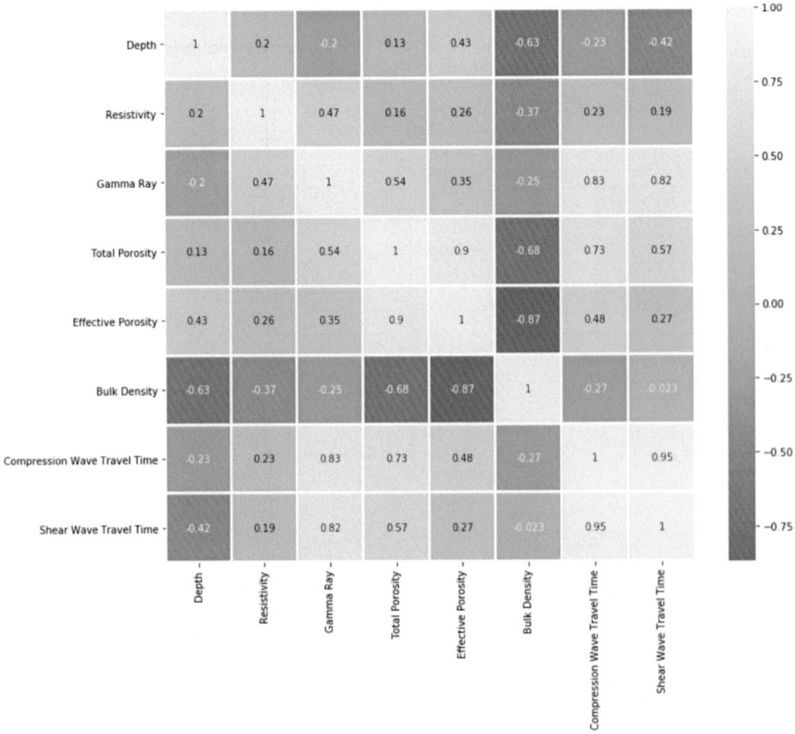

FIGURE 5.35 Pearson correlation coefficient heat map.

Next, use the MinMax scaler to scale all features between 0 and 1 as follows:

```
from sklearn import preprocessing
scaler=preprocessing.MinMaxScaler(feature_range=(0,1))
scaler.fit(df)
df_scaled=scaler.transform(df)
```

Since df_scaled is in an array format now, let's convert to a data frame followed by defining x_scaled and y_scaled features. Please note that the output features are two features of compression and shear wave travel times, and that's why both parameters were included under "y_scaled."

```
df_scaled=pd.DataFrame(df_scaled, columns=['Depth', 'Resistivity',
  'Gamma Ray', 'Total Porosity', 'Bulk Density',
  'Compression Wave Travel Time', 'Shear Wave Travel Time'])
y_scaled=df_scaled[['Compression Wave Travel Time',
  'Shear Wave Travel Time']]
x_scaled=df_scaled.drop(['Compression Wave Travel Time',
  'Shear Wave Travel Time'], axis=1)
```

Next let's import train_test_split and use a 70/30 split to divide the data as follows (make sure to use the same seed number):

```
from sklearn.model_selection import train_test_split
seed=1000
np.random.seed(seed)
X_train,X_test,y_train, y_test=train_test_split(x_scaled,
  y_scaled, test_size=0.30)
```

Next, let's import SVR (support vector regression) "from sklearn.svm." In addition, since this is a multi-output model, let's import "Multi-OutputRegressor" from "sklearn.multioutput" as follows:

```
from sklearn.svm import SVR
from sklearn.multioutput import MultiOutputRegressor
```

Next, let's define the SVR model as follows:

```
np.random.seed(seed)
SVM=MultiOutputRegressor((SVR(kernel='rbf', gamma=1,C=1)))
```

As illustrated, the gamma and C values were assigned as 1 and rbf is used in this model. These parameters are by no means the optimum hyperparameters and grid search is highly recommended to optimize this model. Next, fit the "SVM" to "(X_train,y_train)" and predict "(X_train)" and "(X_test)" as follows:

```
SVM.fit(X_train,y_train)
y_pred_train=SVM.predict(X_train)
y_pred_test=SVM.predict(X_test)
```

Next, let's obtain the R^2 for the 70% of the data that was used as the training set. As shown below, "np.corrcoef" is used to obtain the R value between "y_train['Compression Wave Travel Time']" and "y_pred_train[:,0]." Please note that "y_pred_train[:,0]" are the predicted **training** values for compression wave travel time. In addition, "y_pred_train[:,0]" are the predicted training values for shear wave travel time. The predicted training accuracy is obtained to get a better perspective on the accuracy of the model on the training data set. However, please note that what's truly important when training a ML model is to obtain the **testing** accuracy. If the training accuracy happens to be high but the testing accuracy is low, it indicates that the model has been overfitted. In other words, the model can yield a high accuracy on the training samples but as soon as an unknown data set is introduced to the model, the model fails to predict accurately.

```
corr_train1=np.corrcoef(y_train['Compression Wave Travel Time'],
  y_pred_train[:,0]) [0,1]
corr_train2=np.corrcoef(y_train['Shear Wave Travel Time'],
  y_pred_train[:,1]) [0,1]
print('Compression Wave Travel Time Train Data r^2=',
  round(corr_train1**2,4),'r=', round(corr_train1,4))
print('Shear Wave Travel Time Train Data r^2=',
  round(corr_train2**2,4),'r=', round(corr_train2,4))
Python output=
Compression Wave Travel Time Train Data r^2=0.9257 r=0.9621
Shear Wave Travel Time Train Data r^2=0.9182 r=0.9582
```

Next, let's obtain the testing accuracy as well:

```
corr_test1=np.corrcoef(y_test['Compression Wave Travel Time'],
  y_pred_test[:,0]) [0,1]
corr_test2=np.corrcoef(y_test['Shear Wave Travel Time'],
  y_pred_test[:,1])[0,1]
print('Compression Wave Travel Time Test Data r^2=',
  round(corr_test1**2,4),'r=', round(corr_test1,4))
print('Shear Wave Travel Time Test Data r^2=',
  round(corr_test2**2,4),'r=', round(corr_test2,4))
Python output=
Compression Wave Travel Time Test Data r^2=0.928 r=0.9633
Shear Wave Travel Time Test Data r^2=0.9277 r=0.9632
```

Next, let's obtain MAE, MSE, and RMSE for the compression and shear wave travel times' testing set as follows:

```
from sklearn import metrics
print('Testing Compression MAE:', round(metrics.mean_absolute_error
  (y_test['Compression Wave Travel Time'], y_pred_test[:,0]),4))
print('Testing Compression MSE:', round(metrics.mean_squared_error
  (y_test['Compression Wave Travel Time'], y_pred_test[:,0]),4))
print('Testing Compression RMSE:', round(np.sqrt(metrics.mean_
  squared_error(y_test['Compression Wave Travel Time'], y_pred_test
  [:,0])),4))
```

```
Python output=
Testing Compression MAE: 0.0457
Testing Compression MSE: 0.0036
Testing Compression RMSE: 0.0599
```

```
print('Testing Shear MAE:', round(metrics.mean_absolute_error
  (y_test['Shear Wave Travel Time'], y_pred_test[:,1]),4))
print('Testing Shear MSE:', round(metrics.mean_squared_error
  (y_test['Shear Wave Travel Time'], y_pred_test[:,1]),4))
print('Testing Shear RMSE:', round(np.sqrt(metrics.mean_squared_
  error(y_test['Shear Wave Travel Time'], y_pred_test[:,1])),4))
Python output=
Testing Shear MAE: 0.0437
Testing Shear MSE: 0.0029
Testing Shear RMSE: 0.0542
```

The next step is to plot the training and testing actual versus prediction values as follows:

```
plt.figure(figsize=(10,8))
plt.plot(y_test['Compression Wave Travel Time'],
  y_pred_test[:,0], 'b.')
plt.xlabel('Compression Wave Travel Time Testing Actual')
plt.ylabel('Compression Wave Travel Time Testing Prediction')
plt.title('Compression Wave Travel Time Testing Actual Vs.
  Prediction')
Python output=Fig. 5.36
```

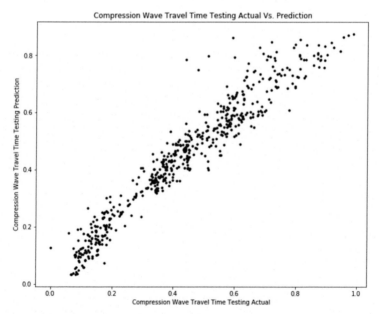

FIGURE 5.36 Compression wave travel time testing actual versus predicted values (normalized values).

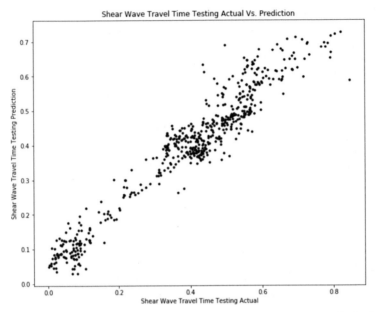

FIGURE 5.37 Shear wave travel time testing actual versus predicted values (normalized values).

```
plt.figure(figsize=(10,8))
plt.plot(y_test['Shear Wave Travel Time'], y_pred_test[:,1], 'b.')
plt.xlabel('Shear Wave Travel Time Testing Actual')
plt.ylabel('Shear Wave Travel Time Testing Prediction')
plt.title('Shear Wave Travel Time Testing Actual Vs. Prediction')
Python output=Fig. 5.37
```

The final step is to apply a few complete blind data sets (blind logs) to see how this trained model performs when applied to new blind wells. Afterward, compare the predicted values from the trained model versus actual shear and compression wave travel time values. If the accuracy is satisfactory with low error, simply proceed with applying the trained ML model in areas with triple combo logs that did not utilize sonic logs. This will eliminate the need for investing in more sonic logs and being able to predict the shear and compression wave travel times with very good accuracy in order to calculate geomechanical properties such as Young's modulus, Poisson's ratio, and minimum horizontal stress.

Decision tree

Decision tree is another supervised ML algorithm that can be used for classification and regression problems. Before discussing random forest, which is heavily related to decision trees, it is crucial to understand how a decision tree

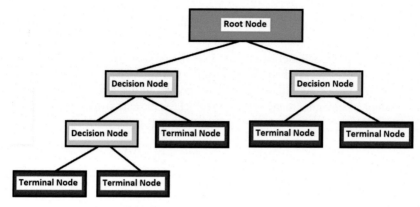

FIGURE 5.38 Decision tree illustration.

works. A decision tree splits the data into subtrees which are then split into other subtrees. As illustrated in Fig. 5.38, a **root node** is at the topmost level which essentially represents the entire population, a **decision node** also referred to as an **internal node** has two or more branches, and finally, a **terminal node** also referred to as a **leaf node** is the lowest node and does not split any longer. Note that the term splitting is defined as splitting a node into two or more subnodes.

There are various decision tree algorithms such as ID3, C4.5, C5.0, and CART. ID3 stands for Iterative Dichotomiser 3 which was developed in 1986 (Quinlan, 1986). This algorithm builds decision trees using a top-down greedy manner on **categorical** features that results in the largest information gain (will be discussed). In addition ID3 uses a multiway tree. C4.5 is another algorithm used in decision trees which removes the restrictions that features must be categorical. C5.0 is Quinlan's latest release which uses less memory and constructs smaller rulesets than C4.5. Finally, CART stands for classification and regression trees and is similar to C4.5 but it supports numerical target variables and does not calculate rulesets. CART creates binary trees that result in the largest information gain at each node. Please note that scikit-learn library uses an optimized version of the CART algorithm (scikit-learn, n.d.).

Attribute selection technique

In a data set with N attributes, deciding the attributes to be at the root or internal node can be complex and challenging. Some of the most important criteria for attribute selection are as follows:

1) **Entropy**: is simply a measure of uncertainty or purity which is calculated as follows. Note that high entropy means low purity.

$$E(S) = \sum_{i=1}^{c} -P_i \log_2 P_i \qquad (5.23)$$

where c is the number of classes, P_i is the probability of a class in a data set. For example, if there are 100 samples of two classes of "Frac" versus "Don't Frac." Assuming 40 samples belong to the "Frac" class and 60 samples belonging to the "Don't Frac" class, the entropy can be calculated as follows:

$$-\frac{40}{100}\log_2\frac{40}{100} - \frac{60}{100}\log_2\frac{60}{100} = 0.97 \qquad (5.24)$$

Mathematically, entropy for multiple attributes can be calculated as follows:

$$E(X,Y) = \sum_{C \in Y} P(c)E(c) \qquad (5.25)$$

where X is the current state and Y is the selected attribute. $P(c)$ is the probability of the attribute and $E(c)$ is the entropy of the attribute. For a table listed below, let's calculate the entropy:

MULTIPLE attribute entropy calculation.

Conditions	Continue	Screen-out	Sum
Flat pressure	4	1	5
Rising pressure	4	2	6
Rapid rise	1	4	5
Sum	9	7	16

Let's go through the entropy calculations as follows:

$Probability(Flat\ Pressure) \times Entropy(4,1) + Probability(Rising\ Pressure)$

$\times Entropy(4,2) + Probability(Rapid\ Rise) \times Entropy(1,4)$

$$= \frac{5}{16} \times Entropy(4,1) + \frac{6}{16} \times Entropy(4,2) + \frac{5}{16} \times Entropy(1,4) \qquad (5.26)$$

Let's calculate the entropy values for each term first and plug in the equation above:

$$Entropy\,(4,1) = -\frac{4}{5}\log_2\frac{4}{5} - \frac{1}{5}\log_2\frac{1}{5} = 0.7219$$

$$Entropy\,(4,2) = -\frac{4}{6}\log_2\frac{4}{6} - \frac{2}{6}\log_2\frac{2}{6} = 0.9183 \qquad (5.27)$$

$$Entropy(1,4) = -\frac{1}{5}\log_2\frac{1}{5} - \frac{4}{5}\log_2\frac{4}{5} = 0.7219$$

Plug back the entropy values in the discussed equation as follows:

$$\frac{5}{16} \times Entropy(4,1) + \frac{6}{16} \times Entropy(4,2) + \frac{5}{16} \times Entropy(1,4)$$

$$= \left(\frac{5}{16} \times 0.7219\right) + \left(\frac{6}{16} \times 0.9183\right) + \left(\frac{5}{16} \times 0.7219\right) = 0.7956$$

$$(5.28)$$

2) **Information gain (IG):** When constructing a decision tree, it is important to find an attribute that results in the highest information gain and smallest entropy. IG refers to how well an attribute separates training examples according to their target classification. Please note that IG favors smaller partitions and it can be calculated as follows:

$$IG(Y,X) = E(Y) - E(Y,X) \qquad (5.29)$$

IG example: Calculate information gain for the previous example.
1) Calculate the entropy of the target variable as follows:

$$Entropy(9,7) = -\frac{9}{16}\log_2\frac{9}{16} - \frac{7}{16}\log_2\frac{7}{16} = 0.9887 \qquad (5.30)$$

2) Calculate IG as follows:

$$IG(outcome,\ pressure\ trend) = 0.9887 - 0.7956 = 0.1931 \qquad (5.31)$$

3) **Gini index:** as opposed to information gain, Gini index favors larger partitions and it is calculated by subtracting 1 from the sum of squared probabilities of each class as illustrated below. Please note that in perfectly classified examples, Gini index would be equal to 0.

$$Gini = 1 - \sum_{i=1}^{c}(P_i)^2 = 1 - \left(P(class\ A)^2 + P(class\ B)^2 + ... + P(class\ N)^2\right)$$

$$(5.32)$$

where P_i is the probability of an element being classified under a particular class. One of the most difficult challenges with using a decision tree is overfitting. One of the approaches of avoiding overfitting is pruning. Pruning is referred to trimming off branches or sections of the tree that do not provide much info in terms of classifying instances or disturbing the overall accuracy. It is also important to use cross-validation when using decision tree to make sure the model is not overfitted. Another approach is to use a random forest algorithm which usually outperforms a decision tree. In the next section of this chapter, random forest is covered in detail. The above discussed criteria such as information gain and gini index are used for attribute selection where these criteria will calculate values for each attribute. Let's assume that information gain is chosen as the criteria for attribute selection. The values are sorted and the attributes with the highest values are placed at the root.

Decision tree using scikit-learn

In this section, a database with the input features listed below is used to tie to an output feature called total organic carbon content. The idea is to build a supervised regression decision tree model to be able to predict TOC.

- The input features are thickness, bulk density, resistivity, effective volume, effective porosity, clay volume, and water saturation.
- The output feature is TOC.

The data set used in this exercise can be found below:
https://www.elsevier.com/books-and-journals/book-companion/9780128219294
Let's start a new Jupyter Notebook and import the main libraries and the df data frame as follows:

```
import numpy as np
import pandas as pd
import matplotlib.pyplot as plt
import seaborn as sns
%matplotlib inline
df=pd.read_excel('Chapter5_TOC_Prediction_DataSet.xlsx')
df.describe().transpose()
Python output=Fig. 5.39
```

	count	mean	std	min	25%	50%	75%	max
Thickness_ft	987.0	150.448933	52.452284	50.218753	123.462354	141.662622	166.707110	475.992627
Bulk Density_gg per cc	987.0	2.423001	0.019059	2.386117	2.409469	2.422639	2.433418	2.540608
Resistivity_ohmsm	987.0	3.892432	1.342193	1.680451	3.120852	3.650354	4.319585	15.970625
Effective Porosity_Fraction	987.0	0.061492	0.014805	0.017432	0.051250	0.061158	0.072289	0.096054
Clay Volume_ Fraction	987.0	0.271257	0.045289	0.153118	0.238607	0.264785	0.303776	0.413083
Water Saturation_Fraction	987.0	0.435876	0.080023	0.230041	0.372234	0.442414	0.490972	0.683304
TOC_Fraction	987.0	0.052630	0.005062	0.030830	0.051026	0.053662	0.056100	0.060907

FIGURE 5.39 df description.

Next, let's visualize the distribution plot of each feature as follows:

```
f, axes=plt.subplots(4, 2, figsize=(12, 12))
sns.distplot(df['Thickness_ft'], color="red", ax=axes[0, 0])
sns.distplot(df['Bulk Density_gg per cc'], color="blue", ax=axes
    [0, 1])
sns.distplot(df['Resistivity_ohmsm'], color="orange",
    ax=axes[1, 0])
sns.distplot(df['Effective Porosity_Fraction'], color="green",
    ax=axes[1, 1])
sns.distplot(df['Clay Volume_ Fraction'], color="cyan",
    ax=axes[2, 0])
sns.distplot(df['Water Saturation_Fraction'], color="brown",
    ax=axes[2, 1])
sns.distplot(df['TOC_Fraction'], color="black", ax=axes[3, 0])
    plt.tight_layout()
```
Python output=Fig. 5.40

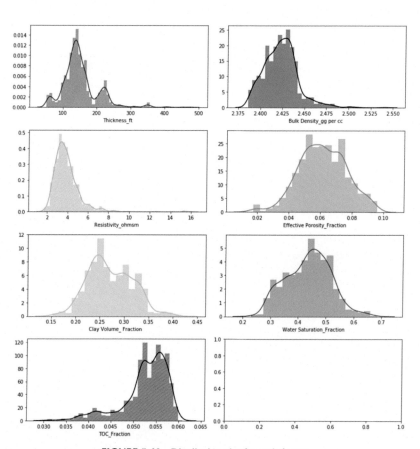

FIGURE 5.40 Distribution plot for each feature.

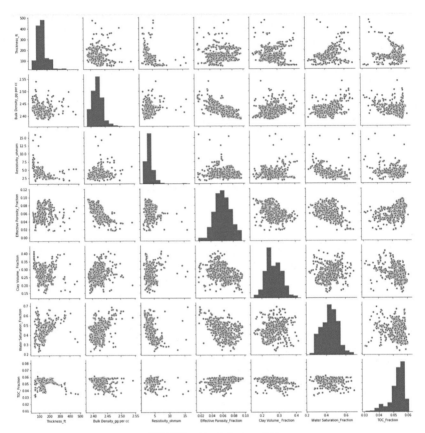

FIGURE 5.41 Pair plot of df.

Also, let's look at the pair plot of all features as follows:

```
sns.pairplot(df)
Python output=Fig. 5.41
```

Let's also look at the heat map of Pearson correlation coefficient as follows:

```
plt.figure(figsize=(12,10))
sns.heatmap(df.corr(), annot=True, linecolor='white',
  linewidths=2, cmap='coolwarm')
Python output=Fig. 5.42
```

Please note that as illustrated in Fig. 5.42, the highest absolute value of Pearson correlation coefficient is between bulk density and effective porosity with a coefficient of −0.72. One of the parameters could potentially be dropped as long as the overall testing accuracy of the model does not decrease.

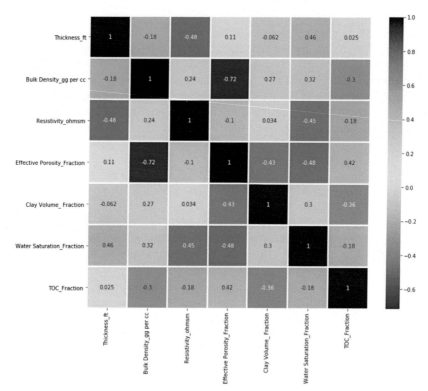

FIGURE 5.42 Heat map of Pearson correlation coefficient.

In this case both parameters were left to be included in the model. Please feel free to drop one and test the model's accuracy.

Tree-based algorithms such as decision tree, random forest, etc., do not require feature normalization or standardization. For example, when applying decision tree, since a tree is being branched down, normalization or standardization will not help. Let's define the x and y variables as follows:

```
x=df.drop(['TOC_Fraction'], axis=1)
y=df['TOC_Fraction']
```

Next, import the train_test_split library and use a 70/30 split as follows:

```
from sklearn.model_selection import train_test_split
  seed=1000
np.random.seed(seed)
X_train,X_test,y_train, y_test=train_test_split\
  (x, y, test_size=0.30)
```

Next, import "DecisionTreeRegressor" from sklearn.tree as follows. For classification problems, simply use "DecisionTreeClassifier."

```
from sklearn.tree import DecisionTreeRegressor
```

Next, let's define the parameters that are used within a decision tree. "Criterion" is used to measure the quality of a split. The options that are provided in scikit-learn when using a "DecisionTreeRegressor" are "mse" standing for mean squared error, "mae" standing for mean absolute error, and "friednman_mse" standing for mean squared error with Friedman's improvement score. In this example, "mse" is chosen; however, this is one of the hyperparameters that can be used when grid search is applied. Please note that in classification problems, "gini" for Gini index or "entropy" for information gain can be used under "criterion."

The next parameter is called **"max_depth"** and according to scikit-learn, it is defined as the "maximum depth of the tree" and the default value is "None" which means "the nodes get expanded" until all leaves are pure or until all leaves contain less than "min_samples_split" samples (will be discussed shortly). In essence, the maximum depth of a decision tree can be the "number of training samples-1," but please note that this length of a depth can lead to severe overfitting. Therefore, if a decision tree model is overfitted, reduce "max_depth" and, as previously discussed, always make sure to use cross-validation. Note that having a low "max_depth" could lead to underfitting. Therefore, compare the training and testing accuracy scores with various values for "max_depth." The higher the "max_depth," the deeper the tree and the more splits that will be associated with that tree. More splits mean capturing more information. Therefore, a higher depth leads to overfitting.

The next parameter is called **"min_sample _split"** and, according to the scikit-learn library, it is defined as "The minimum number of samples required to split an internal node." Please refer to Fig. 5.38 to note that an internal node is the same as a decision node, and it can have children and branch off into other nodes. Please note that high values of "min_sample_split" could lead to underfitting because higher values of "min_sample_split" prevent a model from learning the details. In other words, the higher the "min_sample_split," the more constraint the tree becomes since it must consider more samples at each node. The default value in scikit-learn is 2. Simply increase "min_-sample_split" from the default value of 2−100 and it'll lead to a much lower model accuracy. Therefore, this is one of the parameters used to control overfitting and can be used as a hyper parameter when performing grid search optimization.

"min_sample_leaf" is another parameter used to control overfitting and, according to scikit-learn, this parameter is defined as "The minimum number of samples required to be at a leaf node." The key difference between "min_sample_split" and "min_sample_leaf" is that the first focuses on an internal or decision node while the second focuses on a leaf or terminal node. A high value of "min_sample_leaf" will also lead to severe underfitting while a low value will lead to overfitting. Please make sure to include this parameter as one of the hyperparameters to be optimized when performing grid search optimization (discussed in chapter 7).

The next metric is called **"max_features,"** defined as "The number of features to consider when looking for the best split," in the scikit-learn library. For example, in classification problems, every time there is a split, the decision tree algorithm uses the defined number of features using gini or information gain. One of the main essences of using "max_feature" is to reduce overfitting by choosing a lower number of "max_features." Some of the available options are "auto," "sqrt," "log2," and the default is "None" which sets "max_-features" to number of features. "auto" also sets "max_features" to number of features. "sqrt" sets "max_features" to square root of number of features, and "log2" sets "max_features" to log2 of number of features. If computation time and overfitting are major concerns, consider using "sqrt" or "log2."

The next parameter is called **"ccp_alpha"** which is used for pruning purposes. By default, no pruning will be chosen. Let's define the decision tree parameters as shown below:

```
np.random.seed(seed)
dtree=DecisionTreeRegressor(criterion='mse', splitter='best',
    max_depth=None, min_samples_split=4, min_samples_leaf=2,
    max_features=None,ccp_alpha=0)
```

Let's apply "dtree" to "(X_train,y_train)" as follows:

```
dtree.fit(X_train,y_train)
```

Now that the model has been fit to training inputs and output, let's apply to predict "X_train" and "X_test" as follows. The main reason to apply to "X_train" is to be able to obtain the training accuracy on the model as well as the testing accuracy.

```
y_pred_train=dtree.predict(X_train)
y_pred_test=dtree.predict(X_test)
```

Next, let's obtain the training and testing R^2 as follows:

```
corr_train=np.corrcoef(y_train, y_pred_train) [0,1]
print('Training Data R^2=',round(corr_train**2,4),'R=',
    round(corr_train,4))
```
Python output=Training Data R^2=0.975 R=0.9874

```
corr_test=np.corrcoef(y_test, y_pred_test) [0,1]
print('Testing Data R^2=',round(corr_test**2,4),'R=',
    round(corr_test,4))
```
Python output=Testing Data R^2=0.6833 R=0.8266

As illustrated, the training R^2 is 97.5%. However, the testing R^2 is only 68.33%. Therefore, please spend some time manually adjusting some of the hyperparameters to see whether the testing accuracy can be improved. Grid search optimization will be covered in Chapter 7. For now, just focus on manually adjusting these hyperparameters until the testing accuracy is

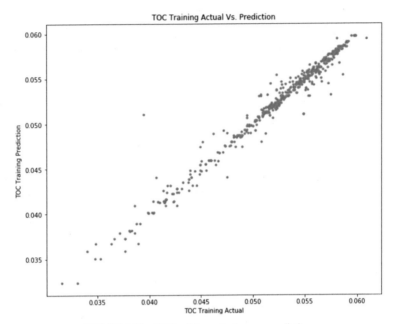

FIGURE 5.43 TOC training actual versus prediction.

optimized. Next, let's visualize the training actual versus prediction and testing actual versus prediction as follows:

```
plt.figure(figsize=(10,8))
plt.plot(y_train, y_pred_train, 'g.')
plt.xlabel('TOC Training Actual')
plt.ylabel('TOC Training Prediction')
plt.title('TOC Training Actual Vs. Prediction')
Python output=Fig. 5.43
```

```
plt.figure(figsize=(10,8))
plt.plot(y_test, y_pred_test, 'g.')
plt.xlabel('TOC Testing Actual')
plt.ylabel('TOC Testing Prediction')
plt.title('TOC Testing Actual Vs. Prediction')
Python output=Fig. 5.44
```

To compare the actual and predicted TOC values for the testing set, the following lines of code can be added:

```
TOC_Actual_Prediction=pd.DataFrame({'Actual':y_test,
  'Predicted':y_pred_test})
TOC_Actual_Prediction
Python output=Fig. 5.45
```

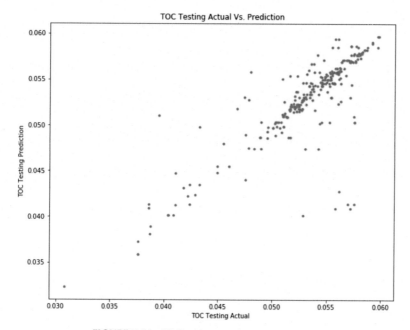

FIGURE 5.44 TOC testing actual versus prediction.

To properly evaluate the model from all aspects, let's also add MAE, MSE, and RMSE as follows:

```
from sklearn import metrics
print('MAE:', round(metrics.mean_absolute_error(y_test,
  y_pred_test),5))
print('MSE:', round(metrics.mean_squared_error(y_test,
  y_pred_test),5))
print('RMSE:',  round(np.sqrt(metrics.mean_squared_error(y_test,
  y_pred_test)),5))
Python output=MAE: 0.00127
MSE: 1e-05
RMSE: 0.00278
```

Decision tree also allows feature ranking by calling "dtree.feature_importances_" as follows:

```
dtree.feature_importances_
Python output=array([0.28651806, 0.11955648, 0.13959011,
  0.24352775, 0.13713719, 0.07367041])
```

Let's place this in a visualization format using a tornado chart as follows:

```
feature_names=df.columns[:-1]
plt.figure(figsize=(10,8))
```

	Actual	Predicted
834	0.051979	0.051979
604	0.053110	0.053445
747	0.057625	0.058708
908	0.057235	0.054457
545	0.054575	0.054347
...	...	...
809	0.056807	0.057144
166	0.054815	0.054572
172	0.058651	0.058527
263	0.058120	0.057927
427	0.051019	0.049853

297 rows × 2 columns

FIGURE 5.45 TOC actual versus predicted values (testing set).

```
feature_imp=pd.Series(dtree.feature_importances_,
  index=feature_names).sort_values(ascending=False)
sns.barplot(x=feature_imp, y=feature_imp.index)
plt.xlabel('Feature Importance Score Using Decision Tree')
plt.ylabel('Features')
plt.title("Feature Importance Ranking")
Python output=Fig. 5.46
```

As illustrated in Fig. 5.46, thickness and effective porosity are two of the most important features impacting the output of the model which was defined

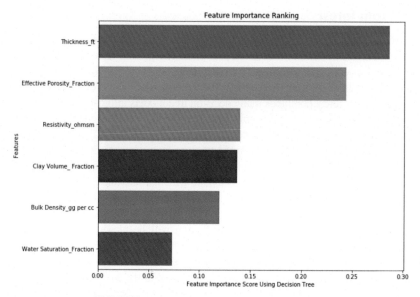

FIGURE 5.46 Feature importance using decision tree.

as TOC. Note that as accuracy of this model improves, the feature ranking will also vary.

Note that the train_test_split was done randomly with 70% of the data used as training and 30% of the data used as testing. Let's also do a five-fold cross-validation to observe the resulting average R^2 as follows:

```
from sklearn.model_selection import cross_val_score
np.random.seed(seed)
scores_R2=cross_val_score(dtree, x, y,cv=5,scoring='r2')
print(" R2_Cross-validation scores: {}". format(scores_R2))
Python output=R2_Cross-validation scores: [0.69672778 0.56657738
   0.51290659 0.60992366 0.76534161]

print(" Average R2_Cross-validation scores: {}".
   format(scores_R2.mean()))
Python output=Average R2_Cross-validation scores:
   0.6302954016341682
```

As illustrated, first import "cross_val_score" library from "sklearn.model_selection." Next, use "dtree" as the model, x and y as input and output variables, cv or cross-validation of 5 (five-fold cross-validation), and use R^2 as the metric to yield. Afterward, print R^2 for each fold, and finally take the average of all 5 R^2 values to yield the average cross-validation R^2 for this example. As can be observed, the R^2 from 5 fold cross-validation appears to be slightly lower than the train_test_split technique.

Random forest

Random forest is another powerful supervised ML algorithm which can be used for both regression and classification problems. The general technique of random decision forests was first proposed by Ho in 1995 (Kam Ho, 1995). Random forest is an ensemble of decision trees or it can be thought of as a forest of decision trees. Since random forest combines many decision tree models into one, it is known as an ensemble algorithm. For example, instead of building a decision tree to predict EUR/1000 ft, using a single tree could result in an erroneous value due to **variance** in predictions. One way to avoid this variance when predicting the EUR/1000 ft is to take predictions from hundreds or thousands of decision trees and using the average of those trees to calculate the final answer. Combining many decision trees into a single model is essentially the fundamental concept behind using random forest. The predictions made by a single decision tree could be inaccurate, but when combined, the prediction will be closer to average. The reason random forest is typically more accurate than a single decision tree is because much more knowledge is incorporated from many predictions. For regression problems, random forest uses the average of the decision trees for final prediction. However, as previously mentioned, classification problems can also be solved using random forest by taking a majority vote of the predicted class. Another type of ensemble learning method is gradient boosting which will be discussed next. Fig. 5.47 illustrates the difference between a single decision tree versus a random forest which consists of an ensemble of decision trees.

There are two main types of combining multiple decision trees into one and they are as follows:

- **Bagging**, also referred to as "bootstrap aggregation," is used in random forest. Bagging was first proposed by Leo Breiman in a technical report in 1994 (Breiman, 1994). In bagging or bootstrap aggregation, an average of many **built independent** models is used. In bagging, the decision trees are trained based on randomly sampling subsets of data and sampling is done

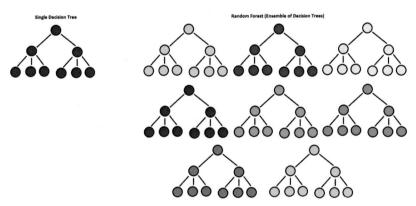

FIGURE 5.47 Decision tree versus random forest.

with replacement. Please note that bootstrapping is a statistical resampling technique which means that samples are drawn with replacement. In other words, the idea behind bootstrapping is that some samples will be used multiple times in one single decision tree (Efron, 1979).

One of the main advantages of using bagging when applying a random forest algorithm is **variance reduction** of the model. For example, when a single decision tree is used, it is very prone to overfitting and can be sensitive to the noise in the data. However, bootstrap aggregation reduces this problem. When feature bagging is used, at each split in the decision tree, not all input features are considered. In other words, only a **random subset** of input features are used. Let's assume that a data set has 10 inputs and 1 output features. Let's also assume that sand/ft and cluster spacing (two out of the 10 features) have the strongest impact on the output feature which is EUR/1000 ft. When applying bagging, since a random subset of input features are used, it will help to reduce the effects of sand/ft and cluster spacing on the target class which is EUR/1000 ft in this example. In essence, selecting a random subset of input features results in lower correlations across decision trees and, as a result, more diversification.

- Boosting, on the other hand, (used in gradient boosting) is also considered to be an ensemble technique where the models are built sequentially as opposed to **independently.** In boosting, more weights are placed on instances with incorrect predictions. Therefore, the focus in boosting is the challenging cases that are being predicted inaccurately. As opposed to bagging where an **equal** weighted average is used, boosting uses **weighted average** where more weight is applied to the models with better performance. In other words, in boosting, the samples that were predicted inaccurately get a higher weight which would then lead to sampling them more often. This is the main reason why bagging can be performed independently while boosting is performed sequentially. As opposed to bagging where variance is reduced, boosting will try to reduce bias. In addition to gradient boosting, AdaBoost (adaptive gradient boosting) and XGBoost (extreme gradient boosting) are examples of boosting algorithms.

Since majority of the terminologies used in random forest have already been covered in decision tree section of this chapter, let's start implementing random forest to the same TOC data set covered under decision tree and compare the accuracy of both models.

Random forest implementation using scikit-learn

In this section, the same TOC data set used under decision tree will be used to apply for random forest regression. Therefore, open a new Jupyter Notebook and follow the exact same codes covered in decision tree (or use the existing Jupyter Notebook to continue). Instead of importing "DecisionTreeRegressor" from sklearn.ensemble, import the "RandomForestRegressor" as follows:

```
from sklearn.ensemble import RandomForestRegressor
```

Next, let's define the parameters inside the "RandomForestRegressor." There are multiple important hyper-tuning parameters within a random forest model such as "n_estimators," "criterion," "max_depth," etc. Some of these parameters were covered under decision tree and the rest will be covered here. "n_estimators" defines the number of trees in the forest. Usually the higher this number, the more accurate the model is without leading to overfitting. In the example below, "n_estimators" is set to be 5000 which means 5000 independent decision trees will be constructed and the average of the 5000 trees will be used as the predicted value for each prediction row. "criterion" of "mse" was chosen for this model which means variance reduction is desired. Since bootstrapping aggregation that was discussed is desired to be chosen for this model, "bootstrap" was set to "True." If "bootstrap" is set to "False," the whole data set is used to build each decision tree. "n_jobs" is set to "-1" in an attempt to use all processors. If this is not desired, simply change from -1 to a different integer value.

```
np.random.seed(seed)
rf=RandomForestRegressor(n_estimators=5000,
  criterion='mse',max_depth=None, min_samples_split=4,
  min_samples_leaf=2, max_features='auto', bootstrap=True,
  n_jobs=-1)
```

Next, let's apply these defined "rf" parameters to the training inputs and output features (X_train,y_train) and obtain the accuracy of both training and testing sets as shown below:

```
rf.fit(X_train,y_train)
y_pred_train=rf.predict(X_train)
y_pred_test=rf.predict(X_test)
corr_train=np.corrcoef(y_train, y_pred_train) [0,1]
print('Training Data R^2=',round(corr_train**2,4),'R=',
  round(corr_train,4))
```
Python output=Training Data R^2=0.9658 R=0.9827

```
corr_test=np.corrcoef(y_test, y_pred_test) [0,1]
print('Testing Data R^2=',round(corr_test**2,4),'R=',
  round(corr_test,4))
```
Python output=Testing Data R^2=0.8182 R=0.9046

As can be observed, the testing R^2 is 81.82% compared to 68.33% of the decision tree. Therefore, without doing further parameter fine-tuning, the random forest algorithm appears to be outperforming the decision tree. Let's also visualize the cross plots of actual versus predicted training and testing data sets as follows:

```
plt.figure(figsize=(10,8))
plt.plot(y_train, y_pred_train, 'r.')
plt.xlabel('TOC Training Actual')
plt.ylabel('TOC Training Prediction')
plt.title('TOC Training Actual Vs. Prediction')
```
Python output=Fig. 5.48

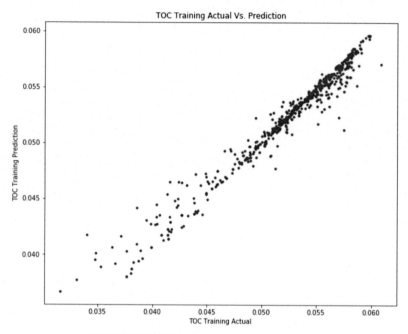

FIGURE 5.48 TOC training actual versus prediction.

```
plt.figure(figsize=(10,8))
plt.plot(y_test, y_pred_test, 'r.')
plt.xlabel('TOC Testing Actual')
plt.ylabel('TOC Testing Prediction')
plt.title('TOC Testing Actual Vs. Prediction')
Python output=Fig. 5.49
```

Next, let's also obtain MAE, MSE, and RMSE for the testing set as follows:

```
from sklearn import metrics
print('MAE:', round(metrics.mean_absolute_error(y_test,
  y_pred_test),5))
print('MSE:', round(metrics.mean_squared_error(y_test,
  y_pred_test),5))
print('RMSE:', round(np.sqrt(metrics.mean_squared_error(y_test,
  y_pred_test)),5))
Python output=MAE: 0.00118
MSE: 0.0
RMSE: 0.00201
```

As illustrated, MAE, MSE, and RMSE values are lower as compared to the decision tree model. Next, let's also obtain the feature ranking using random forest as follows:

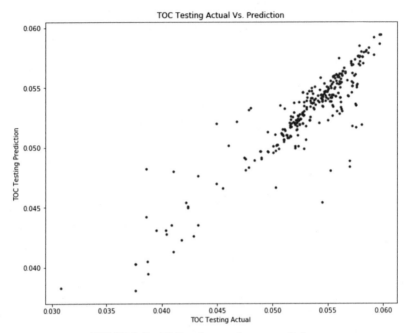

FIGURE 5.49 TOC testing actual versus prediction.

```
feature_names=df.columns[:-1]
plt.figure(figsize=(10,8))
feature_imp=pd.Series(rf.feature_importances_,
  index=feature_names).sort_values(ascending=False)
sns.barplot(x=feature_imp, y=feature_imp.index)
plt.xlabel('Feature Importance Score Using Random Forest')
plt.ylabel('Features')
plt.title("Feature Importance Ranking")
Python output=Fig. 5.50
```

As illustrated in Fig. 5.50, the important features obtained by random forest is different than what was obtained from decision tree. This is primarily attributed to the higher accuracy of the random forest model. The recommendation is to go with the model with higher accuracy which is the random forest model in this particular example. Tree-based algorithms, such as decision tree, random forest, extra trees, etc., use percentage improvement in the purity of the node to naturally rank the input features. As previously discussed, in classification problems, the idea is to minimize Gini impurity (if Gini impurity is selected). Therefore, nodes that lead to the greatest reduction in Gini impurity happen at the start of the trees while nodes with the least amount of reduction occur at the end of the trees. This is how tree-based algorithms perform feature ranking.

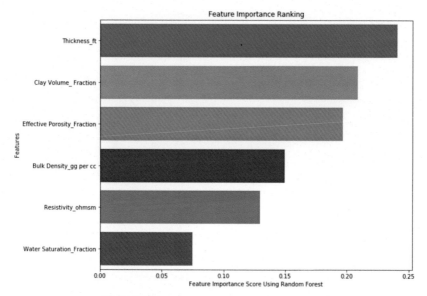

FIGURE 5.50 Feature ranking using random forest.

To be consistent with the decision tree model, let's also do a five-fold cross-validation to observe the resulting average R^2 for the random forest model as follows:

```
from sklearn.model_selection import cross_val_score
np.random.seed(seed)
scores_R2=cross_val_score(rf, x, y,cv=5,scoring='r2')
print(" R2_Cross-validation scores: {}". format(scores_R2))
Python output=R2_Cross-validation scores: [0.76480359 0.78500227
  0.74946927 0.82476465 0.74985886]
print(" Average R2_Cross-validation scores: {}".
  format(scores_R2.mean()))
Average R2_Cross-validation scores: 0.7747797298703516
```

On average, the cross-validation R^2 on random forest is 77.48% as compared to 63.03% of the decision tree model.

Extra trees (extremely randomized trees)

Before proceeding to gradient boosting, extra trees is covered in this section. Extra trees is another form of supervised ML algorithm that is similar to random forest used for both classification and regression problems. As discussed in the previous section, random forest uses bootstrapping which means samples are drawn with replacement. However, extra trees algorithm uses the whole original sample. In the scikit-learn library, the default term for

TABLE 5.3 Primary differences between decision tree, random forest, and extra trees.

Metrics	Decision tree	Random forest (RF)	Extra trees (ET)
Number of trees	1	Many	Many
Bootstrapping	N/A	Yes	No (option for bootstrapping is available in scikit-learn)
# Of features for split at each decision (internal) node	All features	Random subset of features	Random subset of features
Splitting criteria	Best split	Best split	Random split

"bootstrap" is "False" but it does allow for bootstrapping if desired. Another difference between random forest and extra trees is that random forest chooses the optimum split when splitting nodes while extra trees chooses it randomly. Please note that after selecting the split points, the two algorithms choose the best one between all subset of features. From a computational standpoint, extra tree algorithm is faster than random forest because it randomly selects the split points. Please refer to Table 5.3 to get a better understanding between decision tree, random forest, and extra trees.

Extra trees implementation using scikit-learn

Let's apply extra trees to the same TOC data set. Start a new Jupyter Notebook and follow the same codes until importing the extra trees library as follows (or continue with the same Jupyter Notebook). For classification problems, "ExtraTreesClassifier" can be used instead of "ExtraTreesRegressor."

```
from sklearn.ensemble import ExtraTreesRegressor
```

Next, let's define the extra trees parameters, apply to "(X_train, y_train)," and predict "(X_train)" and "X_test" as shown below:

```
np.random.seed(seed)
et=ExtraTreesRegressor(n_estimators=5000,
  criterion='mse',max_depth=None, min_samples_split=4,
  min_samples_leaf=2, max_features='auto', bootstrap=False,
  n_jobs=-1)
et.fit(X_train,y_train)
y_pred_train=et.predict(X_train)
y_pred_test=et.predict(X_test)
```

As shown above, "bootstrap" is set to "False" and the same number of trees of 5000 was chosen for this algorithm as well. Next, obtain the R^2 for training and testing set as follows:

```
corr_train=np.corrcoef(y_train, y_pred_train) [0,1]
print('Training Data R^2=',round(corr_train**2,4),'R=',
  round(corr_train,4))
Python output=Training Data R^2=0.9832 R=0.9916

corr_test=np.corrcoef(y_test, y_pred_test) [0,1]
print('Testing Data R^2=',round(corr_test**2,4),'R=',
  round(corr_test,4))
Python output=Testing Data R^2=0.8663 R=0.9307
```

When comparing the testing R^2 from extra trees to that of random forest, the testing R^2 accuracy appears to be higher by a few percentages. Please note that both models have not been optimized and the accuracy of these models could potentially increase. Next, let's plot the training and testing actual versus predicted values.

```
plt.figure(figsize=(10,8))
plt.plot(y_train, y_pred_train, 'b.')
plt.xlabel('TOC Training Actual')
plt.ylabel('TOC Training Prediction')
plt.title('TOC Training Actual Vs. Prediction')
Python output=Fig. 5.51
```

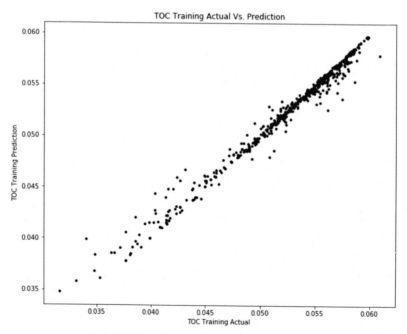

FIGURE 5.51 TOC training actual versus prediction.

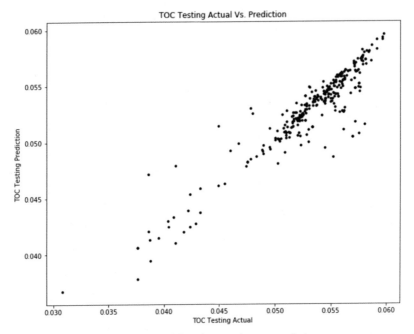

FIGURE 5.52 TOC testing actual versus prediction.

```
plt.figure(figsize=(10,8))
plt.plot(y_test, y_pred_test, 'b.')
plt.xlabel('TOC Testing Actual')
plt.ylabel('TOC Testing Prediction')
plt.title('TOC Testing Actual Vs. Prediction')
```
Python output=Fig. 5.52

Next, let's look at the actual versus predicted values side by side in a table format as follows:

```
TOC_Actual_Prediction=pd.DataFrame({'Actual':y_test,
  'Predicted':y_pred_test})
TOC_Actual_Prediction
```
Python output=Fig. 5.53

Let's also obtain MAE, MSE, and RMSE as follows:

```
from sklearn import metrics
print('MAE:', round(metrics.mean_absolute_error(y_test,
  y_pred_test),5))
print('MSE:', round(metrics.mean_squared_error(y_test,
  y_pred_test),5))
print('RMSE:', round(np.sqrt(metrics.mean_squared_error(y_test,
  y_pred_test)),5))
```
Python output=MAE: 0.00104
MSE: 0.0
RMSE: 0.00176

	Actual	Predicted
834	0.051979	0.051964
604	0.053110	0.053851
747	0.057625	0.054805
908	0.057235	0.056245
545	0.054575	0.053831
...	...	...
809	0.056807	0.054295
166	0.054815	0.054232
172	0.058651	0.058608
263	0.058120	0.057312
427	0.051019	0.050779

297 rows × 2 columns

FIGURE 5.53   TOC actual versus predicted for the test data set.

As illustrated, the extra trees model appears to have the lowest MAE, MSE, and RMSE. This does not imply that extra tree model will outperform the random forest or other algorithms every time. Each problem must be treated separately and independently to find the best ML algorithm that would result in the highest testing accuracy. Let's perform a feature ranking using the trained extra trees model as shown below:

```
feature_names=df.columns[:-1]
plt.figure(figsize=(10,8))
```

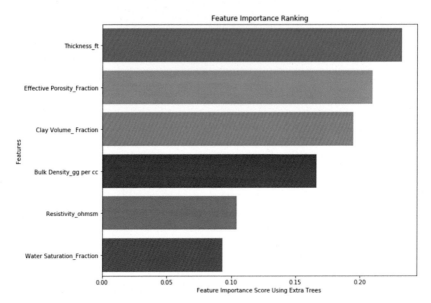

FIGURE 5.54 Feature ranking using extra trees.

```
feature_imp=pd.Series(et.feature_importances_,
  index=feature_names).sort_values(ascending=False)
sns.barplot(x=feature_imp, y=feature_imp.index)
plt.xlabel('Feature Importance Score Using Extra Trees')
plt.ylabel('Features')
plt.title("Feature Importance Ranking")
Python output=Fig. 5.54
```

As illustrated in Fig. 5.54, the overall feature ranking using the extra trees algorithm is similar to that of random forest. The top three features that have the highest impact on TOC appear to be thickness, effective porosity, and clay volume.

To be consistent with decision tree and random forest models, let's also perform a five-fold cross-validation to observe the resulting average R^2 for the extra tree model as follows:

```
from sklearn.model_selection import cross_val_score
np.random.seed(seed)
scores_R2=cross_val_score(et, x, y,cv=5,scoring='r2')
print(" R2_Cross-validation scores: {}". format(scores_R2))
Python output=R2_Cross-validation scores: [0.78910151 0.82774583
  0.78219988 0.84719793 0.80344111]
print(" Average R2_Cross-validation scores: {}".
  format(scores_R2.mean()))
Average R2_Cross-validation scores: 0.8099372535306054
```

As illustrated, the average R^2 value obtained from five-fold cross-validation using the extra tree model is 80.99% as compared to 63.03% and 77.48% in decision tree and random forest models covered previously.

Gradient boosting

Gradient boosting is another type of ensemble supervised ML algorithm that can be used for both classification and regression problems. The main reason why algorithms such as random forest, extra trees, gradient boosting, etc., are called **ensemble** is because a final model is generated based on many individual models. As described under the bagging/boosting comparison section, gradient boosting will train many models sequentially by placing more weights on instances with erroneous predictions. Therefore, challenging cases are the focus during training. A sequential model training using gradient boosting will gradually minimize a loss function. The loss function is minimized similarly to an artificial neural network model (will be discussed in the next chapter) in which weights are optimized. In gradient boosting, after building the weak learners, the predictions are compared with actual values. The difference between prediction and actual values represents the error rate of the model. The error rate of the model can now be used to calculate the gradient, which is essentially the partial derivative of the loss function. The gradient is used to find the direction that the model parameters would have to change to reduce the error in the next round of training. As opposed to a neural network model where the objective function is to minimize a loss function in a **single model**, multiple models' predictions are combined in gradient boosting. Therefore, gradient boosting uses some of the hyperparameters used in random forest/extra trees and other hyperparameters such as learning rate, loss function, etc., that are used in an ANN model. These hyperparameters will be discussed in detail in the next chapter.

Gradient boosting implementation using scikit-learn

In this section, the goal is to use gradient boosting to train a ML model to predict net present value (NPV) per well @ $3/MMBTU gas pricing using the following input features:

- Lateral length (ft), stage length (ft), sand to water ratio or SWR (lb/gal), sand per ft (lb/ft), water per ft (BBL/ft), estimated ultimate recovery or EUR (BCF)

The data set for this example can be found below:
https://www.elsevier.com/books-and-journals/book-companion/9780128219294
Therefore, let's open a new Jupyter Notebook and start importing the following libraries and data set as follows:

	count	mean	std	min	25%	50%	75%	max
Lateral_length	258.0	8911.663566	2995.195633	2439.500000	6433.225000	8089.450000	11582.100000	16918.400000
Stage_length	258.0	223.834920	52.900698	85.600003	178.830607	204.708946	263.308520	386.754534
Sand_to_water_ratio	258.0	1.268741	0.338050	0.119433	1.108849	1.249180	1.437271	2.442476
Sand_per_ft	258.0	2066.112041	903.288312	310.728368	1416.164360	2027.515360	2777.796760	4964.854720
Water_per_ft	258.0	39.069433	15.740073	8.118385	26.593500	40.575506	51.229328	75.340213
EUR_BCF	258.0	9.795406	6.193401	0.173786	5.212174	8.570267	12.465853	37.375668
NPV_at_3.0_MMBTU_Gas Pricing	258.0	-0.634117	3.631055	-8.092757	-3.107133	-1.587870	1.061966	13.828945

FIGURE 5.55 df description.

```
import numpy as np
import pandas as pd
import matplotlib.pyplot as plt
import seaborn as sns
%matplotlib inline
sns.set_style("darkgrid")
df=pd.read_excel('Chapter5_NPV_DataSet.xlsx')
df.describe().transpose()
Python output=Fig. 5.55
```

Next, let's look at the distribution, box, and scatter plots of the following features:

```
f, axes=plt.subplots(4, 2, figsize=(12, 12))
sns.distplot(df['Lateral_length'], color="red", ax=axes[0, 0])
sns.distplot(df['Stage_length'], color="olive", ax=axes[0, 1])
sns.distplot(df['Sand_to_water_ratio'], color="blue",
  ax=axes[1, 0])
sns.distplot(df['Sand_per_ft'], color="orange", ax=axes[1, 1])
sns.distplot(df['Water_per_ft'], color="black", ax=axes[2, 0])
sns.distplot(df['EUR_BCF'], color="green", ax=axes[2, 1])
sns.distplot(df['NPV_at_3.0_MMBTU_Gas Pricing'], color="cyan",
  ax=axes[3, 0])
plt.tight_layout()
Python output=Fig. 5.56
```

As observed, the distribution of majority of the parameters is normal. Stage length appears to have a slight dent between 200 and 250 ft stage length.

```
f, axes=plt.subplots(4, 2, figsize=(12, 12))
sns.boxplot(df['Lateral_length'], color="red", ax=axes[0, 0])
sns.boxplot(df['Stage_length'], color="olive", ax=axes[0, 1])
sns.boxplot(df['Sand_to_water_ratio'], color="blue", ax=axes[1, 0])
sns.boxplot(df['Sand_per_ft'], color="orange", ax=axes[1, 1])
sns.boxplot(df['Water_per_ft'], color="black", ax=axes[2, 0])
sns.boxplot(df['EUR_BCF'], color="green", ax=axes[2, 1])
sns.boxplot(df['NPV_at_3.0_MMBTU_Gas Pricing'], color="cyan",
  ax=axes[3, 0])
plt.tight_layout()
Python output=Fig. 5.57
```

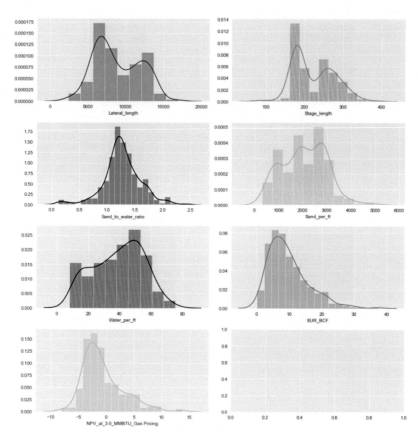

FIGURE 5.56 Distribution plots of df.

```
f, axes=plt.subplots(3, 2, figsize=(12, 12))
sns.scatterplot(df['Lateral_length'],df['NPV_at_3.0_MMBTU_Gas
  Pricing'], color="red", ax=axes[0, 0])
sns.scatterplot(df['Stage_length'],df['NPV_at_3.0_MMBTU_Gas
  Pricing'], color="olive", ax=axes[0, 1])
sns.scatterplot(df['Sand_to_water_ratio'], df['NPV_at_3.0_
  MMBTU_Gas Pricing'],color="blue", ax=axes[1, 0])
sns.scatterplot(df['Sand_per_ft'], df['NPV_at_3.0_MMBTU_Gas
  Pricing'],color="orange", ax=axes[1, 1])
sns.scatterplot(df['Water_per_ft'], df['NPV_at_3.0_MMBTU_Gas
  Pricing'],color="black", ax=axes[2, 0])
sns.scatterplot(df['EUR_BCF'],df['NPV_at_3.0_MMBTU_Gas
  Pricing'], color="green", ax=axes[2, 1])
plt.tight_layout()
```
Python output=Fig. 5.58

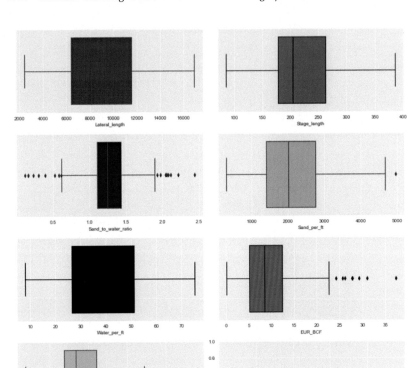

FIGURE 5.57 Box plots of df.

Next, let's visualize the Pearson correlation coefficient in a heat map as follows:

```
plt.figure(figsize=(14,10))
sns.heatmap(df.corr(),linewidths=2,
  linecolor='black',cmap='coolwarm', annot=True)
Python output=Fig. 5.59
```

Since gradient boosting is a tree-based algorithm, no normalization will be needed. Therefore, let's define the x and y variables as follows:

```
x=df.drop(['NPV_at_3.0_MMBTU_Gas Pricing'], axis=1)
y=df['NPV_at_3.0_MMBTU_Gas Pricing']
```

Next, import "train_test_split" and divide the data into 70/30 split (70% for training and 30% for testing) as follows:

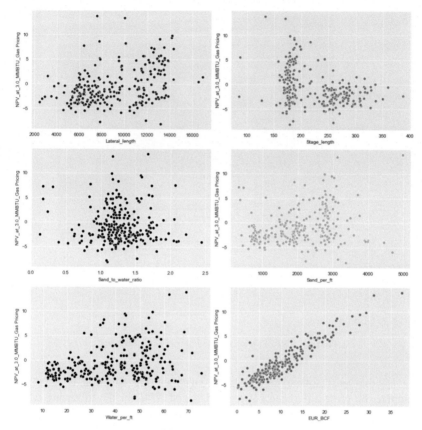

FIGURE 5.58   Scatter plots.

```
from sklearn.model_selection import train_test_split
seed=1000
np.random.seed(seed)
X_train, X_test, y_train, y_test=train_test_split(x, y,
    test_size=0.3)
```

Next, let's import the "GradientBoostingRegressor" library and define the parameters within this algorithm. "n_estimators" is defined as the number of boosting stages or in other words, it is the number of trees in the forest. Please note that a large number of trees usually do not result in overfitting. Therefore, feel free to increase the number of trees to observe the resulting outcome. Learning rate determines the step size at each sequential iteration while moving toward minimizing the loss function. The default value is 0.1. "loss" is referred to the loss function to be optimized. "ls" is the least squares loss function that was used for this exercise. Let's go over the following example to understand the least square regression intuitively.

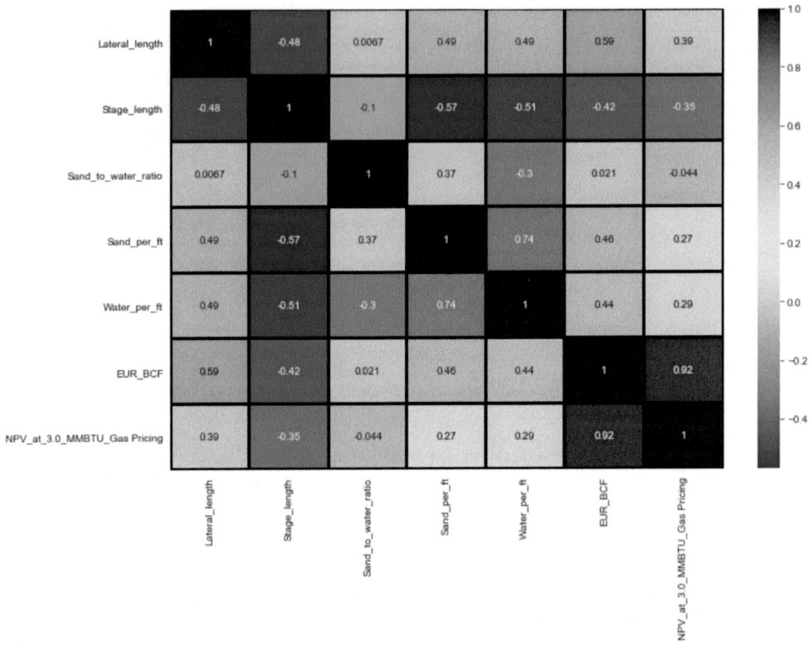

FIGURE 5.59 Pearson correlation coefficient heat map.

Use the least square regression to calculate the equation of line and compute the error difference between EUR and NPV for the following table:

EUR versus NPV example.

EUR (x-variable)	NPV (y-variable)
5	3
2	0.2
8	7
9	10
3	1

Step 1) Calculate x^2, xy, sum of x, sum of y, sum of x^2, and sum of xy as shown in the table below:

Calculations sample.

EUR (x)	NPV (y)	x^2	$x \times y$
5	3	25	15
2	0.2	4	0.4
8	7	64	56
9	10	81	90
3	1	9	3
Total (27)	Total (21.2)	Total (183)	Total (164.4)

Step 2) Calculate the slope (m) as follows:

$$m = \frac{N\sum(xy) - \sum x \sum y}{N\sum(x^2) - (\sum x)^2} = \frac{(5 \times 164.4) - (27 \times 21.2)}{(5 \times 183) - (27^2)} = 1.3419 \quad (5.33)$$

Step 3) Calculate intercept (b) as follows:

$$b = \frac{\sum y - m\sum x}{N} = \frac{21.2 - (1.3419 \times 27)}{5} = -3.00645 \quad (5.34)$$

Step 4) Assemble the equation of the line as follows:

$$y = mx + b \rightarrow y = 1.3419x - 3.00645 \quad (5.35)$$

Step 5) The last step is to compare the y values with the calculated y-values from the line equation calculated in step 4 as shown in the table below:

Difference between y-values and calculated y-values from the line equation.

NPV (y)	Calculated y-values from line equation	Difference (error)
3	3.7032	0.7032
0.2	−0.3226	−0.5226
7	7.7290	0.7290
10	9.0710	−0.9290
1	1.0194	0.0194

Therefore, least square regression was used to minimize the loss function in this exercise. Please note that other loss functions such as "lad," "huber," and "quantile" can also be used. "lad" stands for least absolute deviation that is the same as MAE which essentially minimizes the absolute value of residuals. "huber" is a combination of "ls" and "lad" loss functions developed by Peter Huber (Huber, 1964). Finally, "quantile" allows quantile regression where alpha must be specified.

The default "criterion" parameter is "friedman_mse" which is the MSE with an improvement score by Friedman. Other criterion such as "mse" and "mae" can also be used but "friedman_mse" is pretty robust as it provides great approximations. The rest of the parameters have been previously discussed.

```
from sklearn.ensemble import GradientBoostingRegressor
np.random.seed(seed)
gb=GradientBoostingRegressor(loss='ls', learning_rate=0.1,
  n_estimators=100, criterion='friedman_mse',
  min_samples_split=4, min_samples_leaf=2, max_depth=3,
  max_features=None)
```

Next, let's fit to "(x_train, y_train)" and predict the test data:

```
gb.fit(X_train,y_train)
y_pred_train=gb.predict(X_train)
y_pred_test=gb.predict(X_test)
```

Let's also obtain the R^2 for training and testing data as follows:

```
corr_train=np.corrcoef(y_train, y_pred_train) [0,1]
print('Training Data R^2=',round(corr_train**2,4),'R=',
  round(corr_train,4))
```
Python output=Training Data R^2=0.9903 R=0.9952

```
corr_test=np.corrcoef(y_test, y_pred_test) [0,1]
print('Testing Data R^2=',round(corr_test**2,4),'R=',
  round(corr_test,4))
```
Python output=Testing Data R^2=0.9168 R=0.9575

Let's also visualize training and testing actual versus prediction as follows:

```
plt.figure(figsize=(10,8))
plt.plot(y_train, y_pred_train, 'b.')
plt.xlabel('NPV Training Actual')
plt.ylabel('NPV Training Prediction')
plt.title('NPV Training Actual Vs. Prediction')
```
Python output=Fig. 5.60

```
plt.figure(figsize=(10,8))
plt.plot(y_test, y_pred_test, 'b.')
plt.xlabel('NPV Testing Actual')
plt.ylabel('NPV Testing Prediction')
plt.title('NPV Testing Actual Vs. Prediction')
```
Python output=Fig. 5.61

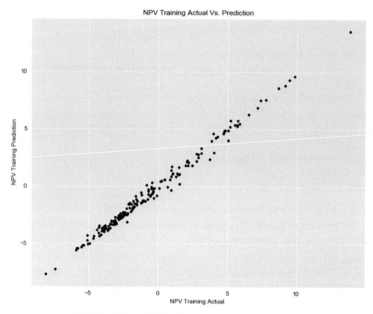

FIGURE 5.60 NPV training actual versus prediction.

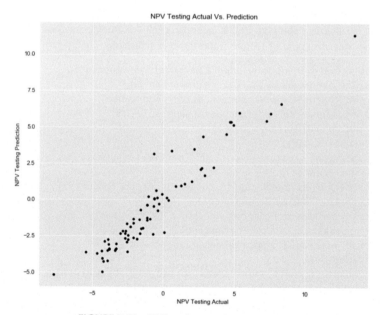

FIGURE 5.61 NPV testing actual versus prediction.

Let's also obtain MAE, MSE, and RMSE for the 30% testing data as follows:

```
from sklearn import metrics
print('MAE:', round(metrics.mean_absolute_error(y_test,
  y_pred_test),5))
print('MSE:', round(metrics.mean_squared_error(y_test,
  y_pred_test),5))
print('RMSE:', round(np.sqrt(metrics.mean_squared_error(y_test,
  y_pred_test)),5))
Python output=MAE: 0.76514
MSE: 1.06659
RMSE: 1.03276
```

Let's also obtain the feature importance of each input variable and **permutation feature importance**. The permutation feature importance is defined as the decrease in a model score when a single feature value is randomly shuffled (Breiman, 2011). Permutation feature importance evaluates input and output features by calculating the drop in the model score. The bigger the drop in the model score, the more impact the input feature has on the model, and the higher it will rank. Permutation importance can be obtained by importing the "permutation_importance" library from "sklearn.inspection" and passing in the trained model (in this case "gb"), testing set (in this case "X_test,y_test"), n_repeats (in this case 10 was used), and random_state (to use a seed number of 1000 as was previously used). Please note that "n_repeats" refers to the number of times a feature is randomly shuffled. In this example, each feature was randomly shuffled 10 times and a sample of feature importance is returned.

```
from sklearn.inspection import permutation_importance
feature_importance=gb.feature_importances_
sorted_features=np.argsort(feature_importance)
pos=np.arange(sorted_features.shape[0]) + .5
fig=plt.figure(figsize=(12, 6))
plt.subplot(1, 2, 1)
plt.barh(pos, feature_importance[sorted_features],
  align='center')
plt.yticks(pos, np.array(df.columns)[sorted_features])
plt.title('Feature Importance')
result=permutation_importance(gb, X_test, y_test,
  n_repeats=10,random_state=seed)
sorted_idx=result.importances_mean.argsort()
plt.subplot(1, 2, 2)
plt.boxplot(result.importances[sorted_idx].T, vert=False,
  labels=np.array(df.columns)[sorted_idx])
plt.title("Permutation Importance (test set)")
fig.tight_layout()
plt.show()
Python output=Fig. 5.62
```

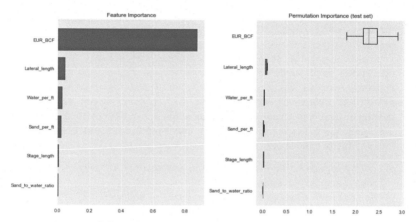

FIGURE 5.62   Feature and permutation importance.

Let's also call "result" which will show the **average** permutation importance score for each feature in the first two lines (based on randomly shuffling
10 times). The next two lines show the **standard deviation** of each feature that
was randomly shuffled 10 times. Finally, the rest of the result shows the list of
10 elements of individual permutation score in each array for each feature.
Since there are 6 input features, there are also 6 arrays with each one including
10 elements based on 10 random shuffling.

```
result
Python output=Fig. 5.63
```

```
{'importances_mean': array([ 5.62673558e-02,  4.21978379e-03, -1.68587786e-03,  1.08985184e-02,
          1.29103662e-02,  2.28967526e+00]),
 'importances_std': array([0.01485314, 0.00449182, 0.00290839, 0.00638356, 0.00391517,
          0.27624977]),
 'importances': array([[ 8.15151462e-02,  3.95292772e-02,  7.02221645e-02,
          3.65059457e-02,  6.03294438e-02,  4.96642294e-02,
          5.30148570e-02,  5.79832600e-02,  3.92243716e-02,
          7.46848629e-02],
        [ 7.44887785e-03,  9.59443016e-03,  2.58807132e-03,
         -1.38147086e-03, -1.05835466e-03,  9.27570663e-03,
          5.14050355e-03,  4.35297088e-03, -2.86902680e-03,
          9.10612984e-03],
        [ 4.48758354e-03,  1.25757307e-03, -2.61008665e-03,
         -1.66488246e-03, -5.10349406e-03, -3.54065341e-03,
         -4.17053339e-03,  1.94360646e-04, -5.02201479e-03,
         -6.86631090e-04],
        [ 2.16017350e-02,  1.38400077e-02,  1.45205740e-02,
          1.05278583e-02,  9.32294255e-04,  1.76246999e-02,
          3.92768661e-02,  1.23552019e-02,  2.21219261e-03,
          1.14429341e-02],
        [ 1.92971526e-02,  1.29535259e-02,  1.81843586e-02,
          1.41525885e-02,  1.00042563e-02,  9.58542022e-03,
          9.02224063e-03,  1.14259918e-02,  7.47400867e-03,
          1.70041187e-02],
        [ 1.78325738e+00,  2.50606649e+00,  2.24622373e+00,
          2.09877842e+00,  2.87685350e+00,  2.45690805e+00,
          2.17434168e+00,  2.27078646e+00,  2.36532335e+00,
          2.11821354e+00]]])}
```

FIGURE 5.63

Just like the previous examples, let's apply a 10-fold cross-validation to see the resultant R^2 as follows:

```
from sklearn.model_selection import cross_val_score
np.random.seed(seed)
scores_R2=cross_val_score(gb, x, y,cv=10,scoring='r2')
print(" R2_Cross-validation scores: {}". format(scores_R2))
Python output=R2_Cross-validation scores: [0.8862245 0.87216255
  0.76550356 0.946582 0.94277734 0.89203469 0.54233081 0.9442973
  0.93938502 0.80444424]

print(" Average R2_Cross-validation scores: {}".
  format(scores_R2.mean()))
Python output=Average R2_Cross-validation scores:
  0.8535741998748938
```

As illustrated, the average R^2 score for the 10-fold cross-validation is roughly 6% lower than the R^2 score obtained from random 70/30 split validation.

Extreme gradient boosting

Extreme gradient boosting, also known as XGBoost, developed by Tianqi Chen (Chen & Guestrin, 2016), is another type of ensemble supervised ML algorithm used for both classification and regression problems. XGBoost is a type of gradient boosting method and is different from a gradient boosting model as follows:

- XGBoost uses L_1 and L_2 regularization which helps with model generalization and overfitting reduction.
- As previously discussed under the gradient boosting section, the error rate of a gradient boosting model is used to calculate the gradient, which is essentially the partial derivative of the loss function. In contrast, XGBoost uses the **second** partial derivative of the loss function. Using the second partial derivative of the loss function will provide more info about the direction of the gradient.
- XGBoost is usually faster than gradient boosting due to the parallelization of tree construction.
- XGBoost can handle missing values within a data set; therefore, the data preparation is not as time-consuming.

Extreme gradient boosting implementation using scikit-learn

Let's also use the previous data set to apply the regression version of XGBoost to see its performance. Start a new Jupyter Notebook and follow the same previous steps shown in the gradient boosting section until after train_-test_split (or continue with the same Jupyter Notebook). Instead of importing "GradientBoostingRegressor," import "XGBRegressor" as shown below. For classification problems, simply import "XGBClassifier" instead. Prior to importing the "XGBRegressor" library shown below, make sure to use the Anaconda command prompt to install this library using the following command: pip install xgboost

```
from xgboost import XGBRegressor
```

Next, let's define the XGBoost parameters. Objective is set to "reg:squarederror" which uses the squared error to minimize the loss function. "n_estimators" specifies the number of trees (as was discussed under multiple sections before). Learning rate defines the step size which means each weight in all trees will be multiplied by this value (in this case 0.1 was used). Learning rate can range between 0 and 1. "reg_alpha" controls L_1 regularization on leaf weights. A high "reg_alpha" leads to more regularization. "reg_lambda" specifies L_2 regularization on leaf weights. "max_depth" specifies the maximum depth of the tree during any boosting run. "gamma" determines whether to proceed with further partitioning on a leaf node of a tree based on the expected loss reduction.

```
np.random.seed(seed)

xgb=XGBRegressor(objective ='reg:squarederror',n_estimators=200,
   reg_lambda=1, gamma=0,max_depth=3, learning_rate=0.1,
   reg_alpha=0.1)
```

Next, let's apply the defined model "xgb" to fit the training data (70% training). In addition, also obtain "y_pred_train" and "y_pred_test" as follows:

```
xgb.fit(X_train,y_train)
y_pred_train=xgb.predict(X_train)
y_pred_test=xgb.predict(X_test)
```

Next, obtain R^2 for both training and testing data sets as follows:

```
corr_train=np.corrcoef(y_train, y_pred_train) [0,1]
print('Training Data R^2=',round(corr_train**2,4),'R=',
   round(corr_train,4))
```
Python output=Training Data R^2=0.9957 R=0.9979

```
corr_test=np.corrcoef(y_test, y_pred_test) [0,1]
print('Testing Data R^2=',round(corr_test**2,4),'R=',
   round(corr_test,4))
```
Python output=Testing Data R^2=0.9138 R=0.9559

Also, visualize training and testing actual versus prediction values as follows:

```
plt.figure(figsize=(10,8))
plt.plot(y_train, y_pred_train, 'm.')
plt.xlabel('NPV Training Actual')
plt.ylabel('NPV Training Prediction')
plt.title('NPV Training Actual Vs. Prediction')
```
Python output=Fig. 5.64

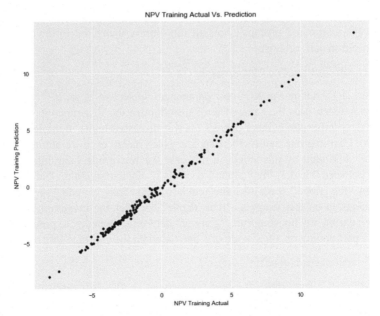

FIGURE 5.64 NPV training actual versus prediction.

```
plt.figure(figsize=(10,8))
plt.plot(y_test, y_pred_test, 'm.')
plt.xlabel('NPV Testing Actual')
plt.ylabel('NPV Testing Prediction')
plt.title('NPV Testing Actual Vs. Prediction')
Python output=Fig. 5.65
```

Next, let's obtain MAE, MSE, and RMSE as follows:

```
from sklearn import metrics
print('MAE:', round(metrics.mean_absolute_error(y_test,
  y_pred_test),5))
print('MSE:', round(metrics.mean_
  squared_error(y_test, y_pred_test),5))
print('RMSE:', round(np.sqrt(metrics.mean_squared_error(y_test,
  y_pred_test)),5))
Python output=MAE: 0.76634
MSE: 1.0761
RMSE: 1.03735
```

Let's also get the feature ranking and permutation importance for this model. Please note that the permutation importance and feature ranking's orders are not the same when using XGBoost algorithm in this particular example. The top four features are very similar in both rankings. Note that permutation importance does have shown the predictive capability of a feature, but it illustrates how essential a feature is for a model.

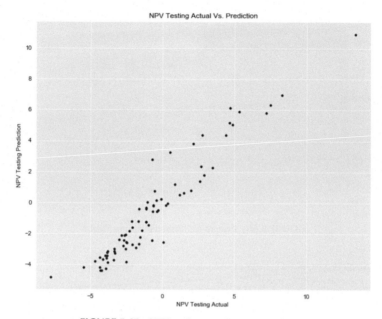

FIGURE 5.65   NPV testing actual versus prediction.

```
from sklearn.inspection import permutation_importance
feature_importance=xgb.feature_importances_
sorted_features=np.argsort(feature_importance)
pos=np.arange(sorted_features.shape[0]) + .5
fig=plt.figure(figsize=(12, 6))
plt.subplot(1, 2, 1)
plt.barh(pos, feature_importance[sorted_features],
  align='center')
plt.yticks(pos, np.array(df.columns)[sorted_features])
plt.title('Feature Importance')
result=permutation_importance(xgb, X_test, y_test,
  n_repeats=10,random_state=seed)
sorted_idx=result.importances_mean.argsort()
plt.subplot(1, 2, 2)
plt.boxplot(result.importances[sorted_idx].T, vert=False,
  labels=np.array(df.columns)[sorted_idx])
plt.title("Permutation Importance (test set)")
fig.tight_layout()
```
Python output=Fig. 5.66

Let's also do a 10-fold cross-validation and obtain the average R^2 value as follows:

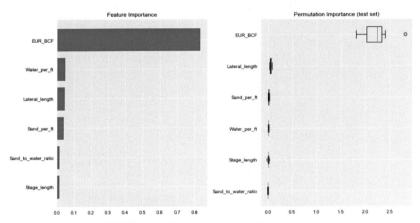

FIGURE 5.66 Feature and permutation importance.

```
from sklearn.model_selection import cross_val_score
np.random.seed(seed)
scores_R2=cross_val_score(xgb, x, y,cv=10,scoring='r2')
print(" R2_Cross-validation scores: {}". format(scores_R2))
```
**Python output=R2_Cross-validation scores: [0.89722326 0.86415599
 0.79374482 0.94312108 0.94706938 0.88140055 0.52043121 0.94923754
 0.92307646 0.81054857]**

```
print(" Average R2_Cross-validation scores: {}".
    format(scores_R2.mean()))
```
**Python output=Average R2_Cross-validation scores:
 0.8530008849657191**

Thus far, we have observed that both gradient boosting and XGBoost algorithms have very similar performance for this particular data set.

Adaptive gradient boosting

Adaptive gradient boosting or AdaBoost is another form of supervised ML boosting algorithm. In adaptive boosting, the weights on the sample observations that are challenging to classify or predict are **increased** by altering the distribution of the training data set. As opposed to AdaBoost where the sample distribution gets modified, in gradient boosting the distribution does **not** get modified and the weak learners train on the remaining errors of the strong learners. Remember that a weak learner is defined as a classifier that has a strong correlation with true classification. In other words, it can correctly classify or predict slightly better than random guessing. On the other hand, a strong learner is a classifier with a **stronger** correlation with true classification.

Adaptive gradient boosting implementation using scikit-learn

Let's open a new Jupyter Notebook, and follow the same procedure as previously discussed until after the train_test_split portion (or continue with the same Jupyter Notebook). Let's import "AdaBoostRegressor" library as follows:

```
from sklearn.ensemble import AdaBoostRegressor
```

Please note that if a classification problem is in question, simply use "AdaBoostClassifier" instead. Next, define the hyperparameters in the adaptive boosting regression algorithm. "base_estimator" defines how the boosted ensemble is built. If "None" is selected, a "DecisionTreeRegressor(max_depth=3)" is the default model estimator that will be used. For this example, the "DecisionTreeRegressor()" was first imported "from sklearn.tree" and the hyperparameters within the decision tree algorithm were adjusted to have a max depth of "None," "min_sample_split" of 4, and "min_sample_leaf" of 2. A loss function of "linear," "square," and "exponential" can be used. A linear loss function was used in the example, but feel free to adjust to the other two and observe the resulting impact. The rest of the parameters such as "n_estimators" and "learning_rate" were previously discussed.

```
from sklearn.tree import DecisionTreeRegressor
np.random.seed(seed)
agb=AdaBoostRegressor(base_estimator=
  DecisionTreeRegressor(max_depth=None,min_samples_split=4,
  min_samples_leaf=2),n_estimators=200,
  learning_rate=0.1,loss='linear')
```

Next, fit "agb" to "(X_train,y_train)" and predict both training and testing data, and obtain R^2 on both training and testing data as follows:

```
agb.fit(X_train,y_train)
y_pred_train=agb.predict(X_train)
y_pred_test=agb.predict(X_test)
corr_train=np.corrcoef(y_train, y_pred_train)[0,1]
print('Training Data R^2=',round(corr_train**2,4),'R=',
  round(corr_train,4))
Python output=Training Data R^2=0.9984 R=0.9992

corr_test=np.corrcoef(y_test, y_pred_test)[0,1]
print('Testing Data R^2=',round(corr_test**2,4),'R=',
  round(corr_test,4))
Testing Data R^2=0.8917 R=0.9443
```

Next, let's visualize the training and testing actual versus predicted NPV values as follows:

```
plt.figure(figsize=(10,8))
plt.plot(y_train, y_pred_train, 'y.')
plt.xlabel('NPV Training Actual')
plt.ylabel('NPV Training Prediction')
plt.title('NPV Training Actual Vs. Prediction')
Python output=Fig. 5.67
```

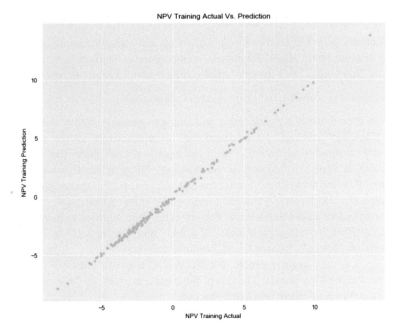

FIGURE 5.67 Training actual versus prediction.

```
plt.figure(figsize=(10,8))
plt.plot(y_test, y_pred_test, 'y.')
plt.xlabel('NPV Testing Actual')
plt.ylabel('NPV Testing Prediction')
plt.title('NPV Testing Actual Vs. Prediction')
```
Python output=Fig. 5.68

Next, let's obtain MAE, MSE, and RMSE as follows:

```
from sklearn import metrics
print('MAE:', round(metrics.mean_absolute_error(y_test,
  y_pred_test),5))
print('MSE:', round(metrics.mean_squared_error(y_test,
  y_pred_test),5))
print('RMSE:', round(np.sqrt(metrics.mean_squared_error(y_test,
  y_pred_test)),5))
```
Python output=MAE: 0.77298
MSE: 1.33434
RMSE: 1.15514

Next, let's get feature and permutation importance as follows:

```
from sklearn.inspection import permutation_importance
feature_importance=agb.feature_importances_
sorted_features=np.argsort(feature_importance)
```

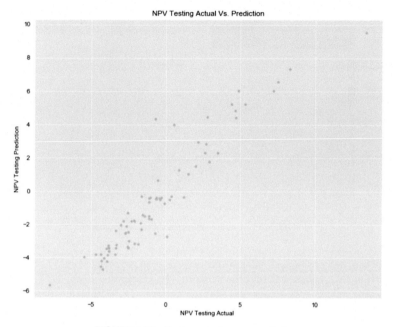

FIGURE 5.68   Testing actual versus prediction.

```
pos=np.arange(sorted_features.shape[0]) + .5
fig=plt.figure(figsize=(12, 6))
plt.subplot(1, 2, 1)
plt.barh(pos, feature_importance[sorted_features],
  align='center')
plt.yticks(pos, np.array(df.columns)[sorted_features])
plt.title('Feature Importance')
result=permutation_importance(agb, X_test, y_test,
  n_repeats=10,random_state=seed)
sorted_idx=result.importances_mean.argsort()
plt.subplot(1, 2, 2)
plt.boxplot(result.importances[sorted_idx].T, vert=False,
  labels=np.array(df.columns)[sorted_idx])
plt.title("Permutation Importance (test set)")
fig.tight_layout()
```
Python output=Fig. 5.69

As illustrated in Fig. 5.69, the top four features obtained from this algorithm are in line with the previously discussed two gradient boosting algorithms. Finally, let's also apply a 10-fold cross-validation to "agb" to observe the resulting R^2:

```
from sklearn.model_selection import cross_val_score
np.random.seed(seed)
```

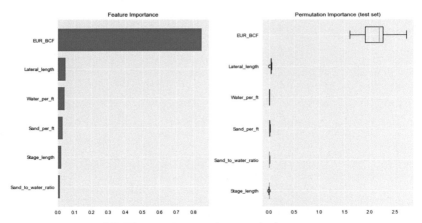

FIGURE 5.69 Feature and permutation importance.

```
scores_R2=cross_val_score(agb, x, y,cv=10,scoring='r2')
print(" R2_Cross-validation scores: {}". format(scores_R2))
```
Python output=R2_Cross-validation scores: [0.92186019 0.88601973
  0.75506971 0.94139885 0.92490025 0.88551602 0.50663126 0.95476168
  0.91861629 0.80053136]

```
print(" Average R2_Cross-validation scores: {}".
   format(scores_R2.mean()))
```
Python output=Average R2_Cross-validation scores:
  0.8495305328469369

Although grid search optimization has not been applied in all three algorithms to further improve the test set accuracy, the average accuracy of all three models appears to be very similar. After reading the "model evaluation" chapter, please make sure to apply grid search optimization to fine-tune the hyperparameters discussed in each model.

Frac intensity classification example

The next example is to go over a classification problem and apply the discussed algorithms to solve the problem using the scikit-learn library. One of the applications of ML in the O&G is predicting hydraulic frac stages that are expected to be challenging to treat. In other words, if these stages can be flagged ahead of time prior to the frac start date, it will provide a lot of insight prior to the frac start date and it will eliminate surprises during frac jobs. This info can also be provided to the frac supervisor and consultant on site. In this exercise, the following input features are available per stage: measured depth in ft, resistivity in ohm-m, Young's modulus over Poisson's ratio (YM/PR in 10^6 psi), gamma ray in gAPI, and minimum horizontal stress in psi. The completions engineer then used the actual frac treatment data to classify the stage as 0 or 1. 0 indicates stages that were not challenging to frac and 1 indicates stages that were difficult to treat which led to using more chemicals to treat the stage.

https://www.elsevier.com/books-and-journals/book-companion/9780128219294

Let's start a new Jupyter Notebook and import the usual libraries as well as the frac stage data set that can be found below:

```python
import numpy as np
import pandas as pd
import matplotlib.pyplot as plt
import seaborn as sns
%matplotlib inline
sns.set_style("darkgrid")
```

Let's look at the description of the frac stage data set. As shown below, there are 1000 stages that are labeled as either 0 or 1.

```python
df=pd.read_excel('Chapter5_Fracability_DataSet.xlsx')
df.describe().transpose()
```
Python output=Fig. 5.70

Next, let's visualize the data using distribution, box, and scatter plots:

```python
f, axes=plt.subplots(3, 2, figsize=(12, 12))
sns.distplot(df['MD_ft'], color="red", ax=axes[0, 0])
sns.distplot(df['Resistivity'], color="olive", ax=axes[0, 1])
sns.distplot(df['YM/PR'], color="blue", ax=axes[1, 0])
sns.distplot(df['GR'], color="orange", ax=axes[1, 1])
sns.distplot(df['Minimum Horizontal Stress Gradient'],
  color="black", ax=axes[2, 0])
plt.tight_layout()
```
Python output=Fig. 5.71

```python
f, axes=plt.subplots(3, 2, figsize=(12, 12))
sns.boxplot(df['MD_ft'], color="red", ax=axes[0, 0])
sns.boxplot(df['Resistivity'], color="olive", ax=axes[0, 1])
sns.boxplot(df['YM/PR'], color="blue", ax=axes[1, 0])
sns.boxplot(df['GR'], color="orange", ax=axes[1, 1])
sns.boxplot(df['Minimum Horizontal Stress Gradient'],
  color="black", ax=axes[2, 0])
plt.tight_layout()
```
Python output=Fig. 5.72

	count	mean	std	min	25%	50%	75%	max
MD_ft	1000.0	14000.000000	4269.740397	2571.139190	10943.876726	13864.280896	17149.039581	25382.097071
Resistivity	1000.0	500.000000	179.604705	-2.624400	364.746085	497.077253	620.324542	970.103975
YM/PR	1000.0	24.000000	8.388762	4.917917	17.708133	23.399724	29.588017	49.567271
GR	1000.0	300.000000	122.443017	-2.296323	210.515488	302.737171	387.137024	627.305824
Minimum Horizontal Stress Gradient	1000.0	0.926801	0.234990	0.292125	0.747472	0.932445	1.106538	1.508552
Fracability	1000.0	0.500000	0.500250	0.000000	0.000000	0.500000	1.000000	1.000000

FIGURE 5.70 Basic statistics of frac stage data set.

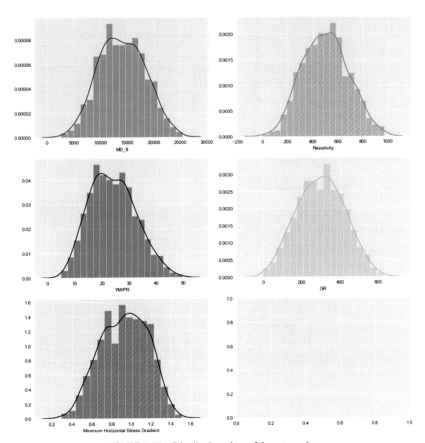

FIGURE 5.71 Distribution plots of frac stage data.

Let's also visualize the pair plot of the frac stage data by the fracability column.

```
sns.pairplot(df,hue='Fracability', palette="viridis")
Python output=Fig. 5.73
```

Let's also visualize the count plot of each class to make sure there is a balanced count between the two categories:

```
plt.figure(figsize=(10,8))
sns.countplot(df['Fracability'])
plt.title('Count Plot of Fracabilty', fontsize=15)
plt.xlabel('Categories', fontsize=15)
plt.xlabel('Count', fontsize=15)
Python output=Fig. 5.74
```

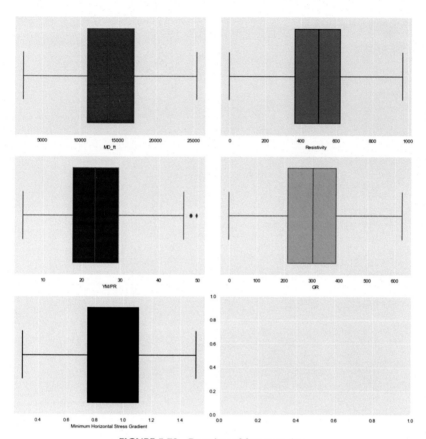

FIGURE 5.72   Box plots of frac stage data.

Next, let's look at the heat map of the Pearson correlation coefficient as follows:

```
plt.figure(figsize=(12,10))
sns.heatmap(df.corr(),linewidths=2,
  linecolor='black',cmap='plasma', annot=True)
Python output=Fig. 5.75
```

Next, let's normalize the input features using MinMax scaler as follows:

```
from sklearn import preprocessing
scaler=preprocessing.MinMaxScaler(feature_range=(0, 1))
y=df['Fracability']
x=df.drop(['Fracability'], axis=1)
x_scaled=scaler.fit(x)
x_scaled=scaler.transform(x)
```

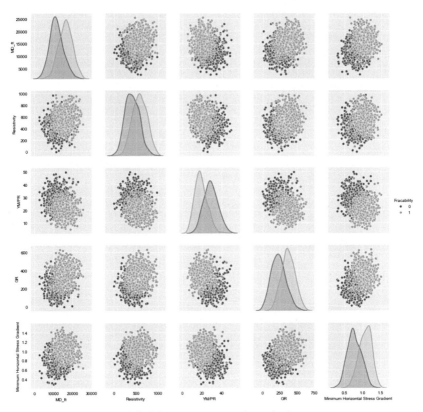

FIGURE 5.73 Frac stage data pair plot

Next, apply a 70/30 train_test_split as follows:

```
seed=50
np.random.seed(seed)
from sklearn.model_selection import train_test_split
x_train, x_test, y_train, y_test=train_test_split(x_scaled,y,
    test_size=0.30)
```

Support vector machine classification model

Let's start off by training a SVM model as follows:

```
from sklearn import svm
np.random.seed(seed)
svm=svm.SVC(C=1.0, kernel='rbf', degree=3, gamma=1, tol=0.001)
svm.fit(x_train, y_train)
from sklearn.metrics import accuracy_score
from sklearn.metrics import confusion_matrix
```

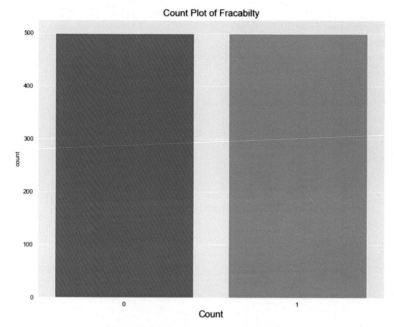

FIGURE 5.74 Count plot of fracability.

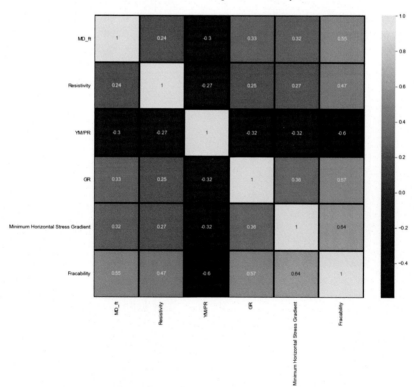

FIGURE 5.75 Heat map of Pearson correlation coefficient.

```
from sklearn.metrics import classification_report
y_pred=svm.predict(x_test)
cm=confusion_matrix(y_test, y_pred)
print('Accuracy Score:',accuracy_score(y_test, y_pred))
print('Confusion Matrix:')
print(cm)
print(classification_report(y_test, y_pred))
sns.heatmap(cm, center=True, annot=True, cmap='viridis',
  linewidths=3, linecolor='black')
plt.show()
```
Python output=Fig. 5.76

```
Accuracy Score: 0.9366666666666666
Confusion Matrix:
[[138  14]
 [  5 143]]
              precision    recall  f1-score   support

           0       0.97      0.91      0.94       152
           1       0.91      0.97      0.94       148

    accuracy                           0.94       300
   macro avg       0.94      0.94      0.94       300
weighted avg       0.94      0.94      0.94       300
```

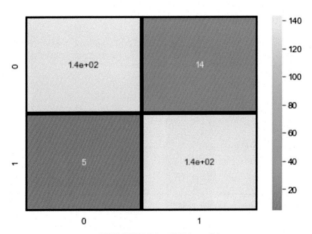

FIGURE 5.76 SVM model summary.

Let's use a 10-fold cross-validation as opposed to train_test_split:

```
from sklearn.model_selection import cross_val_score
np.random.seed(seed)
scores=cross_val_score(svm, x_scaled, y,cv=10,
scoring='accuracy')
print("Cross-validation scores: {}". format(scores))
print("Average cross-validation score: {}". format(scores.mean()))
Python output=Cross-validation scores: [0.94 0.93 0.93 0.93 0.9 0.95
   0.92 0.95 0.97 0.97]
Average cross-validation score: 0.9390000000000003
```

Random forest classification model

Next, let's apply a random forest model in the same Jupyter Notebook:

```
from sklearn.ensemble import RandomForestClassifier
np.random.seed(seed)
rf=RandomForestClassifier(n_estimators=5000, criterion='gini',
  max_depth=None, min_samples_split=2, min_samples_leaf=5,
  max_features='auto')
rf.fit(x_train, y_train)
y_pred=rf.predict(x_test)
cm=confusion_matrix(y_test, y_pred)
print('Accuracy Score:',accuracy_score(y_test, y_pred))
print('Confusion Matrix:')
print(cm)
print(classification_report(y_test, y_pred))
sns.heatmap(cm, center=True, annot=True, cmap='coolwarm',
  linewidths=3, linecolor='black')
plt.show()
Python output=Fig. 5.77
```

Let's also obtain feature ranking and permutation importance for this model. As illustrated, minimum horizontal stress and YM/PR appear to have the highest impact on the model output which was defined as a binary classifier.

```
from sklearn.inspection import permutation_importance
feature_importance=rf.feature_importances_
sorted_features=np.argsort(feature_importance)
pos=np.arange(sorted_features.shape[0]) + .5
fig=plt.figure(figsize=(12, 6))
plt.subplot(1, 2, 1)
plt.barh(pos, feature_importance[sorted_features], align='center',
  color='green')
```

```
Accuracy Score: 0.9366666666666666
Confusion Matrix:
[[139  13]
 [  6 142]]
              precision    recall  f1-score   support

           0       0.96      0.91      0.94       152
           1       0.92      0.96      0.94       148

    accuracy                           0.94       300
   macro avg       0.94      0.94      0.94       300
weighted avg       0.94      0.94      0.94       300
```

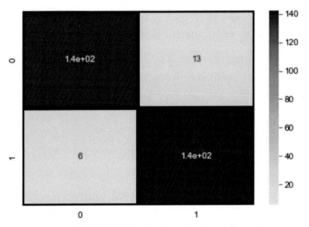

FIGURE 5.77 Random forest model summary.

```
plt.yticks(pos, np.array(df.columns)[sorted_features])
plt.title('Feature Importance')
result=permutation_importance(rf, x_test, y_test,
  n_repeats=10,random_state=seed)
sorted_idx=result.importances_mean.argsort()
plt.subplot(1, 2, 2)
plt.boxplot(result.importances[sorted_idx].T, vert=False,
  labels=np.array(df.columns)[sorted_idx])
plt.title("Permutation Importance (test set)")
fig.tight_layout()
Python output=Fig. 5.78
```

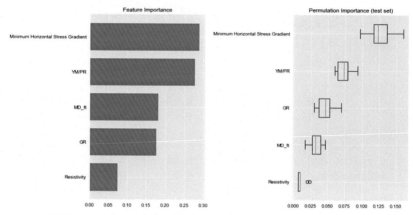

FIGURE 5.78 Feature ranking and permutation importance using random forest.

Let's also get the average R^2 accuracy for the 10-fold cross-validation as follows:

```
np.random.seed(seed)
scores=cross_val_score(rf, x_scaled, y,cv=10,scoring='accuracy')
print("Cross-validation scores: {}". format(scores))
print("Average cross-validation score: {}". format(scores.mean()))
Python output=Cross-validation scores: [0.94 0.93 0.92 0.91 0.91
   0.95 0.92 0.93 0.93 0.95]
Average cross-validation score: 0.9289999999999999
```

Extra trees classification model

Next, let's apply the extra tree algorithm as follows:

```
from sklearn.ensemble import ExtraTreesClassifier
np.random.seed(seed)
et=ExtraTreesClassifier(n_estimators=5000, criterion='gini',
   max_depth=None, min_samples_split=2, min_samples_leaf=5,
   max_features='auto')
et.fit(x_train, y_train)
y_pred=et.predict(x_test)
cm=confusion_matrix(y_test, y_pred)
print('Accuracy Score:',accuracy_score(y_test, y_pred))
print('Confusion Matrix:')
print(cm)
print(classification_report(y_test, y_pred))
sns.heatmap(cm, center=True, annot=True, cmap='Accent',
   linewidths=3, linecolor='black')
plt.show()
Python output=Fig. 5.79
```

```
Accuracy Score: 0.9333333333333333
Confusion Matrix:
[[137  15]
 [  5 143]]
```

	precision	recall	f1-score	support
0	0.96	0.90	0.93	152
1	0.91	0.97	0.93	148
accuracy			0.93	300
macro avg	0.93	0.93	0.93	300
weighted avg	0.94	0.93	0.93	300

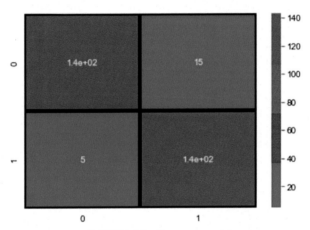

FIGURE 5.79 Extra trees model summary.

In addition, let's visualize the feature ranking and permutation importance as follows:

```python
from sklearn.inspection import permutation_importance
feature_importance=et.feature_importances_
sorted_features=np.argsort(feature_importance)
pos=np.arange(sorted_features.shape[0]) + .5
fig=plt.figure(figsize=(12, 6))
plt.subplot(1, 2, 1)
plt.barh(pos, feature_importance[sorted_features],
  align='center', color='brown')
plt.yticks(pos, np.array(df.columns)[sorted_features])
plt.title('Feature Importance')
result=permutation_importance(et, x_test, y_test,
  n_repeats=10,random_state=seed)
sorted_idx=result.importances_mean.argsort()
```

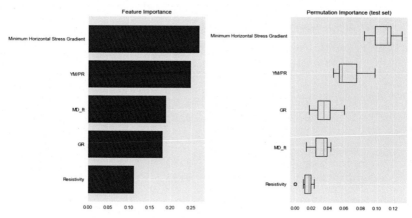

FIGURE 5.80 Feature ranking and permutation importance using extra trees.

```
plt.subplot(1, 2, 2)
plt.boxplot(result.importances[sorted_idx].T, vert=False,
    labels=np.array(df.columns)[sorted_idx])
plt.title("Permutation Importance (test set)")
fig.tight_layout()
Python output=Fig. 5.80
```

Let's also obtain the average R^2 accuracy for the 10-fold cross-validation as follows:

```
np.random.seed(seed)
scores=cross_val_score(et, x_scaled, y,cv=10,scoring='accuracy')
print("Cross-validation scores: {}". format(scores))
print("Average cross-validation score: {}". format(scores.mean()))
Python output=Cross-validation scores: [0.94 0.92 0.91 0.92 0.9 0.94
    0.93 0.93 0.93 0.97]
Average cross-validation score: 0.929
```

Gradient boosting classification model

Next, let's also apply gradient boosting as follows:

```
from sklearn.ensemble import GradientBoostingClassifier
np.random.seed(seed)
gb =GradientBoostingClassifier(loss='deviance',
    learning_rate=0.1, n_estimators=2000, criterion='friedman_mse',
    min_samples_split=2, min_samples_leaf=1, max_depth=3,
    max_features=None)
gb.fit(x_train, y_train)
y_pred=gb.predict(x_test)
cm=confusion_matrix(y_test, y_pred)
```

```
Accuracy Score: 0.9266666666666666
Confusion Matrix:
[[137  15]
 [  7 141]]
               precision    recall  f1-score   support

           0       0.95      0.90      0.93       152
           1       0.90      0.95      0.93       148

    accuracy                           0.93       300
   macro avg       0.93      0.93      0.93       300
weighted avg       0.93      0.93      0.93       300
```

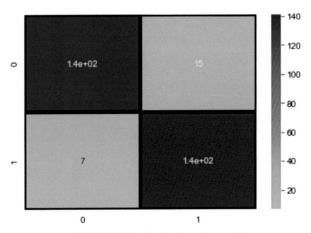

FIGURE 5.81   Gradient boosting model summary.

```
print('Accuracy Score:',accuracy_score(y_test, y_pred))
print('Confusion Matrix:')
print(cm)
print(classification_report(y_test, y_pred))
sns.heatmap(cm, center=True, annot=True, cmap='Greens',
  linewidths=3, linecolor='black')
plt.show()
```
Python output=Fig. 5.81

Also, obtain the feature ranking and permutation importance for this model as follows:

```
from sklearn.inspection import permutation_importance
feature_importance=gb.feature_importances_
sorted_features=np.argsort(feature_importance)
pos=np.arange(sorted_features.shape[0]) + .5
```

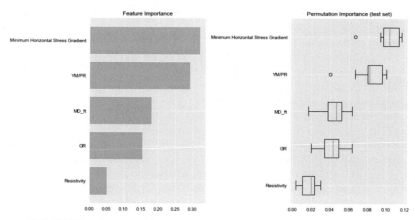

FIGURE 5.82 Feature ranking and permutation importance using gradient boosting.

```
fig=plt.figure(figsize=(12, 6))
plt.subplot(1, 2, 1)
plt.barh(pos, feature_importance[sorted_features], align='center',
  color='orange')
plt.yticks(pos, np.array(df.columns)[sorted_features])
plt.title('Feature Importance')
result=permutation_importance(gb, x_test, y_test, n_repeats=10,
  random_state=seed)
sorted_idx=result.importances_mean.argsort()
plt.subplot(1, 2, 2)
plt.boxplot(result.importances[sorted_idx].T, vert=False,
  labels=np.array(df.columns)[sorted_idx])
plt.title("Permutation Importance (test set)")
fig.tight_layout()
```
Python output=Fig. 5.82

Next, obtain the average R^2 accuracy for the 10-fold cross-validation as follows:

```
np.random.seed(seed)
scores=cross_val_score(gb, x_scaled, y,cv=10,scoring='accuracy')
print("Cross-validation scores: {}". format(scores))
print("Average cross-validation score: {}". format(scores.mean()))
```
**Python output=Cross-validation scores: [0.95 0.89 0.89 0.91 0.89
  0.92 0.9 0.94 0.95 0.94]**
Average cross-validation score: 0.9179999999999998

Extreme gradient boosting classification model

The next step is to apply XGBoost algorithm to the frac stage data as follows:

```
from xgboost import XGBClassifier
np.random.seed(seed)
xgb=XGBClassifier(objective ='binary:logistic',
  n_estimators=5000, reg_lambda=1, gamma=0,max_depth=3,
  learning_rate=0.1, alpha=0.5)
xgb.fit(x_train, y_train)
y_pred=xgb.predict(x_test)
cm=confusion_matrix(y_test, y_pred)
print('Accuracy Score:',accuracy_score(y_test, y_pred))
print('Confusion Matrix:')
print(cm)
print(classification_report(y_test, y_pred))
sns.heatmap(cm, center=True, annot=True, cmap='ocean',
  linewidths=3, linecolor='black')
plt.show()
```
Python output=Fig. 5.83

```
Accuracy Score: 0.9266666666666666
Confusion Matrix:
[[135  17]
 [  5 143]]
              precision    recall  f1-score   support

           0       0.96      0.89      0.92       152
           1       0.89      0.97      0.93       148

    accuracy                           0.93       300
   macro avg       0.93      0.93      0.93       300
weighted avg       0.93      0.93      0.93       300
```

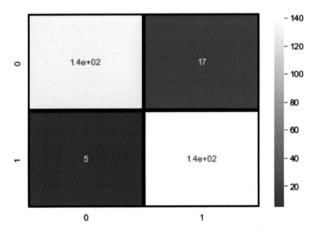

FIGURE 5.83 Extreme gradient boosting model summary.

Next, plot feature ranking and permutation importance using the XGBoost model as follows:

```
from sklearn.inspection import permutation_importance
feature_importance=xgb.feature_importances_
sorted_features=np.argsort(feature_importance)
pos=np.arange(sorted_features.shape[0]) + .5
fig=plt.figure(figsize=(12, 6))
plt.subplot(1, 2, 1)
plt.barh(pos, feature_importance[sorted_features], align='center',
    color='purple')
plt.yticks(pos, np.array(df.columns)[sorted_features])
plt.title('Feature Importance')
result=permutation_importance(xgb, x_test, y_test, n_repeats=10,
    random_state=seed)
sorted_idx=result.importances_mean.argsort()
plt.subplot(1, 2, 2)
plt.boxplot(result.importances[sorted_idx].T, vert=False,
    labels=np.array(df.columns)[sorted_idx])
plt.title("Permutation Importance (test set)")
fig.tight_layout()
Python output=Fig. 5.84
```

Finally, let's apply 10-fold cross-validation using XGBoost to obtain the average R^2 as follows:

```
np.random.seed(seed)
scores=cross_val_score(xgb, x_scaled, y,cv=10,
    scoring='accuracy')
print("Cross-validation scores: {}". format(scores))
print("Average cross-validation score: {}". format(scores.mean()))
Python output=Cross-validation scores: [0.93 0.92 0.91 0.92 0.89
    0.91 0.92 0.93 0.97 0.94]
Average cross-validation score: 0.924
```

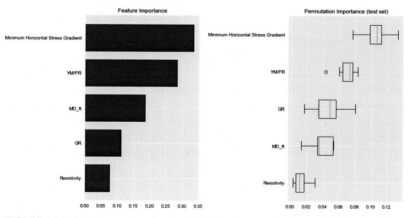

FIGURE 5.84 Feature ranking and permutation importance using extreme gradient boosting.

Handling missing data (imputation techniques)

This chapter has primarily focused on supervised algorithms that can be applied to solve various problems. Another important concept that was briefly discussed in Chapter 1 is handling missing data. The basic types of dealing with missing data were previously discussed. Below are various common approaches to missing data:

1) Simply remove any samples with N/A values: This is the simplest method of treating missing data, and the disadvantage of this approach is that the number of samples could be drastically reduced due to the removal of missing info. For example, if 1000 out of 2000 wells miss one important feature, removing wells with missing features will result in only 1000 wells. This is 50% less wells that could be included in the ML model.

2) Replacing missing values with the mean, median, most frequent, first value after missing, last value before missing of the existing data: Although this approach is overly simplistic and easy to perform, it is not typically recommended since there are other powerful algorithms that can be used instead.

3) Some algorithms such as k-nearest neighbor (KNN) and **MICE** can be used to fill the missing data. These algorithms do a great job only when the following two conditions exist:

   - There is an inherent relationship between the parameters that are missing and the existing parameters. For example, if sand/ft for one well is missing, but water per foot for the well is available, these algorithms might produce meaningful answers.

   - Most of the features with some inherent relationships must be present. One challenge that could be encountered when using these algorithms is that when a well misses a feature, it could potentially miss other important features. For example, when gathering competitor data, when a well misses sand loading, it usually misses water loading, stage spacing, etc.

The fancyimpute package in Python is a powerful package when one or two parameters in a data set are sporadically missing while there are other parameters that have inherent relationships between those parameters. When most of the parameters are simply missing from a well's data, it is not recommended to use these packages because they could quite possibly produce erroneous results. When using these packages, be sure to normalize or standardize the data prior to applying algorithms such as KNN and MICE. K-nearest neighbor uses feature similarity to predict the missing values. KNN essentially assigns a new value based on how closely it resembles points in the data set. The advantage of KNN is that it is more accurate than simply using mean, median, or most frequent methods. The disadvantage when using KNN is that it is computationally expensive, meaning it works by storing the whole training data set in memory. In addition, KNN is sensitive to outliers in the data set. Therefore, make sure the outliers are removed and the data are standardized prior to usage.

Multivariate imputation by chained equations

Multivariate imputation involves filling in the missing values multiple times, creating multiple "complete" data sets. This is described in detail by Schafer and Graham (2002). The missing values are imputed based on the observed values for a given individual and the relations observed in the data for other participants, assuming the observed variables are included in the imputation model. Multiple imputation procedures, particularly MICE, are very flexible and can be used in a broad range of settings. The following steps are taken when using MICE:

Step 1) A simple imputation, such as mean imputation is performed for every missing value in the data set. These mean imputations can be elaborated as "place holders" (Azur et al., 2011).

Step 2) The "place holder" mean imputations for one variable ("var") are set back to missing (Azur et al., 2011).

Step 3) "var" would then become the dependent variable in a regression model and all the other variables become independent variables in the regression model (Azur et al., 2011).

Step 4) The missing values for "var" are then replaced with imputation predictions from the regression model (Azur et al., 2011).

Step 5) Steps 2–4 are then repeated for each variable with missing data. The cycling through each of the variables constitutes one iteration or "cycle." At the end of one cycle, all the missing values have been substituted with predictions from regressions that reflect the observed data relationships (Azur et al., 2011).

Step 6) Steps 2 through 4 are repeated for several cycles, with the imputations being updated at each cycle. The number of cycles to be performed can be specified by the user. At the end of these cycles, the final imputations are retained, resulting in one imputed data set (Azur et al., 2011).

Fancy impute implementation in Python

As discussed, there are various applications for filling in missing values. One of the main applications of imputing missing values is in log analysis. Portions of the data that are believed to be washed out could be imputed using various techniques such as KNN, MICE, iterative imputer, etc. In addition, well data (drilling, completions, and production) could also be imputed if there are important features that could be used to get the missing data. In this section, the implementation of KNN and MICE to impute missing values for one second hydraulic frac data of a stage in the Marcellus Shale is covered. These data were obtained from the MSEEL publicly available data set and a portion of the surface treating pressure data was removed for this exercise. The file can be found at the link below:

https://www.elsevier.com/books-and-journals/book-companion/
9780128219294

Open a new Jupyter Notebook and start importing the following libraries:

```
import numpy as np
import pandas as pd
import matplotlib.pyplot as plt
import seaborn as sns
import missingno as msno
%matplotlib inline
df=pd.read_excel('Chapter5_Imputation_DataSet.xlsx')
df.describe()
Python output=Fig. 5.85
```

It is important to visualize the missing data in each feature as follows:

```
plt.figure(figsize=(10,8))
sns.heatmap(df.isnull(),yticklabels=False,cbar=False,
  cmap='viridis')
Python output=Fig. 5.86
```

Next, let's visualize the missing data using a slightly different approach using "missingno" library as follows. Please note that this library is used for exploratory visualization of missing data.

	Surface Treating Pressure	Slurry Rate	Proppant Concentration
count	4588.000000	4752.000000	4752.000000
mean	8142.469922	99.710795	1.422033
std	85.671249	3.308136	0.703032
min	7913.000000	64.400000	0.000000
25%	8088.000000	99.900000	1.000000
50%	8116.000000	100.300000	1.500000
75%	8193.000000	100.700000	2.000000
max	8708.000000	101.600000	3.500000

FIGURE 5.85 Missing data set description.

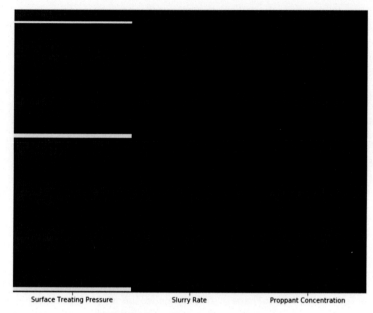

FIGURE 5.86 Missing data visualization.

```
missingdata_df=df.columns[df.isnull().any()].tolist()
msno.matrix(df[Chapter5_Imputation_DataSet_df])
Python output=Fig. 5.87
```

Next, obtain the number of missing values for each feature (in this exercise, only surface treating pressure has missing data) as follows:

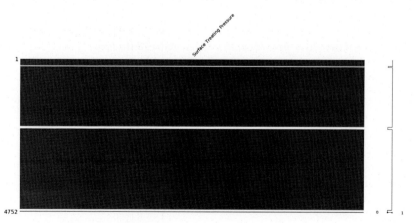

FIGURE 5.87 Missing data visualization using "missingno" library.

```
missing_val_count_by_column=(df.isnull().sum())
print(missing_val_count_by_column
  [missing_val_count_by_column > 0])
```
Python output=Surface Treating Pressure 164
dtype: int64

Next, standardize the data as follows:

```
from sklearn.preprocessing import StandardScaler
scaler=StandardScaler()
scaler.fit(df)
df_scaled=scaler.transform(df)
df_scaled=pd.DataFrame(df_scaled, columns=['Surface Treating\
  Pressure', 'Slurry Rate', 'Proppant Concentration'])
```

First, in "Anaconda prompt," use "pip install fancyimpute" before importing the library as shown below. Next, use two nearest neighbors and the KNN algorithm to impute the missing values:

```
from fancyimpute import KNN
X_filled_knn=KNN(k=2).fit_transform(df_scaled)
X_filled_knn=pd.DataFrame(X_filled_knn,columns=['Surface Treating\
  Pressure', 'Slurry Rate', 'Proppant Concentration'])
X_filled_knn
```
Python output=Fig. 5.88

Next, since the data were standardized prior to feeding the data into the KNN imputation algorithm, convert back to its original form and export into a csv file.

```
X_filled_knn['Surface Treating Pressure']=(X_filled_knn['Surface\
  Treating Pressure']*df['Surface Treating Pressure'].std())+df
  ['Surface Treating Pressure'].mean()
X_filled_knn['Slurry Rate']=(X_filled_knn['Slurry Rate']*df['Slurry\
  Rate'].std())+df['Slurry\ Rate'].mean()
X_filled_knn['Proppant Concentration']=(X_filled_knn['Proppant\
  Concentration']*df['Proppant\ Concentration'].std())+df
  ['Proppant Concentration'].mean()
X_filled_knn.to_csv('KNN_Imputation.csv', index=False)
```

Next, import the MICE library and apply the MICE algorithm to fill in the missing data in average treating pressure column as follows:

```
from impyute.imputation.cs import mice
imputed_training=mice(df_scaled.values)
imputed_training=pd.DataFrame(imputed_training,columns=
  ['Surface Treating Pressure', 'Slurry Rate', 'Proppant
  Concentration'])
```

```
Imputing row 1/4752 with 0 missing, elapsed time: 6.403
Imputing row 101/4752 with 0 missing, elapsed time: 6.403
Imputing row 201/4752 with 0 missing, elapsed time: 6.403
Imputing row 301/4752 with 0 missing, elapsed time: 6.428
Imputing row 401/4752 with 0 missing, elapsed time: 6.432
Imputing row 501/4752 with 0 missing, elapsed time: 6.432
Imputing row 601/4752 with 0 missing, elapsed time: 6.432
Imputing row 701/4752 with 0 missing, elapsed time: 6.436
Imputing row 801/4752 with 0 missing, elapsed time: 6.436
Imputing row 901/4752 with 0 missing, elapsed time: 6.436
Imputing row 1001/4752 with 0 missing, elapsed time: 6.440
Imputing row 1101/4752 with 0 missing, elapsed time: 6.440
Imputing row 1201/4752 with 0 missing, elapsed time: 6.440
Imputing row 1301/4752 with 0 missing, elapsed time: 6.444
Imputing row 1401/4752 with 0 missing, elapsed time: 6.444
Imputing row 1501/4752 with 0 missing, elapsed time: 6.444
Imputing row 1601/4752 with 0 missing, elapsed time: 6.448
Imputing row 1701/4752 with 0 missing, elapsed time: 6.448
Imputing row 1801/4752 with 0 missing, elapsed time: 6.448
Imputing row 1901/4752 with 0 missing, elapsed time: 6.448
Imputing row 2001/4752 with 0 missing, elapsed time: 6.452
Imputing row 2101/4752 with 1 missing, elapsed time: 6.452
Imputing row 2201/4752 with 0 missing, elapsed time: 6.460
Imputing row 2301/4752 with 0 missing, elapsed time: 6.460
Imputing row 2401/4752 with 0 missing, elapsed time: 6.464
Imputing row 2501/4752 with 0 missing, elapsed time: 6.464
Imputing row 2601/4752 with 0 missing, elapsed time: 6.464
Imputing row 2701/4752 with 0 missing, elapsed time: 6.468
Imputing row 2801/4752 with 0 missing, elapsed time: 6.468
Imputing row 2901/4752 with 0 missing, elapsed time: 6.468
Imputing row 3001/4752 with 0 missing, elapsed time: 6.472
Imputing row 3101/4752 with 0 missing, elapsed time: 6.472
Imputing row 3201/4752 with 0 missing, elapsed time: 6.472
Imputing row 3301/4752 with 0 missing, elapsed time: 6.476
Imputing row 3401/4752 with 0 missing, elapsed time: 6.476
Imputing row 3501/4752 with 0 missing, elapsed time: 6.476
Imputing row 3601/4752 with 0 missing, elapsed time: 6.476
Imputing row 3701/4752 with 0 missing, elapsed time: 6.480
Imputing row 3801/4752 with 0 missing, elapsed time: 6.480
Imputing row 3901/4752 with 0 missing, elapsed time: 6.480
Imputing row 4001/4752 with 0 missing, elapsed time: 6.484
Imputing row 4101/4752 with 0 missing, elapsed time: 6.484
Imputing row 4201/4752 with 0 missing, elapsed time: 6.484
Imputing row 4301/4752 with 0 missing, elapsed time: 6.488
Imputing row 4401/4752 with 0 missing, elapsed time: 6.488
Imputing row 4501/4752 with 0 missing, elapsed time: 6.488
Imputing row 4601/4752 with 0 missing, elapsed time: 6.492
Imputing row 4701/4752 with 1 missing, elapsed time: 6.496
```

	Surface Treating Pressure	Slurry Rate	Proppant Concentration
0	-2.678786	-10.675047	-2.022926
1	-1.663165	-10.554120	-2.022926
2	-1.091149	-10.433193	-2.022926
3	-0.717588	-10.282035	-2.022926
4	-0.122224	-10.070413	-2.022926
...	...	...	...
4747	-0.682566	-0.003264	-2.022926
4748	-0.659219	0.026968	-2.022926
4749	-0.495785	0.057200	-2.022926
4750	-0.554154	0.057200	-2.022926
4751	-0.519133	0.057200	-2.022926

FIGURE 5.88   KNN Python output.

Next, convert the data back to its original form and export into an excel as follows:

```
imputed_training['Surface Treating Pressure']=(imputed_training
  ['Surface Treating Pressure']*df['Surface Treating
  Pressure'].std())+df['Surface Treating Pressure'].mean()
imputed_training['Slurry Rate']=(imputed_training['Slurry
  Rate']*df['Slurry Rate'].std())+df['Slurry Rate'].mean()
imputed_training['Proppant Concentration']=(imputed_training
  ['Proppant Concentration']*df['Proppant Concentration'].std())+
  df['Proppant Concentration'].mean()
imputed_training.to_csv('MICE_Imputation.csv', index=False)
```

Iterative imputer imputes the missing values in a **round-robin fashion** where at each step, a feature column is selected as the model output and the remaining columns are treated as inputs. Afterward, a regression model is built based on the x and y variables and the trained regression model is then used to predict the missing y values. This is accomplished for each feature in an iterative fashion and it is repeated for "max_iter" imputation rounds. Let's apply iterative imputer to the standardized data set as follows:

```
from fancyimpute import IterativeImputer
X_filled_ii=IterativeImputer(max_iter=10).fit_transform
  (df_scaled)
X_filled_ii=pd.DataFrame(X_filled_ii,columns=['Surface    Treating
  Pressure', 'Slurry Rate', 'Proppant Concentration'])
X_filled_ii['Surface    Treating    Pressure']=(X_filled_ii['Surface
  Treating Pressure']*df['Surface Treating Pressure'].std())+df
  ['Surface Treating Pressure'].mean()
X_filled_ii['Slurry  Rate']=(X_filled_ii['Slurry  Rate']*df['Slurry
  Rate'].std())+df['Slurry Rate'].mean()
X_filled_ii['Proppant Concentration']=(X_filled_ii['Proppant
  Concentration']*df['Proppant Concentration'].std())+df['Proppant
  Concentration'].mean()
X_filled_ii.to_csv('Iterative_Imputer_Imputation.csv',
  index=False)
```

Let's use the iterative imputer with the exception that it will run 5 times and the average imputation will be used as the final output.

```
from fancyimpute import IterativeImputer
XY_incomplete=df_scaled
n_imputations=5
XY_completed=[]
for i in range(n_imputations):
    imputer=IterativeImputer(sample_posterior=True,
      random_state=i)
    XY_completed.append(imputer.fit_transform
      (XY_incomplete))
XY_completed_mean=np.mean(XY_completed, 0)
```

```
XY_completed_std=np.std(XY_completed, 0)
XY_completed=np.array(XY_completed)
XY_completed_mean=pd.DataFrame(XY_completed_mean,columns=
  ['Surface Treating Pressure', 'Slurry Rate', 'Proppant
  Concentration'])
XY_completed_mean['Surface Treating Pressure']=(XY_completed_mean
  ['Surface Treating Pressure']*df['Surface Treating
  Pressure'].std())+df['Surface Treating Pressure'].mean()
XY_completed_mean['Slurry Rate']=(XY_completed_mean['Slurry Rate']*
  df['Slurry Rate'].std())+df['Slurry Rate'].mean()
XY_completed_mean['Proppant Concentration']=(XY_completed_mean
  ['Proppant Concentration']*df['Proppant Concentration'].std())+
  df['Proppant Concentration'].mean()
XY_completed_mean.to_csv('Mean_Iterative_Imputer.csv',
  index=False)
```

Please feel free to apply the same workflow and methodology to impute missing values on any other problems. Also, make sure to first apply this to full data sets where a portion of the data is intentionally removed for the purpose of testing the model. Afterward, compare the imputed values from different algorithms discussed to the actual values to comprehend the accuracy of different models.

Rate of penetration (ROP) optimization example

Another regression example that can be illustrated is drilling rate of penetration (ROP) prediction and optimization. There are various important features that are captured when drilling a well. These features include but are not limited to hook load, rpm, torque, weight on bit (WOB), differential pressure, gamma, and ROP. An important output feature that must be predicted or optimized is referred to as ROP. Maximizing ROP is the absolute goal in drilling. Therefore, the key to ROP maximization is building a supervised regression machine learning model where ROP can be used as the output of the model. Once a satisfactory model has been trained, ROP can be easily predicted for new wells. In addition, the last chapter of this book (evolutionary optimization) can be used to find the set of input features that result in maximizing the ROP.

In this section, MIP3H and 5H from the MSEEL project were used as shown in the link below:

http://www.mseel.org/research/research_MIP.html

The data have been cleaned up and can be accessed using the portal below:

https://www.elsevier.com/books-and-journals/book-companion/
9780128219294

One well was used as the training data set and the other well was used as a complete blind set to test the model accuracy and capabilities. The depth-based data in the curve and lateral portion of each well were used in this analysis. This is because we are interested in the ROP optimization in the curve and lateral portions only. It is highly recommended to use all the drilling data (10s or 100s of wells) across a field of interest to build a more comprehensive model. This example is for illustration purposes of the workflow, and the accuracy of the model predictive capabilities can certainly be improved when incorporating more wells and data. In this example, both curve and lateral portions were combined, however, when including more wells, a better approach could be evaluating the curve and lateral portions independently by building two separate models for curve and lateral portions individually.

The input features used in this exercise are as follows:

- Hole depth (is the measured depth or MD in ft)
- Hook load (Klbs)
- Rotary rpm
- Rotary torque (Klbs-ft)
- Weight on bit (WOB in Klbs)
- Differential pressure (psi)
- Gamma ray at bit (gAPI)

The output of the model is ROP which has a unit of ft/hr.

The first step is to import the necessary libraries and the training data set as follows:

```
import numpy as np
import pandas as pd
import matplotlib.pyplot as plt
import seaborn as sns
%matplotlib inline
df=pd.read_csv('Training_Data_Set_One_Well.csv')
```

Obtain the basic statistics of each feature and verify the range of each feature to ensure outlier removal prior to proceeding:

```
df.describe().transpose()
```
Python output=Fig. 5.89

	count	mean	std	min	25%	50%	75%	max
Hole Depth	7934.0	10487.930804	2289.973781	6524.000	8500.250	10487.500	12469.750	14454.00
Hook Load	7934.0	129.672107	7.724432	107.200	123.800	129.500	134.400	156.40
Rotary RPM	7934.0	65.961054	24.280720	9.000	49.000	70.000	90.000	101.00
Rotary Torque	7934.0	11.459822	3.386672	2.701	9.096	11.373	14.198	20.05
Weight on Bit	7934.0	19.826758	5.611785	0.000	16.300	20.400	23.900	39.40
Differential Pressure	7934.0	520.255067	142.477894	2.900	429.350	565.900	627.700	783.30
Gamma at Bit	7934.0	211.783237	81.536917	54.120	148.240	204.710	235.290	600.00
Rate Of Penetration	7934.0	143.107280	55.738087	1.610	100.160	161.160	185.240	259.29

FIGURE 5.89 df basic statistics.

Next, let's visualize the distribution plot for each feature as follows:

```
f, axes =plt.subplots(4, 2, figsize=(12, 12))
sns.distplot(df['Hole Depth'], color="red", ax=axes[0, 0],
    axlabel='Measured Depth (ft)')
sns.distplot(df['Hook Load'], color="olive", ax=axes[0, 1],
    axlabel='Hook Load (Klbs)')
sns.distplot(df['Rate Of Penetration'], color="blue", ax=axes
    [1, 0],axlabel='Rate of Penetration (ft/hr)')
sns.distplot(df['Rotary RPM'], color="orange", ax=axes[1, 1],
    axlabel='Rotary rpm')
sns.distplot(df['Rotary Torque'], color="black",
    ax=axes[2, 0],axlabel='Rotary Torque (Klbs-ft)')
sns.distplot(df['Weight on Bit'], color="green",
    ax=axes[2, 1],axlabel='Weight on bit (Klbs)')
sns.distplot(df['Differential Pressure'], color="brown",
    ax=axes[3, 0],axlabel='Differential Pressure (psi)')
sns.distplot(df['Gamma at Bit'], color="gray", ax=axes[3, 1],
    axlabel='Gamma Ray at Bit (gAPI)')
plt.tight_layout()
```
Python output=Fig. 5.90

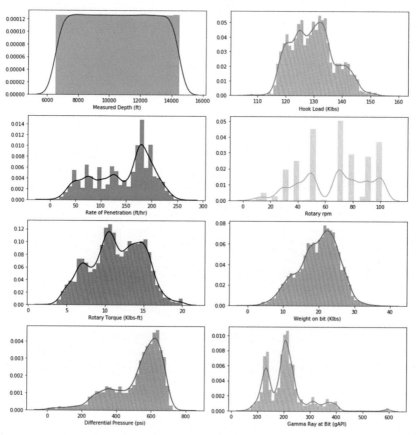

FIGURE 5.90 Distribution plots of drilling features.

As observed, each parameter has a good distribution, and this is simply based on the training data for one well and the distribution of each feature can be improved by including more wells. Let's also visualize the box plots as follows:

```
f, axes =plt.subplots(4, 2, figsize=(12, 12))
sns.boxplot(df['Hole Depth'], color="red", ax=axes[0, 0])
sns.boxplot(df['Hook Load'], color="olive", ax=axes[0, 1])
sns.boxplot(df['Rate Of Penetration'], color="blue",
  ax=axes[1, 0])
sns.boxplot(df['Rotary RPM'], color="orange", ax=axes[1, 1])
sns.boxplot(df['Rotary Torque'], color="black", ax=axes[2, 0])
sns.boxplot(df['Weight on Bit'], color="green", ax=axes[2, 1])
sns.boxplot(df['Differential Pressure'], color="brown",
  ax=axes[3, 0])
sns.boxplot(df['Gamma at Bit'], color="gray", ax=axes[3, 1])
plt.tight_layout()
```
Python output=Fig. 5.91

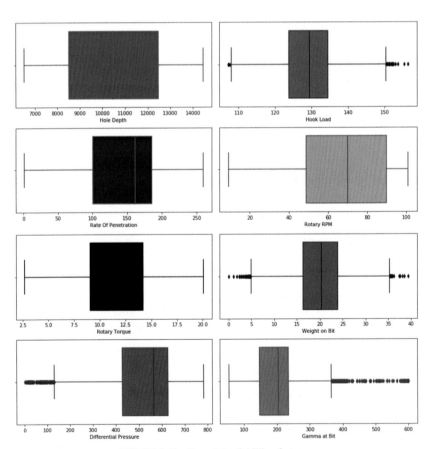

FIGURE 5.91 Box plots of drilling features.

Pearson correlation coefficient heatmap can be generated as follows:

```
plt.figure(figsize=(12,10))
sns.heatmap(df.corr(), annot=True, linecolor='white',
  linewidths=2, cmap='coolwarm')
```
Python output=Fig. 5.92

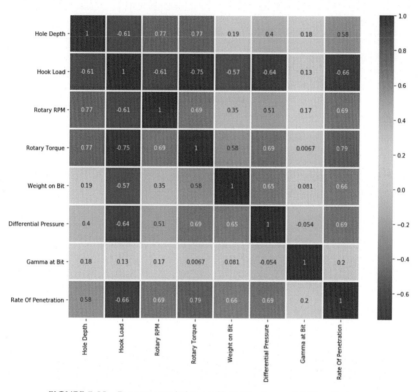

FIGURE 5.92 Pearson correlation coefficient heatmap of drilling features.

As illustrated, the majority of the features such as rpm, torque, WOB, differential pressure, etc., have a strong correlation with ROP. This is very encouraging to see as this could potentially indicate that these features are important in relation to ROP. Next, let's use the MinMax scaler to normalize the data as follows:

```
from sklearn import preprocessing
scaler=preprocessing.MinMaxScaler(feature_range=(0,1))
  scaler.fit(df)
df_scaled=scaler.transform(df)
```

Let's also define the scaled X and y variables as follows:

```
df_scaled=pd.DataFrame(df_scaled, columns=['Hole Depth', 'Hook\
  Load', 'Rotary RPM', 'Rotary Torque', 'Weight on Bit',
  'Differential\
Pressure', 'Gamma at Bit', 'Rate Of Penetration'])
y_scaled=df_scaled[['Rate Of Penetration']]
x_scaled=df_scaled.drop(['Rate Of Penetration'], axis=1)
```

Next, split the data into training and testing set (70/30 split):

```
from sklearn.model_selection import train_test_split
  seed=1000
np.random.seed(seed)
X_train, X_test,y_train, y_test=train_test_split(x_scaled,
  y_scaled, test_size=0.30)
```

Let's start off by using a support vector machine model to examine the model's performance:

```
from sklearn.svm import SVR
np.random.seed(seed)
SVM =SVR(kernel='rbf', gamma=1.5,C=5)
SVM.fit(X_train,np.ravel(y_train))
```

Also, obtain the R^2 for training and testing set as follows:

```
y_pred_train=SVM.predict(X_train)
y_pred_test=SVM.predict(X_test)
corr_train=np.corrcoef(y_train['Rate Of Penetration'],
  y_pred_train) [0,1]
print('ROP Train Data r^2=',round(corr_train**2,4),'r=',
  round(corr_train,4))
Python output=
ROP Train Data r^2=0.8967 r=0.947
corr_test=np.corrcoef(y_test['Rate Of Penetration'], y_pred_test)
  [0,1]
print('ROP Test Data r^2= ',round(corr_test**2,4),'r=',
  round(corr_test,4))
Python output=
ROP Test Data r^2=0.901 r=0.9492
```

The result looks reasonable considering that only one well was used for train_test_split. Please note that the fine-tuning parameters such as gamma, C, etc., have not been optimized in this model and chapter 7 can be used to perform this step. Next, plot the ROP testing actual versus prediction as follows:

```
plt.figure(figsize=(10,8))
plt.plot(y_test['Rate Of Penetration'], y_pred_test, 'b.')
plt.xlabel('ROP Testing Actual')
plt.ylabel('ROP Testing Prediction')
plt.title('ROP Testing Actual vs. Prediction')
```
Python output=Fig. 5.93

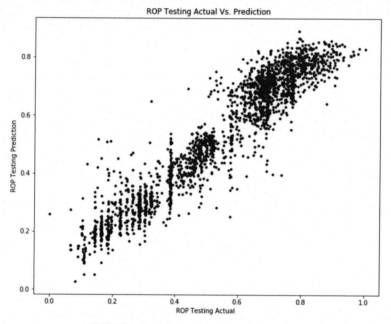

FIGURE 5.93 ROP testing actual vs. prediction.

Let's also obtain MAE, MSE, and RMSE:

```
from sklearn import metrics
print('Testing ROP MAE:', round(metrics.mean_absolute_error(y_test
  ['Rate Of Penetration'], y_pred_test),4))
print('Testing ROP MSE:', round(metrics.mean_squared_error(y_test
  ['Rate Of Penetration'], y_pred_test),4))
print('Testing ROP RMSE:', round(np.sqrt
  (metrics.mean_squared_error(y_test['Rate Of Penetration'],
y_pred_test)),4))
```
Python output=
Testing ROP MAE: 0.0514
Testing ROP MSE: 0.0048
Testing ROP RMSE: 0.0689

Next, the second well is the blind well that will be used to test the accuracy of the model. Since only one well was used to train a model and not enough variations of parameters were seen during training, the expected accuracy outcome on this complete blind set is expected to be low. This file is named "Blind_Data_Set" and is imported as follows:

```
df_blind=pd.read_csv('Chapter5_ROP_Blind_DataSet.csv')
```

Before defining the X and y variables on the blind data set, first normalize the data using "scaler.transform" as shown below. This ensures the data are normalized based on the training data set minimum and maximum values for each feature.

```
scaled_blind=scaler.transform(df_blind)
```

After applying "scaler.transform" to the blind data set, the data were converted into a numpy array. Therefore, let's convert the data from an array to a data frame and define "x_scaled_blind" and "y_scaled_blind."

```
scaled_blind=pd.DataFrame(scaled_blind, columns=['Hole Depth',
    'Hook Load', 'Rotary RPM', 'Rotary Torque','Weight on Bit',
    'Differential Pressure', 'Gamma at Bit','Rate Of Penetration'])
y_scaled_blind=scaled_blind['Rate Of Penetration']
x_scaled_blind=scaled_blind.drop(['Rate Of Penetration'],axis=1)
```

Next, use the trained "SVM" model to predict the output variable (ROP) on the blind data set as follows:

```
y_pred_blind=SVM.predict(x_scaled_blind)
```

Also obtain the R^2 for the blind data set as follows:

```
corr_test=np.corrcoef(y_scaled_blind, y_pred_blind) [0,1]
print('ROP Blind Data r^2=',round(corr_test**2,4),'r=',
    round(corr_test,4))
Python output=
ROP Blind Data r^2=0.6823 r=0.826
```

As illustrated, the R^2 value for the blind set is low (as expected). This is simply due to the size of the training and lack of training variations during the training process. This example illustrates the importance of testing various blind sets to make sure the model remains accurate prior to productionizing or commercializing a model. Next, let's plot ROP blind actual versus prediction as follows:

```
plt.figure(figsize=(10,8))
plt.plot(y_scaled_blind, y_pred_blind, 'g.')
plt.xlabel('ROP Blind Actual')
plt.ylabel('ROP Blind Prediction')
plt.title('ROP Blind Actual vs. Prediction')
Python output=Fig. 5.94
```

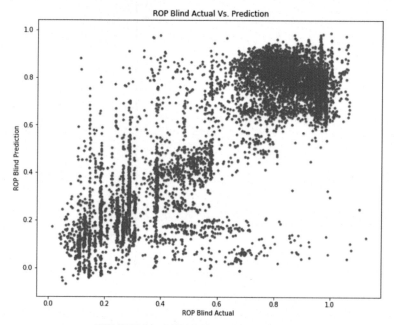

FIGURE 5.94   ROP blind actual vs. prediction.

Next, to visualize the actual blind and predicted ROP data set, the following lines of codes can be written:

```
plt.figure(figsize=(6,12))
sns.scatterplot(y_scaled_blind, df_blind['Hole Depth'],
   label='Actual Blind Data', color='blue')
sns.scatterplot(y_pred_blind, df_blind['Hole Depth'],
   label='Predicted Blind Data', color='green')
Python output=Fig. 5.95
```

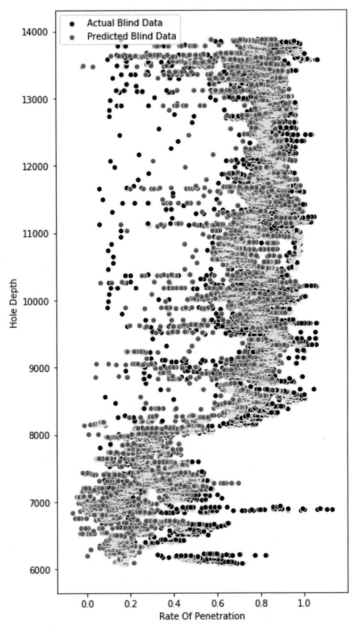

FIGURE 5.95 ROP blind vs. prediction comparison plot.

As illustrated in this figure, the overall trend appears to be ok and other algorithms should be tested to see their performances versus SVM. Let's also train an extra tree model in the same Jupyter Notebook file to see how it performs as follows:

```
from sklearn.ensemble import ExtraTreesRegressor
np.random.seed(seed)
ET=ExtraTreesRegressor(n_estimators=100,criterion='mse',
  max_depth=None, min_samples_split=2,
    min_samples_leaf=1)
ET.fit(X_train,np.ravel(y_train))
y_pred_train=ET.predict(X_train)
y_pred_test=ET.predict(X_test)
corr_train=np.corrcoef(y_train['Rate Of Penetration'],
  y_pred_train) [0,1]
print('ROP Train Data r^2=',round(corr_train**2,4),'r=',
  round(corr_train,4))
```
Python output=ROP Train Data r^2=1.0 r=1.0

```
corr_test=np.corrcoef(y_test['Rate Of Penetration'], y_pred_test)\
  [0,1]
print('ROP Test Data r^2=',round(corr_test**2,4),'r=',
  round(corr_test,4))
```
Python output=
ROP Test Data r^2=0.9503 r=0.9748

Let's also visualize the resulting scatter plot of ROP testing actual and prediction using the extra trees model as follows:

```
plt.figure(figsize=(10,8))
plt.plot(y_test['Rate Of Penetration'], y_pred_test, 'b.')
plt.xlabel('ROP Testing Actual')
plt.ylabel('ROP Testing Prediction')
plt.title('ROP Testing Actual vs. Prediction Using Extra Trees Model')
```
Python output=Fig. 5.96

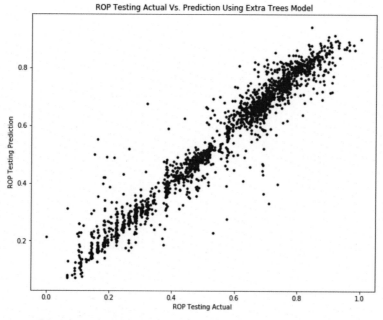

FIGURE 5.96 ROP testing actual vs. prediction using extra trees model.

As illustrated, the testing accuracy when using extra trees is higher than the previous model (SVM). Let's also apply this trained extra trees model to the blind data set that was previously imported as follows:

```
y_pred_blind=ET.predict(x_scaled_blind)
corr_test=np.corrcoef(y_scaled_blind, y_pred_blind) [0,1]
print('ROP Blind Data r^2=',round(corr_test**2,4),'r=',
  round(corr_test,4))
```

Python output=
ROP Blind Data r^2=0.7476 r=0.8646

As shown above, the blind set accuracy is also higher than the blind set accuracy of the SVM model.

Next, plot the ROP blind actual versus prediction as follows:

```
plt.figure(figsize=(10,8))
plt.plot(y_scaled_blind, y_pred_blind, 'g.')
plt.xlabel('ROP Blind Actual')
plt.ylabel('ROP Blind Prediction')
plt.title('ROP Blind Actual vs. Prediction Using Extra Trees Model')
```

Python output=Fig. 5.97

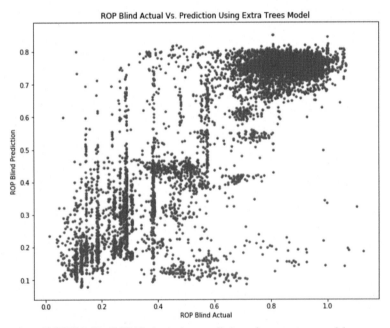

FIGURE 5.97   ROP blind actual vs. prediction using extra trees model.

Finally, let's also plot this as a function of depth as follows:

```
plt.figure(figsize=(6,12))
sns.scatterplot(y_scaled_blind, df_blind['Hole Depth'],
    label='Actual Blind Data', color='blue')
sns.scatterplot(y_pred_blind, df_blind['Hole Depth'], label='\
    Predicted Blind Data', color='green')
```
Python output=Fig. 5.98

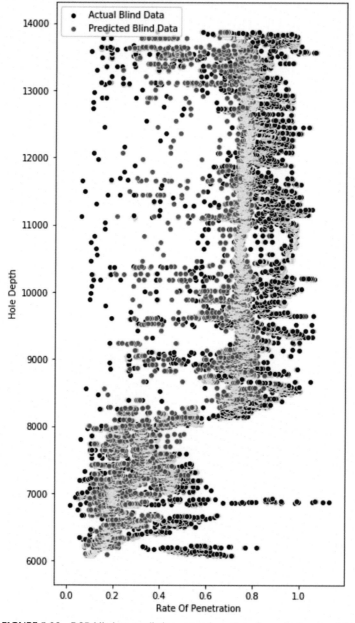

FIGURE 5.98 ROP blind vs. prediction comparison plot using extra trees model.

As illustrated, the extra trees model appears to be underperforming on a high level when applied to the blind data set. Please spend some time applying other covered algorithms to see whether any of the discussed algorithms can beat the accuracy of the SVM or extra trees. This example was to show a proof of concept of training a model to predict ROP based on training one well and testing another well on the same pad and the result expected to be ok due to lack of the adequate amount of data for training purposes. The next step is to build a model based on 10s or 100s of wells followed by testing the accuracy to make sure the predictive capabilities are available. Although the selected features in this problem appear to be the correct features, this does not indicate that other important features should not be considered. The public availability of data is extremely limited. Therefore, please use your domain expertise to make sure to use all the available features to build a more robust and powerful machine learning model within your organization, incorporating 10s or 100s of wells into your training portion. Let's also obtain the feature importance using the extra tree algorithm as follows:

```
feature_names =df.columns[:-1]
plt.figure(figsize=(10,8))
feature_imp =pd.Series(ET.feature_importances_,
  index=feature_names).sort_values(ascending=False)
sns.barplot(x=feature_imp, y=feature_imp.index)
plt.xlabel('Feature Importance Score Using Extra Trees')
plt.ylabel('Features')
plt.title("Feature Importance Ranking")
```
Python output=Fig. 5.99

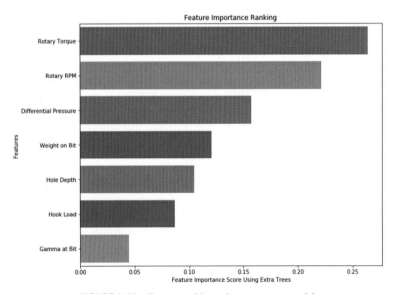

FIGURE 5.99 Feature ranking using extra trees model.

Please make sure to also apply grid search optimization for fine tuning the parameters for each selected ML model and also use the last chapter of this book (Evolutionary Optimization) to find the best set of input features that result in maximizing the ROP.

References

Avriel, M. (2003). *Nonlinear programming: Analysis and methods.* Dover Publishing.

Azur, M., Stuart, E., Frangakis, C., & Philip, L. (2011). Multiple imputation by chained equations: what is it and how does it work? *International Journal of Methods in Psychiatric Research, 20*(1), 40−49.

Belyadi, H., Fathi, E., & Belyadi, F. (2019). *Hydraulic fracturing in unconventional reservoirs* (2nd ed.). Elsevier, 9780128176658.

Breiman, L. (1994). *Bagging predictors.* Department of Statistics, University of California at Berkeley. https://www.stat.berkeley.edu/ ~ breiman/bagging.pdf.

Breiman, L. (2011). *Random forests.* Springer. https://doi.org/10.1023/A:1010933404324.

Chen, T., & Guestrin, C. (2016). *XGBoost: A scalable tree boosting system.* Cornell University. https://doi.org/10.1145/2939672.2939785.

Cortes, C., & Vapnik, V. (1995). *Support-vector networks.* https://doi.org/10.1007/BF00994018.

Cover, T., & Hart, P. (1967). Nearest neighbor pattern classification. *IEEE Transactions on Information Theory, 13,* 21−27. https://pdfs.semanticscholar.org/a3c7/50febe8e72a1e377fbae1a723768b2 33e9e9.pdf?_ga=2.114560259.5041978.1593729195-1056142191.1593729195.

Efron, B. (1979). Bootstrap methods: Another look at the jackknife. *The Annals of Statistics, 7*(1), 1−26. https://doi.org/10.1214/aos/1176344552.

Huber, P. (1964). Robust estimation of a location parameter. *Annals of Mathematical Statistics, 35*(1), 73−101. https://doi.org/10.1214/aoms/1177703732.

Kam Ho, T. (1995). *Random decision forests* (pp. 278−282). https://web.archive.org/web/20160417030218/http://ect.bell-labs.com/who/tkh/publications/papers/odt.pdf.

Quinlan, R. (1986). *Induction of decision trees. 1* pp. 81−106). https://hunch.net/ ~ coms-4771/quinlan.pdf.

Raschka, S. (n.d.). Why is nearest neighbor a lazy algorithm? Retrieved July 3, 2020, from https://sebastianraschka.com/faq/docs/lazy-knn.html.

Schafer, J. L., & Graham, J. W. (2002). Missing data: our view of the state of the art. *Psychological Methods, 7,* 147−177. https://doi.org/10.1111/1467-9574.00218.

scikit-learn. (n.d.). Retrieved July 17, 2020, from https://scikit-learn.org/stable/modules/tree.html#tree-algorithms-id3-c4-5-c5-0-and-cart.

Statsmodels. (2020). https://pypi.org/project/statsmodels/.

Tolles, J., & Meurer, W. J. (2016). Logistic regression: Relating patient characteristics to outcomes. *Journal of the American Medical Association.* https://doi.org/10.1001/jama.2016.7653.

Wright, S. (2015). *Coordinate descent algorithms.* Springer. https://doi.org/10.1007/s10107-015-0892-3.

Chapter 6

Neural networks and Deep Learning

Introduction and basic architecture of neural network

Artificial Intelligence (AI) may be defined as the collection of analytical and numerical tools that try to learn and imitate a process. When the learning process is accomplished, AI is capable of handling and responding to new situations. Neural networks, genetic algorithms, and fuzzy logic are the building blocks of AI (Mohaghegh, 2000).

The neural network claims its artificial information processing correlates closely with biological neural networks. Neurons are the main elements comprising artificial neural networks. They pass signals between each other, like those of the human brain. The artificial neurons connect several inputs to one or multiple outputs by associated weighs and nonlinear activation functions. When data are provided to neural networks, they undergo a learning process, by specific algorithms, to find the appropriate weights that describe the behavior of the output with respect to multiple inputs.

Neural networks provide a high potential for exploring and analyzing large historical databases that don't seem to be used in conventional modeling (Mohaghegh, 2000). In other words, neural networks should be applied in cases where mathematical modeling is impractical. The nature of the problems in the O&G industry is complex. These issues may be solved by unconventional methods such as AI.

The artificial neural network was originally developed from the behavior of biological neurons in the brain by McCulloch and Pitts (Mohaghegh, 2000). The information processing in neural networks mimics mechanisms of neurons in biological systems. A schematic of the nervous system block body, which is also called a neuron, is shown in Fig. 6.1. Generally, a neuron consists of a cell body, axon, and dendrite connecting to another neuron with a synaptic connection. Information communication occurs through electrochemical signals that enter the cells through dendrites. Based on the characteristics of input, the neuron is stimulated and releases an output signal that passes through the axons. The action is referred to as "firing a signal" if a threshold is reached by a great enough electrical potential (Mohaghegh, 2000). The output

Machine Learning Guide for Oil and Gas Using Python. https://doi.org/10.1016/B978-0-12-821929-4.00008-1
297

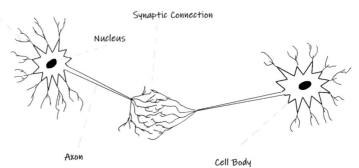

FIGURE 6.1 Sketch of a biological neuron.

signal from one neuron is an input for another neuron, which generates a new electrochemical pulse as the output. Each module in the brain may have more than 100,000 neurons connecting to thousands of other neurons and forming the complex architecture for neural networks; this neural network architecture is the basis for the learning process in the human brain.

Artificial neural networks are mathematically modeled based on the functionality of biological neural networks. From now on, neural networks and artificial neural networks will be used interchangeably. There are two main applications for neural networks: clustering (unsupervised classification) and finding a correlation between a set of numeric inputs (attributes) and outputs (targets). Neural networks generally consist of a series of "Neurons" in a sequential layering structure. Each input and output variable can be assigned to a node that is like a biological neuron. Nodes are gathered into layers, including an input layer, a hidden layer(s), and an output layer with the connected input and output layers as shown in Fig. 6.2. The topology of neural networks defines the number of hidden layers and the number of nodes in each hidden layer which are used to connect the input to the output layer. The connection between each of the two nodes (synapse from consecutive layers) is represented by weights w_{ij} in which i and j represent nodes in the input and output layers, respectively. The weight, as a given value, is like the strengths of the electrochemical signal. Excitatory reflections are designated by the weight's positive values, while inhibitory ones are identified by negative values (Kriesel, 2005).

Each node in the hidden layer receives signals from the input layer nodes, considers bias, and finally uses an activation function to produce an output signal. To explain this process (Fig. 6.2), assume "n" input nodes. Let's represent the signal from the ith input node, I_i, signal weight from node i to node j, w_{ij}, and bias as θj. The action potential of the node j in the hidden layer (h_j) is calculated as below-weighted summation.

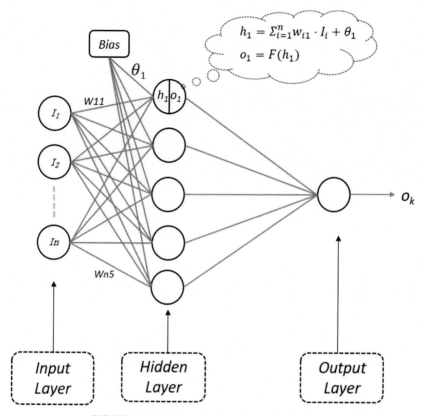

$$h_1 = \Sigma_{i=1}^{n} w_{i1} \cdot I_i + \theta_1$$
$$o_1 = F(h_1)$$

FIGURE 6.2 Topology of a three-layer neural network.

$$h_j = \sum_{i=1}^{N} w_{ij}.I_i + \theta_j \tag{6.1}$$

The bias or θj is a pseudo node with an output value of 1 and is used if the input value is 0. The final step is to calculate the node's output (oj). The output is calculated using an activation function (F) that determines the amplitude of the output signal based on the action potential of the node (h_j). Based on the function type and a bias value, the activation function generates a value between 0 and 1, or -1 to 1. Generally, four types of activation functions (F) have been used in neural network models: (i) F_Step, (ii) F_ReLU, (iii) F_sigmoid, and (iv) F_tanh. Activation functions help neural networks to understand non-linear and complex patterns in the data.

For F_{step}, activation function takes the value of 0 if action potential (h_j) is less than 0, and 1 *when it is* equal or more than 0.

$$oj = F_{step}(hj) = \begin{cases} 0, & hj < 0 \\ 1, & hj \geq 0 \end{cases} \qquad (6.2)$$

Rectified linear unit (ReLU) is the most widely used activation function. If action potential is less than 0, the activation function would be 0. For action potential values greater than 0, activation function and action potential values are the same. In other words, ReLU is actually a maximum function in terms of $F_{ReLU} = Max\ (0, hj)$ or:

$$oj = F_{ReLU}(hj) \begin{cases} 0, & hj < 0 \\ hj, & hj \geq 0 \end{cases} \qquad (6.3)$$

The sigmoid function uses the following relation and ranges from 0 to 1. For large negative numbers, the sigmoid function returns 0 and for large positive numbers, it returns 1. The sigmoid activation function has been practiced less recently (compared with other activation functions). It is due to the fact that this function's gradient would not be appropriate for weight optimization (will be discussed later) and also sigmoid outputs are not centered at 0.

$$oj = F_{sigmoid}(hj) = \frac{1}{1 + e^{-hj}} \qquad (6.4)$$

Tanh activation function is similar to sigmoid function with a primary difference of returning values from -1 to 1. Tanh is preferred over the sigmoid activation function since it is zero centered with better gradient performance. All the mentioned activation functions are shown in Fig. 6.3.

$$oj = F_{tanh}(hj) = \tanh(hj) = \frac{1 - exp(-2hj)}{1 + exp(-2hj)} \qquad (6.5)$$

In order to solve the system of equations containing all the nodes in the neural network, linear algebra can be utilized by rewriting all the mentioned equations in a matrix form.

For example, by assuming no bias, Formula 6.1 can be written as:

$$H = W.I$$

where W is the weights matrix, I is the input vector, and H is the action potential vector. To use the activation function and calculate the output vector for each hidden layer, it is possible to apply sigmoid functions to all the elements of vector H as follows:

$$o = F(H)$$

Neural network topologies are determined by connection patterns between neurons and propagation of data. The mentioned process can be repeated for each hidden layer considering that the output of the first hidden layer nodes

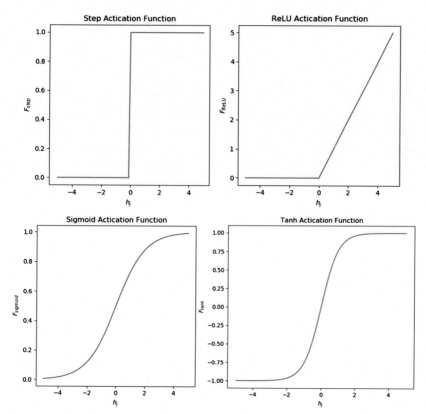

FIGURE 6.3 Four types of activation functions (top left: step, top right: ReLU, bottom left: Sigmoid, bottom right: Tanh).

would be the input for the second hidden layer. This consecutive process continues until the output of the last or final hidden layer is found. This process is called "feedforward," when there is no feedback from the output results into the layers and units. The process of how these weights are optimized is explained in the next section.

Backpropagation technique

The objective of the training procedure in neural networks is to adjust the calculated weights until the model estimates the output values correctly. The backpropagation method requires updating the weights to minimize the error which is the difference between network outputs and the target values. These errors are filtered back through the network. The backpropagation process includes a feedforward calculation to find the initial output, an error calculation that propagates backward into the network, and finally, weight adjustments based on the magnitude of errors.

The "Gradient Descent" method, a popular way to find the minimum of complex functions, is used to calculate weight changes. This method finds the steepest path to reach the minimum of the function. The steepness is defined as the slope (first derivative) of the function which is used to take steps towards the minimum. Therefore, considering the error values as a function of the network weights, the gradient descend technique is used to minimize the error subject to the weights, by taking steps of the size of the error gradient "$\frac{\partial e}{\partial w}$". To minimize the error function, the output error may be modified by multiplying it by the activation function gradient (Chain Rule). Depending on whether the error is positive or negative, the gradient of the activation function may move up or down (Fausett, 1993).

The backpropagation procedure is straightforward for the last or output layer because the error can be calculated using the output and target values ($e = t_k - o_k$). In the last layer, to adjust the weights based on error, error gradient (δk) at each neuron k of the output layer is calculated by the following formula:

$$\frac{\partial e_k}{\partial w_k} = \delta_k = (t_k - o_k)\frac{\partial F(O_k)}{\partial w_k} \tag{6.6}$$

where o_k and t_k are output and target values at neuron k of the output layer and F is activation function. If the target value is not available in the hidden layer, the weight adjustment parameter can be calculated as follows:

$$\delta_j = \frac{\partial F(Oj)}{\partial W_j}\sum_{k=1}^{n}w_{jk\delta_k} \tag{6.7}$$

The new weights $\left(W_{jk}^n\right)$ are then manipulated into the following equation:

$$w_{jk}^n = w_{jk}^{n-1} + \Delta w_{jk}, \quad \Delta w_{jk} = \alpha.\delta_k \tag{6.8}$$

where α represents the learning rate and takes the value between 0 and 1; this value determines the rate of the weight adjustments and learning speed. A small value of α causes low learning rates and high values may lead to network instability or stuck in local optima. To overcome this problem, "Momentum" term (μ) is recommended. It increases step sizes towards the global minimum and bypasses local minima. Therefore, new weights can be calculated using the following equation:

$$\Delta_{j,\,k}^n = \alpha.\delta_k + \mu.\Delta w_{j,\,k}^{n-1} \tag{6.9}$$

To start the training process, it is necessary to assign initial random values to weights. These values can be randomly selected within the range of -1 to 1 with a uniform probability distribution. However, some specific random weight initialization might be required, depending on the network topology and the type of activation function. One common rule for the weight initialization of a node with "n" number of incoming links is random sampling in the range of $\pm\frac{1}{\sqrt{n}}$ (Rashid, 2016).

Before the training process, the data set should be prepared to prevent the activation function from flattening towards 1 which would make the gradient descend method ineffective. This occurs with large input values (Fig. 6.3). To do so, it is recommended to scale or normalize all the inputs in the range of 0.01 (instead of zero) and 1. The same rescaling procedure can also be applied for the output values considering range 0–1.

Data partitioning

The main objective of building a neural network model is to use its prediction capabilities. It means a model should predict reliable output(s) when it is exposed to new data set that was not used during the training phase. To make sure that the model's predictions are reliable, the whole data set would be partitioned into three segments consisting of training, test, and validation sets. Training data sets are only used for model training, the weight modification procedure that was discussed earlier. After training a model, its reliability can be verified by using a portion of the data that is new to the network and was not used during the training process. This portion of the data, set aside to verify the prediction precision (capability) of the model, is called the test set. It is worth mentioning that neural network topology and parameters can be tuned (i.e., number of layers, nodes per layer, learning rate, etc.) to improve the accuracy of the model on the test data set. Therefore, one must measure the accuracy of the model for the second time on the portion of the data set that was not seen during training and testing. This portion of the data is called the validation set.

The most commonly used partitioning technique is random sampling. For example, it is common to have 70% of data for training, 15% for testing, and 15% for validation. There are also some other techniques that use oversampling that helps to populate new data points around low-frequency intervals. If any of the input attributes represents nonuniformity, the training data set should contain some data points from the less represented intervals. Otherwise, neural networks would not be capable of predicting reliable outputs in the range within such low populated intervals.

The last step of the neural network training process is to decide when to end. During neural network training, "Epoch" indicates one complete cycle of the whole training process (feedforward calculation, finding errors, and backpropagation). In most cases, it is necessary to go through a number of iterations or epochs to complete the training process. The main objective during neural network training is to minimize the global error or "Loss Function," defined as follows (this formula is actually Mean Squared Error; MSE):

$$E = \frac{1}{n} \sum_{i=1}^{n} (t_{ik} - o_{ik})^2 \tag{6.10}$$

As the number of epochs increases, the error of training and test data sets generally decreases. The training process should continue if the error of the test data set decreases. When the test error starts increasing and the training error continues to decrease, the neural network starts memorizing the training pattern, an activity that does not improve the generalization of the model. At this point, it is recommended to stop the training process (Fig. 6.4). In some instances, iteration through the epochs continues a number of times (i.e., 1000 times) or for a time period (i.e., 24 h), depending on computational capacity. At the end, it is important to select and memorize the information of the epoch at which the minimum error of the test data set is achieved.

The steps required to train a neural network model are summarized as follows:

1. Normalize or rescale the input and output data
2. Partition the data into training, validation, and testing
3. Define the topology of the neural network, describing the number of hidden layers, the number on nodes (neurons) per hidden layer, the activation function, learning rate, momentum, solver, etc.
4. Initialize all the weights and bias values

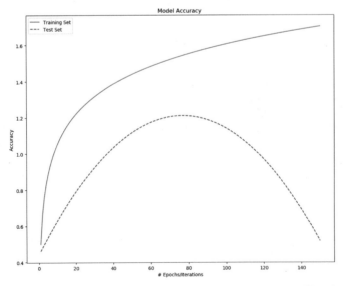

FIGURE 6.4 Training and validation data set accuracy versus number of iterations.

5. For each node in the hidden layer, calculate the node's output using Formula 6.5 (Formula 6.2, 6.3, and 6.4 can also be used depending on the type of activation function) and continue the feedforward process until the final output from the output layer is calculated
6. Calculate the output layer error relative to the actual output and update the weights for nodes in the prior layers using Formula 6.6, 6.8, and 6.9.
7. For the rest of the hidden layers, update the nodes, weights using Formula 6.7, 6.8 and 6.9
8. Repeat Step 5 and calculate new output using the updated weights
9. Calculate the global error using Formula 6.10. Stop the training based on the termination criteria (number of epochs or minimum error validation data set)

Neural network applications in oil and gas industry

Neural networks have shown a wide variety of applications in different E&P disciplines, but the implementation of neural networks is not recommended if conventional methods provide firm solutions. Neural networks have been able to perform accurate analyses in large historical data sets that cannot reveal explicit information by conventional modeling (Mohaghegh, 2000).

The early use of neural networks in the oil and gas industry was mostly related to reservoir characterization, specifically obtaining porosity, permeability, and fluid saturation from well logs. Well logs were generally used as the inputs for neural networks while core results such as porosity and permeability were considered outputs. It was possible to predict reservoir characteristics using well logs when core data were unavailable (Mohaghegh, 2000). Additionally, it was possible to train neural networks to generate synthetic magnetic resonance logs with conventional wire line logs such as SP, gamma ray, density, and induction logs (Mohaghegh et al., 2000).

Another well-known implication of neural networks is oil and gas PVT property estimation. There are many empirical correlations that estimate some PVT properties. Gharbi et al. (Gharbi & Elsharkawy, 1997) trained a universal neural network using 5200 PVT data points gathered from all around the world to predict bubble point pressure (P_b) and oil formation volume factor (B_{ob}) as a function of solution gas ratio (R_s), gas specific gravity (γ_g), oil specific gravity (γ_o), and reservoir temperature. The results were more accurate than the existing PVT correlations.

Neural networks were also used to predict the conditions for wax precipitation in the pipelines. One important parameter that determines wax formations is temperature. A neural network was trained by Adeymi et al. (Adeyemi & Sulaimon, 2012) with a combination of data including molecular weight, density, and activation energy, providing very good estimation for wax appearance temperature (WAT).

Forecasting the post-fracture deliverability of wells in the Clinton Sandstone gas storage field was done by Mohaghegh et al. (Mohaghegh et al., 1995) using a backpropagation neural network. The input parameters in this work included well data such as date of completion, well type, sand thickness, flow test values, number of stages, fracturing fluid type, total water used, total sand, acid volume, injection rate, injection pressure, etc. The output variable used was the maximum flow rate after fracture stimulation. Post-fracture well deliverability was forecasted by neural networks with very high accuracy. Additionally, neural networks were used with genetic algorithms to optimize the fracturing operation and select the best candidate for well restimulation.

The other application of neural networks is in the development of the Surrogate Reservoir Model (SRM, recently it is also called "Smart Proxy")—a replica of the reservoir simulation models that reproduces simulation results with high accuracy and short computation time. SRM can be a good substitute for a reservoir simulation model, especially when numerous simulation runs are needed. Risk assessment, uncertainty analysis, optimization, and history matching are the typical analyses requiring many simulation runs. Amini et al. (Amini et al., 2012) developed an SRM for the reservoir simulation model of the CO_2 storage site in Australia which included 100,000 grid blocks. The grid-based SRM was developed by 12 simulation runs to generate an inclusive database for the neural network, including well data, static, and dynamic data for all grid blocks in the reservoir.

The Top-Down Model (TDM), invented by Mohaghegh (Mohaghegh et al., 2012), is another neural network application that incorporates field measurements such as drilling data, well logs, cores, well tests, production history, etc. to build a comprehensive, full field reservoir model. Haghighat et al. (Haghighat et al., 2014) analyzed the production behavior of 145 wells located in an unconventional asset in Wattenberg Field-Niobrara using wells' static (reservoir properties, well completion information) and dynamic data (operational information like days of production per month).

Apart from the aforementioned applications, neural networks were used in the oil and gas industry for drill-bit diagnoses (Arehart, 1990), inversion of seismic waveforms (Gunter & Albert, 1992), seismic attribute calibration (Johnston, 1993), lithology prediction from well logs (Ford & Kelly, 2001), pitting potential prediction (Gartland et al., 1999), reservoir facies classification (Tang, 2008), EOR method evaluation and screening (Surguchev & Li, 2000), stuck pipe prediction (Siruvuri et al., 2006), assessment of formation damage (Kalam et al., 1996), water flooding analysis (Nakutnyy et al., 2008), conductive fracture identification (Thomas & La Pointe, 1995), bit bounce detection (Vassallo & Bernasconi, 2004), and calibration of quarts transducers (Schultz & Chen, 2003).

TABLE 6.1 List of input attributes for dry gas wells.

Inputs		
Geology	Drilling	Completion
Porosity (%)	Lateral length (ft)	Proppant/foot (#/ft)
Thickness (ft)	Updip/downdip	Stage spacing (ft)
Water saturation (%)	Well spacing (ft)	Fluid/foot (bbl/ft)
Pressure gradient (psi/ft)		Injection rate (bbl/minute)
		Instantaneous shut-in pressure-ISIP (psi)
		Percentage of linear gel (%)

Example 1: estimated ultimate recovery prediction in shale reservoirs

In this example, we are going to use a neural network as a nonlinear regression estimator to find a correlation between 20 years gas EUR[1] (estimated ultimate recovery) and multiple well/reservoir attributes like geology, drilling, and completion attributes. Our data set consists of 507 dry gas horizontal multi-stage fractured wells in a shale asset. It is worth mentioning that this data is NOT an actual data set. You can find the "Shale Gas Wells" data set in the following location https://www.elsevier.com/books-and-journals/book-companion/9780128219294. The list of all the attributes is shown in Table 6.1.

Let us clarify some of the input attributes that will be used during neural network training. The producing geologic formation height indicates "Thickness." Updip/Downdip attribute is a categorical input. To use the learning algorithm, these data are transformed into numbers. 1 is assigned to Updip wells and 0 to Downdip wells. Well spacing is the distance between two adjacent wells. There are multiple ways to calculate well spacing that are beyond the scope of this chapter. For completion parameters, refer to the book *Hydraulic Fracturing in Unconventional Reservoirs* (Belyadi et al. 2019).

Descriptive statistics

As mentioned in the first chapter, to load the data into Python as a data set, one must upload "Chapter6_Shale Gas Wells.csv" into Jupyter notebook. Next, the data should be loaded in Python using pandas.

---

1. It is a common practice to normalize the output variable which is EUR by lateral length and removing lateral length as an input from the model.

```
import pandas as pd
dataset=pd.read_csv('Chapter6_Shale Gas Wells.csv')
```

The Shale Gas Wells data set includes 14 columns and 506 rows (the first row is data header). Before starting data preparation, it is important to review the data characteristics or simple descriptive statistics by running the following snippet. Some of the statistics summary parameters like minimum, maximum, and average will be used in generating trends between output and attributes.

```
print(dataset.describe())
Python output=Fig. 6.5
```

Date preprocessing

As mentioned earlier in this chapter, to prevent activation function convergence towards 1, one must normalize or rescale all the input and output data.

	Stage Spacing	bbl/ft	Well Spacing	Dip	Thickness	\
count	506.000000	506.000000	506.000000	506.000000	506.000000	
mean	147.640316	35.134387	820.158103	0.069170	162.365613	
std	18.392128	10.533197	135.736986	0.253994	15.471044	
min	140.000000	30.000000	650.000000	0.000000	120.000000	
25%	140.000000	30.000000	700.000000	0.000000	153.000000	
50%	141.000000	30.000000	800.000000	0.000000	165.000000	
75%	148.000000	36.000000	900.000000	0.000000	176.000000	
max	330.000000	75.000000	1350.000000	1.000000	185.000000	

	Lateral Length	Injection Rate	Porosity	ISIP	\
count	506.000000	506.000000	506.000000	506.000000	
mean	8153.086957	63.079051	7.337549	7010.490119	
std	942.393981	7.250106	0.749451	1211.452205	
min	4500.000000	55.000000	5.500000	5000.000000	
25%	7617.750000	57.000000	6.600000	5000.000000	
50%	8051.000000	61.000000	7.500000	7643.000000	
75%	8608.000000	69.000000	8.000000	7783.000000	
max	11500.000000	80.000000	8.500000	8200.000000	

	Water Saturation	Percentage of LG	Pressure Gradient	\
count	506.000000	506.000000	506.000000	
mean	19.213439	64.845455	0.930257	
std	3.198579	18.427813	0.046507	
min	15.000000	15.000000	0.750000	
25%	16.800000	55.900000	0.940000	
50%	17.700000	69.900000	0.950000	
75%	24.100000	79.700000	0.950000	
max	25.000000	95.000000	0.950000	

	Proppant Loading	EUR
count	506.000000	506.000000
mean	2567.065217	12.845455
std	413.792220	3.067064
min	1100.000000	7.000000
25%	2317.500000	11.000000
50%	2642.000000	12.400000
75%	2897.750000	13.700000
max	3200.000000	22.000000

FIGURE 6.5 Statistics summary of the Shale Gas Wells data set.

In this example, input and output data are normalized using the MinMaxScaler from sklearn library as follows. "X" was used as the matrix of input attributes and "y" was used as the output of the model. The normalized data are then stored into "Xnorm" and "ynorm".

```
X=dataset.iloc[:,0:13]
y=dataset.iloc[:,13].values
from sklearn.preprocessing import MinMaxScaler
sc=MinMaxScaler()
Xnorm=pd.DataFrame(data=sc.fit_transform(X))
yshape=pd.DataFrame(data=y.reshape(-1,1))
ynorm=pd.DataFrame(data=sc.fit_transform(yshape))
print(Xnorm)
Python output=Fig. 6.6
```

The objective of the first example is to train a neural network to detect patterns between each well's EUR and all the attributes or input features. Sometimes, when the number of attributes is high (not specifically in this case), it is recommended to determine the relative degree of influence of each attribute on EUR (output). This process is also known as feature ranking. By having the most important features, it would be possible to disregard low ranked attributes to improve the training speed and performance of the trained model.

There are multiple feature ranking algorithms in the Python libraries. Each algorithm may lead to a different ranking hierarchy depending on the utilized math and formulations. To decrease the uncertainty in the ranking order, using different methods and compare their rankings is recommended. In this example, two methodologies were used and the reader has the opportunity to test other algorithms as well. Some of these ranking techniques were also discussed in chapter 5.

Before starting to feature rank, it is useful to visualize the data trends for each pair of attributes. "Pairs Plots" is a simple, practical graphing tool in seaborn visualization library that creates histograms and scatter plots of all the attributes with the following command:

```
import seaborn as sns
sns.pairplot(dataset)
Python output=Fig. 6.7
```

	0	1	2	3	4	5	6	7 \
0	0.0	0.177778	0.285714	0.0	0.692308	0.577571	0.36	0.933333
1	0.0	0.000000	0.357143	0.0	0.830769	0.548000	0.20	0.766667
2	0.0	0.000000	0.357143	0.0	0.830769	0.694429	0.40	0.766667
3	0.0	0.000000	0.428571	0.0	0.846154	0.658571	0.56	0.933333
4	0.0	0.000000	0.428571	0.0	0.846154	0.687143	0.48	0.933333
..	...	...	...	...	...	...	...	...
501	0.0	0.000000	0.142857	0.0	0.615385	0.581000	0.32	0.566667
502	0.0	0.000000	0.071429	0.0	0.615385	0.490286	0.24	0.566667
503	0.0	0.000000	0.071429	0.0	0.615385	0.654286	0.08	0.566667
504	0.0	0.000000	0.142857	0.0	0.615385	0.619429	0.12	0.566667
505	0.0	0.000000	0.142857	0.0	0.615385	0.473143	0.20	0.566667

FIGURE 6.6 Normalized data for the first seven attributes.

Pair plots for the Shale Gas Wells data set are shown in Fig. 6.7. Since there are 14 variables (13 attributes and 1 target), pair plots include 14×14 plots. On the diagonal, a histogram on each corresponding variable demonstrates the distribution of each input feature. Scatter plots show the relationship between each of the two variables. By using cross plots, correlated attributes can be determined. It is extremely important to remove collinear input parameters from the model since those parameters provide the same info. Therefore, to enhance the training performance, removing the collinear parameters is recommended. Moreover, the overall relationship between target (EUR) and other attributes can be visualized. For example, in Fig. 6.7, the relationship between lateral length and EUR can be clearly seen. On the other hand, there is no significant relationship between EUR and Dip.

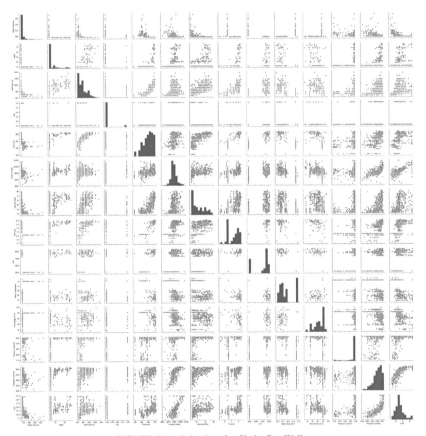

FIGURE 6.7 Pair plots for Shale Gas Wells.

The first technique to calculate correlations between EUR and all the attributes is called "Spearman's rank correlation." This is a nonparametric measurement that shows how two variables are statistically dependent in a monotonic way (increasing or decreasing trends). In this method, Spearman's correlation coefficient or ρ is calculated to measure the degree of influence between two variables. ρ is between -1 and 1. As ρ gets closer to 1 or -1, the two variables become heavily correlated. Positive ρ values show increasing trend and negative ρ values show decreasing trends between values. "Spearmanr" in "stats" library of Python returns a covariance or ρ matrix and P-value matrix (P-value is a hypothesis test to see whether a pair of variables are correlated). ρ matrix includes correlation coefficients between each pair of variables. Spearman technique can be used to rank the features impacting EUR in the Shale Gas Wells data set using the following lines of code:

```
from scipy import stats
import matplotlib.pyplot as plt
datanorm=sc.fit_transform(dataset)
stats.spearmanr(dataset)
rho, pval=stats.spearmanr(datanorm)
corr=pd.Series(rho[:13,13], index=X.columns)
corr.plot(kind='barh')
plt.show()
Python output=Fig. 6.8
```

Please note that "rho" is a 14×14 matrix representing correlation coefficients between each pair of variables. To find the relationship between EURs and other attributes, one can use a bar plot of the ρ values at the last column of rho matrix (rho[:13,13]) versus column headers (X.columns). Feature importance of the Shale Gas Wells using "Spearman's rank correlation" is shown in Fig. 6.8.

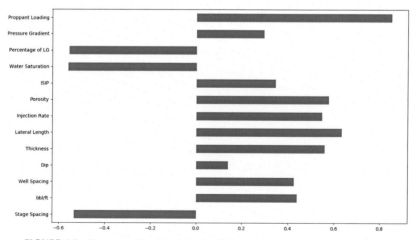

FIGURE 6.8 Feature ranking for Shale Gas Wells using Spearman's rank correlation.

Fig. 6.8 shows that stage spacing, water saturation, and percentage of linear gel (LG) have a negative impact on EUR. If those parameters increase, EUR will decrease. Other attributes show a positive correlation with EUR. By observing the magnitude of the correlation coefficient, it is evident that Proppant loading, lateral length, stage spacing, water saturation, and percentage of LG have the highest impact on EUR.

The second technique that can be used for feature ranking is the "random forest" model (which was discussed in detail in chapter 5). This method works based on dividing the data into subsets—at each level of decision tree—with respect to the attributes with a minimum variance estimator as the main objective (minimum entropy for categorical data). To alleviate the high variance and overfitting of a single decision tree, random forest injects randomness by creating multiple trees and an average estimator. Consequently, the amount of variance reduction at each level of the tree (based on different attributes' values) can be an index of the importance of that attribute. Let's call "RandomForestRegressor" from "sklearn" to see which attributes contribute more to gas production or EUR. In this example, the maximum depth of each tree is assumed to be 10 and the rest of the parameters for RandomForestRegressor are set as the sklearn default values. Note that scaling and normalization can affect the correlation between different features and should be avoided for random forest data.

```
import matplotlib.pyplot as plt
from sklearn.ensemble import RandomForestRegressor
model=RandomForestRegressor(max_depth=10, random_state=0)
model.fit(X,y)
feat_importances=pd.Series(model.feature_importances_,
  index=X.columns)
feat_importances.nlargest(13).plot(kind='barh')
plt.show()
Python output=Fig. 6.9
```

Feature importance of the Shale Gas Wells is shown in Fig. 6.9. For this particular data set, it appears that drilling and completion parameters contribute more to production when compared to geologic parameters. As shown, lateral length has the highest impact. Next, the amount of proppant loading per ft showed the second-highest degree of influence on EUR. The only geological parameter with any sort of importance is formation Thickness. On the other side, porosity, Dip, and liquid per foot (bbl/ft) represent the lowest degree of importance on the EUR. Please note that eliminating lateral length as an input feature is recommended. Instead, normalize the EUR by lateral length and use EUR/1000 ft as the output of the model. This example

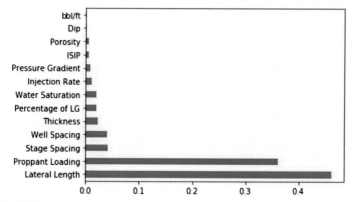

FIGURE 6.9 Feature ranking for Shale Gas Wells using random forest regression.

demonstrates the importance of lateral length on EUR performance if the reserve auditors demand to view it.

The last step prior to the neural network training process is data partitioning. "train_test_split" command from sklearn library to partition Shale Gas Wells data set. "train_test_split" partitions the data into train and test randomly (there is no partition for validation here, but we can run the same command on the test data to divide it into test and validation). In this command, one must define the proportion of the data set to be included in training, between 0 and 1. The following command will randomly split the data into 70% training denoted as X-train and y_train and 30% testing denoted as X_test and y_test. It is worth mentioning that the stochastic nature of data splitting (randomness) will result in a nonunique performance on the trained neural network over the same database with the same neural network topology. However, a seed number can be added to obtain the same result.

```
import numpy as np
seed=50
np.random.seed(seed)
from sklearn.model_selection import train_test_split
X_train, X_test, y_train, y_test=train_test_split(Xnorm, ynorm,
    test_size=0.3)
```

The original data sample (Xnorm) included 506 rows or wells. After splitting the data, X_train includes 354 wells as shown in Fig. 6.10 which is 70% of 506 wells.

Neural network training

Now that the data set for the neural network training is ready, let's build a neural network to regress between gas EUR and the 13 input attributes. To do

	0	1	2	3	4	5	6	7	8	9	10	11	12
251	0.000000	0.222222	0.571429	0.0	0.907692	0.551286	0.92	0.800000	0.739063	0.27	0.69125	0.95	0.948571
3	0.000000	0.000000	0.428571	0.0	0.846154	0.658571	0.56	0.933333	0.913125	0.07	0.64875	1.00	0.966667
257	0.005263	0.200000	0.071429	0.0	0.461538	0.985429	0.12	0.866667	0.825938	0.15	0.04250	1.00	0.906667
35	0.000000	0.000000	0.214286	0.0	0.769231	0.454429	0.32	0.800000	0.825938	0.18	0.70250	1.00	0.780476
339	0.000000	0.000000	0.357143	0.0	0.738462	0.464429	0.56	0.833333	0.825938	0.07	0.80875	1.00	0.779048
...	...	...	...	...	...	...	...	...	...	...	...	...	...
289	0.000000	0.533333	0.571429	0.0	0.953846	0.575571	0.80	0.833333	0.782500	0.20	0.42500	0.95	0.785238
109	0.005263	0.000000	0.142857	0.0	0.723077	0.511143	0.08	0.700000	0.825938	0.38	0.88250	1.00	0.618571
395	0.100000	0.000000	0.071429	0.0	0.369231	0.557571	0.00	0.366667	0.000000	0.91	0.80875	1.00	0.575238
480	0.063158	0.000000	0.214286	0.0	0.692308	0.513714	0.36	0.366667	0.000000	0.91	0.80875	1.00	0.751429
176	0.000000	0.000000	0.214286	0.0	0.738462	0.471143	0.56	0.866667	0.825938	0.21	0.42500	1.00	0.768571

FIGURE 6.10 Training data set.

so, one must use "Dense" and "Sequential" commands from "Keras" library. "Sequential" defines the main characteristics of the neural network configuration such as loss function type or global error (Formula 6.10), optimizer method (one of the gradient descent algorithms) is used to minimize the error and criteria (metric) to use in classification problems. Dense determines layer properties such as the number of nodes found in each layer (input, output, and hidden), the activation function type, and the number of hidden layers. The "fit" command starts the training process by specifying inputs, outputs, and stopping criteria. As mentioned earlier, the training process can be stopped once a specific number of iterations is reached, and the error in the test data does not decrease. Please use "pip install keras" and "pip install tensorflow" in the command prompt to make sure those libraries are installed. Anaconda does not have those libraries installed by default. If these libraries are not installed, the illustrated codes below will return an error. The following lines of codes are used to train the model:

```
from keras.models import Sequential
from keras.layers import Dense
from keras.callbacks import EarlyStopping
model=Sequential()
model.add(Dense(13, activation='relu',input_dim=13))
model.add(Dense(13, activation='relu'))
model.add(Dense(1))
np.random.seed(seed)
model.compile(optimizer='adam', loss='mean_squared_error')
early_stopping_monitor=EarlyStopping(patience=3)
history=model.fit(X_train,y_train,epochs=100,
    validation_data=(X_test, y_test),
    callbacks=[early_stopping_monitor])
```

In this example, a neural network with two hidden layers was used. There are 13 input nodes that are equal to the number of attributes and 1 output node which is "gas EUR." There are 13 nodes in the two hidden layers. The activation function for all the layers is "Relu" indicating a rectified linear unit (Fig. 6.3, Piecewise-Linear). The optimization algorithm is "adam," which calculates the learning rates based on the momentum of changing gradients (Kingma & Ba, 2015). The loss function is "mean squared error," defined in Formula 6.10. The schematic of the neural network topology is shown in Fig. 6.11.

The number of epochs, in this case, is specified as 100. When the training process starts, loss value for both training and test data sets are typically high. As the number of epochs increases, loss values for both data sets decrease until test data loss or accuracy does not improve. The progress of the neural network for the first nine epochs is shown in Fig. 6.12.

In the progress report, epoch number, runtime per epoch, and loss values of training and test are depicted. It is best if both loss values of training and test data decrease while the number of iterations (epochs) increases. In specific situations, training loss values decrease but test loss values start increasing. At this point, networks will start memorizing the training data and their corresponding weights. This will lead to overfitting. To prevent overfitting, an "early-stopping monitor" was used in the code. For this command, a criteria of "patience = 3" is defined, meaning if after three number of epochs, loss value, or performance of test data set is not improved, stop the model from training. This occurrence may happen far before the target number of epochs is reached (in this case it happened at epoch 50 while the total number of epochs

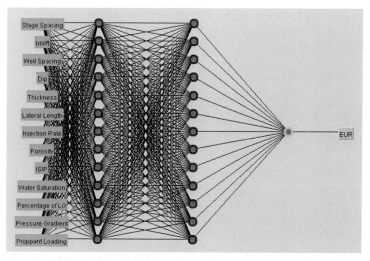

FIGURE 6.11   Neural Network topology for Shale Gas Wells.

```
Epoch 1/100
354/354 [==============================] - 0s 411us/step - loss: 0.0344 - val_loss: 0.0372
Epoch 2/100
354/354 [==============================] - 0s 51us/step - loss: 0.0254 - val_loss: 0.0322
Epoch 3/100
354/354 [==============================] - 0s 64us/step - loss: 0.0215 - val_loss: 0.0301
Epoch 4/100
354/354 [==============================] - 0s 56us/step - loss: 0.0189 - val_loss: 0.0277
Epoch 5/100
354/354 [==============================] - 0s 54us/step - loss: 0.0167 - val_loss: 0.0269
Epoch 6/100
354/354 [==============================] - 0s 62us/step - loss: 0.0157 - val_loss: 0.0263
Epoch 7/100
354/354 [==============================] - 0s 42us/step - loss: 0.0144 - val_loss: 0.0245
Epoch 8/100
354/354 [==============================] - 0s 70us/step - loss: 0.0139 - val_loss: 0.0241
Epoch 9/100
354/354 [==============================] - 0s 56us/step - loss: 0.0129 - val_loss: 0.0230
```

FIGURE 6.12 Epochs progress during neural network training.

is set to be 100). Training and test loss histories are shown in Fig. 6.13 using the following code:

```
plt.plot(history.history['loss'])
plt.plot(history.history["val_loss"],"r--")
plt.title('Model Loss')
plt.ylabel('loss value')
plt.xlabel('# epochs')
plt.legend(['train', 'test'], loc='upper right')
plt.show()
Python output=Fig. 6.13
```

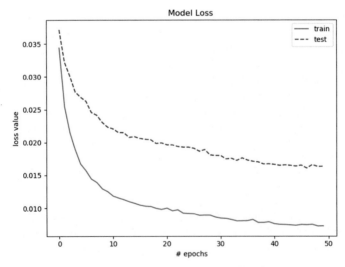

FIGURE 6.13 Training and test loss histories.

When the training process is completed, it is possible to use a trained model to make a prediction. To get the output results of the model for new or unseen data (test), the "model.predict" function is used. One must provide a matrix of inputs with all 13 attributes in the model to get the corresponding EURs. In this example, one hopes to compare the actual or measured gas EURs with model's predictions. As mentioned before, all attributes and target values (EURs) were rescaled to values between 0 and 1. To scale the output from the neural network back to the actual range, the following formula can be used:

$$EUR = EUR.Min + Ynorm \times (EUR.Max - EUR.Min)$$

Y norm can be prediction from the trained model. For comparing the measured with the predicted values, it is useful to show R^2 of the trends on the plots. To calculate R^2, "r2_score" from "sklearn.metrics" library is used. The following codes will plot predicted gas EURs versus measured EURs for both train and test data set with R^2 on the graph. The plots are shown in Fig. 6.14.

```
from sklearn.metrics import r2_score
EUR_test=y_test*(y.max()-y.min())+y.min()
EUR_train=y_train*(y.max()-y.min())+y.min()
EUR_test_prediction=model.predict(X_test)*(y.max()-y.min())+\
    y.min()
EUR_train_prediction=model.predict(X_train)*(y.max()-y.min())+\
    y.min()
r2_test=r2_score(EUR_test, EUR_test_prediction)
r2_train=r2_score(EUR_train, EUR_train_prediction)
fig, ax=plt.subplots()
ax.scatter(EUR_test, EUR_test_prediction)
ax.plot([EUR_test.min(), EUR_test.max()], [EUR_test.min(),
    EUR_test.max()], 'k--', lw=4)
ax.set_xlabel('EUR_Test_Measured(Bcf)')
```

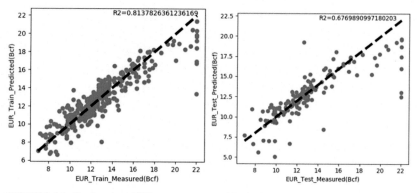

FIGURE 6.14   Predicted gas EURs versus measured for both train (right) and test data set (left).

```
ax.set_ylabel('EUR_Test_Predicted(Bcf)')
plt.text(14,22,"R2="+str(r2_test).format("%.2f"))
plt.show()
fig, ax=plt.subplots()
ax.scatter(EUR_train, EUR_train_prediction)
ax.plot([EUR_train.min(), EUR_train.max()], [EUR_train.min(),
    EUR_train.max()], 'k--', lw=4)
ax.set_xlabel('EUR_Train_Measured(Bcf)')
ax.set_ylabel('EUR_Train_Predicted(Bcf)')
plt.text(14,22,"R2="+str(r2_train).format("%.2f"))
plt.show()
Python output=Fig. 6.14
```

Example 2: develop PVT correlation for crude oils

Petroleum/reservoir engineering calculations are heavily dependent on phase behavior and PVT data. These calculations cover in-place estimation, pressure loss along pipes, fluid flow in porous media, rate transient analysis, and well testing. The most common approach to acquire PVT data is through lab testing. In this method, to calculate PVT properties, the oil sample is tested via multiple methods. Some of these tests are very complex, timely, and expensive. On the other hand, some properties like oil API or gas specific gravity can be easily achieved. Properties like solution gas/oil ratio, bubble/dew point, and liquid drop out (condensate reservoirs) require sophisticated lab tests. In the oil and gas industry, it has been a common practice to derive mathematical and empirical correlations to determine one PVT parameter as function of some easily obtained parameters. Statistical regression techniques are mainly used to derive empirical correlations for different PVT properties. These correlations can be found in *Reservoir Engineering Handbooks* (Ahmed, 2006) or other hydrocarbon properties reference books (McCain, 1999).

"Bubble point" is defined as a pressure at which the first bubble of gas is formed within the liquid oil phase. As pressure falls below the bubble point, more gas is liberated from oil, and single liquid phase becomes two phases. Bubble point (bp) is one of the PVT properties that cannot be measured in the lab easily. Therefore, several correlations like "Standing, Vasque/Beggs, Glaso, Marhoun, and Petrosky/Farshad" were proposed by several authors (Ahmed, 2006). Besides statistical regression techniques, it is also possible to use neural networks to develop correlations between PVT properties. The second example uses a PVT data set, including 249 PVT[2] samples consisting of temperature, solution gas to oil ratio (RS), gas gravity, oil API, and bubble point pressure (Pbp). "PVT Data" can be found in the following location https://www.elsevier.com/books-and-journals/book-companion/9780128219294.

---

2. This example is similar to work done by Osman et al. (2001).

The first step is to import the data and study descriptive statistics. It is necessary to upload "PVT Data.csv" into Jupyter notebook and then load the data in Python using pandas. Descriptive statistics of the PVT data are shown in Fig. 6.15.

```
import pandas as pd
dataset=pd.read_csv('Chapter6_PVT Data.csv')
print(dataset.describe())
Python output=Fig. 6.15
```

After uploading the data in "dataset", the input and output for neural networks should be determined and normalized. The input in this example includes the first four columns as temperature, Rs, gas gravity, and oil API and output is Pbp. "MinMaxScaler" is used to normalize the data as "Xnorm" for input and "ynorm" for output.

```
X=dataset.iloc[:,0:4]
y=dataset.iloc[:,4].values
from sklearn.preprocessing import MinMaxScaler
sc=MinMaxScaler()
Xnorm=pd.DataFrame(data=sc.fit_transform(X))
yshape=pd.DataFrame(data=y.reshape(-1,1))
ynorm=pd.DataFrame(data=sc.fit_transform(yshape))
```

In this example, feature ranking is done to see the relative importance of each parameter on the output data. Like previous examples, "Spearman's rank correlation" and "random forest" are used to rank all the input parameters based on their impact on Pbp. The codes for both methods are as follows.

```
# "Spearman's Rank Correlation" feature ranking
from scipy import stats
import matplotlib.pyplot as plt
datanorm=sc.fit_transform(dataset)
stats.spearmanr(dataset)
rho, pval=stats.spearmanr(datanorm)
corr=pd.Series(rho[:4,4], index=X.columns)
corr.plot(kind='barh')
```

	Temperature	Rs	Gas Gravity	Oil API	Pbp
count	249.000000	249.000000	249.000000	249.000000	249.000000
mean	147.796271	411.145756	1.012870	31.658760	1424.602150
std	41.936641	291.829082	0.162192	5.036067	908.669973
min	70.447234	27.832416	0.797048	17.854225	131.484967
25%	116.365357	183.034832	0.885612	28.469070	697.772696
50%	142.257643	348.735793	0.978741	32.604282	1252.802615
75%	176.074354	587.055416	1.113760	35.875571	1937.635034
max	282.911419	1471.094081	1.632588	39.714096	4306.643567

FIGURE 6.15 Descriptive statistics for PVT data.

```
plt.show()
# "Random Forest" feature ranking
from sklearn.ensemble import RandomForestRegressor
model=RandomForestRegressor(max_depth=10, random_state=0)
model.fit(X,y)
feat_importances=pd.Series(model.feature_importances_,
   index=X.columns)
feat_importances.nlargest(4).plot(kind='barh')
plt.show()
Python output=Fig. 6.16
```

Results for both feature ranking methods are shown in Fig. 6.16. Both methods depict that Rs has the highest impact on bubble point pressure. Spearman's rank correlation method illustrates the negative effect of gas gravity and oil API on Pbp. This means that as gas gravity and oil API increase, Pbp decreases. Based on "random forest" method, oil API, gas gravity, and temperature demonstrate relative impact on Pbp in decreasing order.

In Shale Gas Wells examples "keras" library was used for neural network training. In this example, "MLPregressor" from "sklearn.neural_network" library is used in order to define the neural network topology. First, PVT data set should be partitioned into train and test segments. 70% of data is assigned to train and 30% to testing. As mentioned before, the output for neural networks is Pbp and the inputs are four PVT parameters. Therefore, there are three layers in the neural network as input, hidden, and output layers. Seven neurons are assigned to the hidden layer. Neural network topology for this example is shown in Fig. 6.17. Note that adding one more hidden layer with 5 neurons can be achieved with "hidden_layer_sizes=(7,5)" command within MLPRegressor.

The activation function for the hidden layer is "tanh" or hyperbolic tangent function. The optimizer to utilize gradient descent method is in the family of "Quasi-Newton" technique or "lbfgs." This method is an algorithm that searches for the local minimum by assuming that function around the optimum point is quadratic. Parameter "alpha" is used to regulate the network to overcome overfitting by penalizing larger weights. The learning rate that controls weight updating is 0.1. The maximum number of epochs is 200.

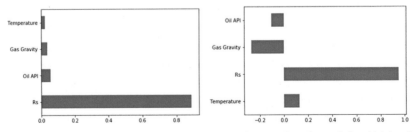

FIGURE 6.16 Feature ranking for PVT data set using Spearman's rank correlation (right) and random forest (left).

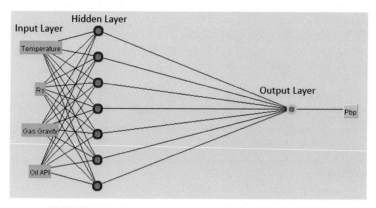

FIGURE 6.17   Neural network topography for Shale Gas Wells.

Parameter "tol" represents tolerance for optimization. If the loss after several iterations does not improve by a larger amount than tol, the training process would stop. Model training is completed with the following codes:

```
import numpy as np
seed=50
np.random.seed(seed)
from sklearn.model_selection import train_test_split
X_train, X_test, y_train, y_test=train_test_split(Xnorm, ynorm,
    test_size=0.3)
from sklearn.neural_network import MLPRegressor
np.random.seed(seed)
clf=MLPRegressor(hidden_layer_sizes=(7), activation='tanh',
    solver='lbfgs', alpha=1,
learning_rate_init=0.1, max_iter=200,
random_state=None, tol=0.01)
y_train_Ravel=y_train.values.ravel()
clf.fit(X_train,y_train_Ravel)
```

When the training process is done, it is necessary to plot the predictions and compare them with actual values to evaluate model's performance. In this example, input data from both the train and test data sets are used to predict bubble point pressure. Then, predicted values are plotted against actual values with the trend line and R^2 to see the model's performance. The following codes are used to plot the results in Fig. 6.18.

```
from sklearn.metrics import r2_score
Pbp_test=y_test*(y.max()-y.min())+y.min()
Ppb_train=y_train*(y.max()-y.min())+y.min()
Pbp_test_prediction=clf.predict(X_test)*(y.max()-y.min())+\
    y.min()
Pbp_train_prediction=clf.predict(X_train)*(y.max()-y.min())+\
    y.min()
```

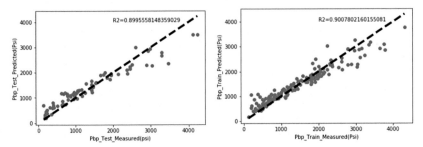

FIGURE 6.18 Predicted bubble point pressures versus measured for both train (right) and test data set (left).

```
r2_test=r2_score(Pbp_test, Pbp_test_prediction)
r2_train=r2_score(Ppb_train, Pbp_train_prediction)
fig, ax=plt.subplots()
ax.scatter(Pbp_test, Pbp_test_prediction)
ax.plot([Pbp_test.min(), Pbp_test.max()], [Pbp_test.min(),
    Pbp_test.max()], 'k--', lw=4)
ax.set_xlabel('Pbp_Test_Measured(psi)')
ax.set_ylabel('Pbp_Test_Predicted(Psi)')
plt.text(2000,4000,"R2="+str(r2_test).format("%.2f"))
plt.show()
fig, ax=plt.subplots()
ax.scatter(Ppb_train, Pbp_train_prediction)
ax.plot([Ppb_train.min(), Ppb_train.max()], [Ppb_train.min(),
    Ppb_train.max()], 'k--', lw=4)
ax.set_xlabel('Pbp_Train_Measured(Psi)')
ax.set_ylabel('Pbp_Train_Predicted(Psi)')
plt.text(2000,4000,"R2="+str(r2_train).format("%.2f"))
plt.show()
Python output=Fig. 6.18
```

The last step of this example compares neural network performance with a PVT correlation. In 1988, "Marhoun" determined a correlation to estimate bubble pressure from some PVT properties (Ahmed, 2006). He used 160 measurements from 69 Middle eastern oil samples and proposed the following correlation:

$$Pb = aR_s^b \gamma_g^c \gamma_o^d T^e \qquad (6.11)$$

where

$T =$ Temperature (°R)
$\gamma_o =$ Oil specific gravity
$\gamma_g =$ Gas specific gravity
Correlation coefficients $a-e$ have the following values:
$a = 5.38\text{E-}03, b = 0.715082, c = -1.877784, d = 3.1437,$ and $e = 1.32657$

In order to use "PVT Data Set" in Marhoun correlation, it is necessary to make two changes. The temperature unit should be converted from °F to °R by adding 460 to °F. Oil API gravity (API) should also be converted to oil specific gravity (γ_o) using following formula:

$$\gamma_o = \frac{145}{API + 135} \tag{6.12}$$

These changes should be applied to the data set that stores the PVT data. To keep all the data in the data set intact, a deep copy is generated as "DatasetTempGamma." Next, the bubble point pressure is calculated using the Marhoun correlation (PBP_Marhoun) and compared with predicted values by clf neural network.

```
DatasetTempGamma=dataset.copy()
DatasetTempGamma['Temperature'] +=460
DatasetTempGamma['Oil API']=145/(DatasetTempGamma['Oil API']+135)
PBP_Marhoun=0.00538088*(pow(DatasetTempGamma['Rs'],0.715082))*\
    (pow(DatasetTempGamma['Gas Gravity'],-1.877784))*\
    (pow(DatasetTempGamma['Oil API'],3.1437))*\
    (pow(DatasetTempGamma['Temperature'],1.32657))
Pbp_NN=clf.predict(Xnorm)*(y.max()-y.min())+y.min()
r2_NN=r2_score(dataset['Pbp'], Pbp_NN)
r2_Marhoun=r2_score(dataset['Pbp'], PBP_Marhoun)
fig, ax=plt.subplots()
ax.scatter(dataset['Pbp'], Pbp_NN)
ax.plot([dataset['Pbp'].min(), dataset['Pbp'].max()],
    [dataset['Pbp'].min(), dataset['Pbp'].max()], 'k--', lw=4)
ax.set_xlabel('Pbp_Measured(psi)')
ax.set_ylabel('Pbp_Neural Network(Psi)')
plt.text(2000,4000,"R2="+str(r2_NN).format("%.2f"))
fig, ax = plt.subplots()
ax.scatter(dataset['Pbp'], PBP_Marhoun)
ax.plot([dataset['Pbp'].min(), dataset['Pbp'].max()],
[dataset['Pbp'].min(), dataset['Pbp'].max()], 'k--', lw=4)
ax.set_xlabel('Pbp_Measured(Psi)')
ax.set_ylabel('Pbp_Marhoun(Psi)')
plt.text(2000,4000,"R2="+str(r2_Marhoun).format("%.2f"))
plt.show()
Python output=Figs. 6.19 and 6.20
```

The results of Marhoun correlation and neural network versus actual data are shown in Figs. 6.19 and 6.20. R^2 for Marhoun correlation versus actual bubble point pressure is 0.87 while R^2 for neural network results versus actual bubble point pressures is 0.9. This shows a better performance of neural networks as compared to the Marhoun correlation. Neural network is a powerful tool to model complicated correlations between data; however, conventional models have many limitations and may not capture all underlying complexities.

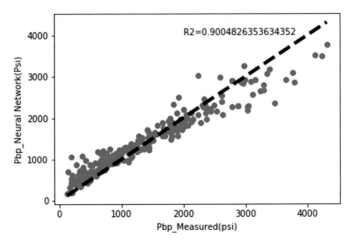

FIGURE 6.19 Predicted bubble point pressures from neural network versus measured values.

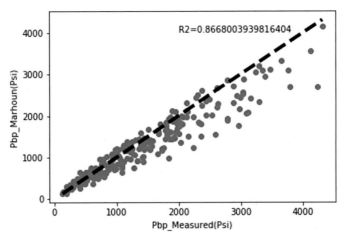

FIGURE 6.20 Predicted bubble point pressures from Marhoun correlation versus measured values.

Deep learning

As explained in Chapter 1, artificial intelligence (AI) is defined as using computers to perform some tasks that are generally handled by human intelligence like learning, reasoning, and discovering. Machine learning (ML) is a subset of AI which mainly focuses on the algorithms that help the machine to interpret patterns in the data. Depending on whether the data are labeled or not, the learning process is supervised or unsupervised.

Due to advancements in technology in the past few decades, the quantity of data that can be generated and stored has tremendously increased. Currently, more than 2 quadrillion bytes of data are produced (Forbes.Com) in the forms of text, image, voice, and video. This value is estimated to significantly increase in the future. Therefore, if the ML algorithms can be exposed to much more labeled data, this exposure can improve the performance of ML models. These algorithms are massively parallelizable, meaning they can benefit from modern graphics processing units (GPUs) architecture. These architectures simply were not available when deep learning (DL) was first introduced decades ago. Supercomputers and cloud computing make parallel processing faster and less expensive. Finally, due to open-source toolboxes such as TensorFlow (which will be discussed in the example), building these models has become extremely straightforward. We will use Keras, a high-level interface that uses TensorFlow for its back end to build a long short-term memory (LSTM) model for one-second frac data prediction later in this chapter.

Fig. 6.21 illustrates the difference between AI, ML, and DL. AI simply uses machine intelligence as opposed to human intelligence; ML is a subset of AI and refers to various algorithms (some of which have been discussed) without being explicitly programmed. Finally, DL is a subset of ML that uses different forms of complex neural networks as the learning algorithm to

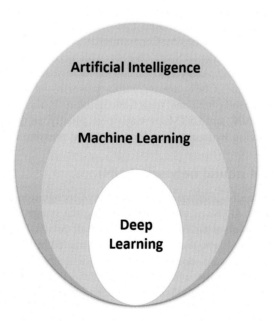

FIGURE 6.21   Artificial intelligence (AI), machine learning (ML), and deep learning (DL).

discover the patterns in big data sets. So, the main building blocks of the DL are neural network. One of the key parameters that define the structure of a neural network is the number of hidden layers that connects input and output nodes. The number of hidden layers determines the depth of a neural network. In deep neural networks, the number of hidden layers can vary from 5 to hundreds or thousands. The other characteristic of a deep neural network is about neurons' connectivity. In this chapter, for the theory and example, it was assumed that every neuron is connected to all the neurons in the next layer. In deep neural networks, it is possible to not have full connectivity between the neurons in consecutive layers. Therefore, the architecture of the neural network may change by partial neuron/layers connectivity, backward connections, and sharing weights. These neural network architectures are covered later in this chapter.

DL has many applications especially in the fields where enough labeled data are available. Speech recognition uses DL in many appliances such as cellphones to recognize voice patterns. Image recognition is another widely used application of DL in automotive, gaming, e-commerce, medicine, manufacturing, and education industries. Another important usage of DL is in natural language processing (NLP) for text classification/extraction/summarization, autocorrection, translation, and market intelligence. DL is also used for self-driving cars, fraud detection, movie making, and product recommendation systems.

There are more than 10 types of DL algorithms. Each algorithm can be suitable to handle specific applications that were mentioned earlier. The most widely used DL algorithms are Convolutional Neural Networks (CNNs), Recurrent Neural Networks (RNNs), LSTMs (which is a subset of RNN), Self-Organizing Map (SOM), Autoencoders, and Generative Adversarial Networks (GANs). For example, CNNs are incredibly powerful in image recognition while RNNs can handle text and speech recognition. In this section, the theories for CNN, RNN, and LSTM are explained. Finally, an example of using LSTM for predicting hydraulic fracturing treatment pressure is illustrated.

Convolutional neural network (CNN)

CNNs were originally developed to handle image recognition specifically for handwritten digits. CNNs work by decomposing the image into three scales from low order to high order. In the low order, small details or local features of the image like lines, edges, and curves can be extracted by early layers of the neural network. Afterward, deeper layers would assemble these features, and final layers reconstruct the whole image (Kelleher). The whole process of CNN training for image classification includes different operations such as convolution, activation function, pooling, and fully connected layers. Each operation is explained in detail below.

Convolution

In the image recognition and classification, the first step is to discretize the image into pixels. Each pixel can have a value depending on the shape and the color. Let us start with a simple example and discretize a plus sign image into 7 by 7 pixels. Black pixels can be represented by 1 and white pixels by zero (Fig. 6.22).

The convolution operation, the main part of the CNN, applies specific filters or kernel functions to a selected region of the image to detect local features. In other words, by convolution, it is possible to focus on a specific feature of the image at a time by applying specific filters. The filter moves over the image to detect specific patterns related to each feature (like line, edge, curve, etc.). Where the pattern is found in a selected region, kernel function returns large positive values. If a 2D matrix is assigned for each specific filter, the convolution operation can be considered as the sum of all elements of the dot product between the selected region (a 2D matrix) and the filter. Let us consider a 3 by 3 matrix as a filter for line detection. A 3×3 matrix with all elements in the middle column equaling to 1 is for vertical line detection, and a 3×3 matrix with all elements in the middle row equaling to 1 is for horizontal line detection (all the other elements are equal to zero). The filter starts at the 3×3 pixel region at the first column and the first row of the image. The sum of all elements of the dot product result is then stored in the first element of the feature map which is a 5×5 matrix. The filter slides one column to right and repeats the same operation until it reaches to the last column. Next, it starts all over from the first column and slides one row down and the same process is applied. It then produces element 2, 1 of the 5×5 feature map. This process repeats until the whole image is covered. The convolution process using vertical and horizontal line filters is illustrated in Fig. 6.23. Multiple filters (kernel functions) can be used to extract a variety of feature map of the image. The array of all these feature maps constitutes the convolutional layer of the network.

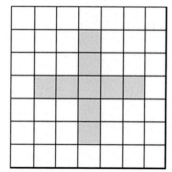

FIGURE 6.22 Digitizing plus sign image into a 7 × 7 binary matrix.

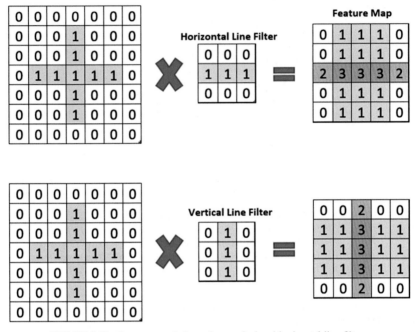

FIGURE 6.23 Image convolution using vertical and horizontal line filters.

Activation function

In most of the pattern recognition practices, the relationship between input and output is nonlinear. To capture nonlinear patterns, the activation function, a nonlinear transformation is used. Activation functions were discussed in this chapter. The most common activation function for CNNs is rectified linear unit (ReLU) (Fig. 6.3). After convolution process, ReLu activation function is applied to each element of the feature map. As an example, the resultant feature map for vertical line feature after applying the ReLu activation function is shown in Fig. 6.24. Please note that ReLU function returns zero for any negative value. Since there is no negative value in the feature map in this example, applying ReLU activation function does not change the initial feature map.

Pooling layer

The next step in the CNN includes pooling, reducing the size of convoluted feature maps by extracting the dominant features. This dimensionality reduction of feature maps retains important information while reducing required computational power especially for processing large images. In pooling operation, a section of the image is covered by a filter or kernel

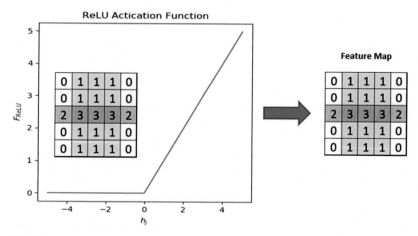

FIGURE 6.24   Applying rectified linear unit (ReLU) activation function on the feature map.

function. Next, the filter returns either maximum (max pooling) or average (average pooling) values of the section that is covered by the filter. The max pooling examples are shown in Fig. 6.25.

Fully connected layers

After the pooled feature maps are generated, they go through a flattening process to generate a one-dimensional matrix that will be considered as the input layer for a deep neural network. A flattening example is shown in Fig. 6.26. Next, the neural network goes through a training process to classify the images. The whole workflow is shown in Fig. 6.27.

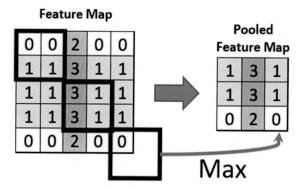

FIGURE 6.25   Applying max pooling on the feature map.

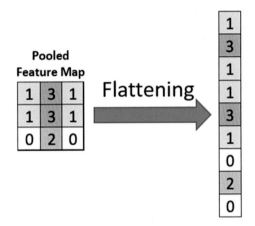

FIGURE 6.26 Flattening the feature map into 1D array.

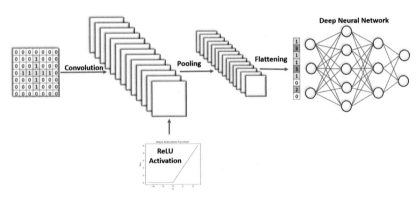

FIGURE 6.27 The whole workflow of convolutional neural network for image classification.

Recurrent neural networks

In many data analytics problems, it is necessary to deal with sequential data such as time series or text sentences. When time or sequence is involved, neural network should be able to deal with the importance of the data at each time step. This can be illustrated in the oil and gas industry by forecasting oil production in the future. When oil production from a well is declining, the oil rate at each time step is not only a function of geology, completion, drilling, and surface facilities but is also heavily related to oil rate at previous time steps. RNNs are designed to handle consecutive data. The main specification of the RNN is that the output value of the previous time step is used to calculate the output at the current time step. The schematic of a RNN is shown in Fig. 6.28.

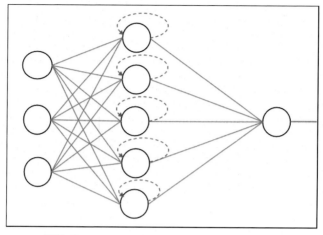

FIGURE 6.28 Schematic of recurrent neural network.

In RNN, the hidden layer state of previous time step (A^{t-1}) which is considered the memory is carried forward and multiplied by a weight vector (W_{aa}). Then, the current hidden layer state (A^t) and output (O^t) can be calculated with the following equation:

$$A^t = F\left(W_{ai}I^t + W_{aa}A^{t-1} + \theta_a\right)$$

$$O^t = F(W_{ao}A^t + \theta_o)$$

where F is activation function, W_{ai} is the weight matrix between input layer and hidden layer, W_{ao} is the weight matrix between hidden layer and output, θ_a is the hidden layer bias, and θ_o is output layer bias. The schematic of a hidden layer in RNN is shown in Fig. 6.29.

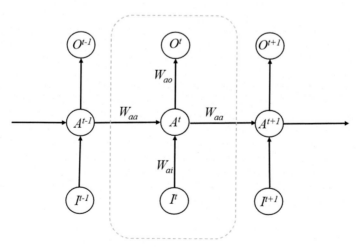

FIGURE 6.29 Sequence of layers in the recurrent neural networks.

For the training of the RNNs, backpropagation technique is used. The overall process is similar to what was explained early in this chapter. However, since there are multiple sequences or hidden layers in RNNs, using gradient descent technique can be challenging. For example, the error gradient for the early time steps becomes very small referred to as diminishing gradient problem. It this case, earlier layers, weights do not change, and neural network cannot learn from the data during the training process. To handle diminishing gradient problem, Hochreiter and Schmidhuber introduced LSTM in 1997 (Raschka & Mirjalili).

LSTM consist of connected memory cells that control the flow of information during the training process for sequential data. Each cell includes 3 operating units that interact with each other to preserve the error through sequential backpropagation. These units that act similar to computer memories are: input gate, forget gate, and output gate. The job of these gates is to control when data should enter, be stored, and/or leave the cells. The schematic of connected cells and cell components are shown in Fig. 6.30.

One important feature of each cell, considered as cell's memory, is cell state (C). The cell states move through all the cells sequentially and are responsible for either retaining or forgetting the previous information. At each time step, cell state from previous time (C^{t-1}), hidden layer state from previous time (A^{t-1}), and input data (I^t) enter the cell. Initially, input data (I^t) and hidden layer state from previous time (A^{t-1}) go through the "forget gate." The

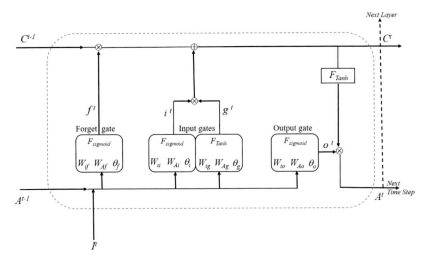

FIGURE 6.30 Schematic of long short-term memory (LSTM) cell and units (Raschka & Mirjalili).

output of the forget gate (f^t) determines which information remains and which is forgotten from the previous step. Forget gates' output is calculated as:

$$f^t = F_{sigmoid}(W_{if}I^t + W_{Af}A^t + \theta_f)$$

where W_{if} is the weight matrix for input data in the forget gate, W_{Af} is weight matrix for hidden layer state in the forget gate, and θ_f is bias in the forget gate.

Then, hidden layer state from previous time (A^{t-1}) and input data (I^t) goes through "input gate" which is responsible for updating the cells state by the following equations:

$$i^t = F_{sigmoid}(W_{ii}I^t + W_{Ai}A^t + \theta_i)$$

$$g^t = F_{tanh}(W_{ig}I^t + W_{Ag}A^t + \theta_g)$$

$$C^t = (C^{t-1} \times f^t) + (i^t \times g^t)$$

where W_{ii}, W_{Ai}, W_{ig}, W_{Ag} are weight matrix for input data and hidden layer state in two segments of the input gate (sigmoid and tanh activation).

Finally, in the "output gate," updating the hidden layer state (A^t) by considering the updated cell state (C^t) is decided using the following equations:

$$O^t = F_{sigmoid}(W_{io}I^t + W_{Ao}A^{t-1} + \theta_o)$$

$$A^t = O^t \times F_{tanh}(C^t)$$

where W_{io} and W_{Ao} are weight matrix for input data and hidden layer state in the output gate (Raschka & Mirjalili).

Deep learning applications in oil and gas industry

The oil and gas industry can be considered one of the main producers of big data. Fiber optics, satellite images, seismic data, downhole gauges/flowmeter records, well logs, and other sources of information are used every day for assessment, monitoring, and decision-making. Many oil and gas companies are now collaborating with IT giants like Amazon, Google, and Microsoft to store all the data on the cloud and use high computational capacities. This resulted in a sharp growth in the application of DL in the oil and gas industry, and few of them would be discussed.

Crnkovic et al. (Crnkovic-Friis & Erlandson, 2015) used DL to estimate EUR from geological data from a region in the Eagle Ford Shale. The data set included 800 wells with 200,000 data points. Seven geological parameters such as porosity, thickness, water saturation, etc. were populated into geological grids and averaged on a well level. They used stacked denoising autoencoder as DL method and used EUR as the target feature. They observed mean absolute percentage error of 28% for the validation set. Omrani et al. (Shoeibi Omrani et al., 2019) assessed the performance of different methods (physical models, decline curve analysis, DL, and hybrid models) for short-term (6 weeks) and long-term (several years) production forecasting for different gas fields. They used wellhead pressure, cumulative production, and choke size as inputs and production rate as the output of the neural network. The forecasted production rate was added to cumulative production to predict new production. They observed that DL models can predict short-term forecast with high precision. Gupta et al. (Gupta et al., 2019) utilized deep neural networks to analyze borehole images. They used fully convolutional networks (FCN) which are a form of CNN but without densely connected layers to detect geological features like induced fractures, natural fractures, and sedimentary surfaces. They used labeled borehole images from four wells to train local and global models that would be used for automated borehole image analysis. Alakeely et al. (Alakeely & Horne, 2020) modeled reservoir behavior with CNNs and RNNs. In their study, reservoir simulation was used to generate data sets based on different well patterns. The data set included pressure, injection, and production data as inputs and pressure or rate as the target feature. They compared the performance of CNN and RNN models in predicting reservoir performance.

Dramsch et al. (Dramsch & Lüthje, 2018) used CNNs for 2D seismic interpretation. 2D in-line seismic slice is labeled by different features such as steep dipping reflectors, continuous high amplitude regions, salt intrusions, etc. Afterward, three different CNN networks (Waldeland CNN, VGG16, and ResNet 50) were used for training. They compared the performance of these CNN networks for seismic facies classification. Zhange et al. (Zhang et al., 2019) trained a deep CNN model to pick the faults in the seismic images and determined fault dips and azimuth. They generated 100,000 3D synthetic images (cubes) with one linear fault for training and 10,000 images for validating the CNN model. The trained model was tested on various seismic images from different regions such as the Gulf of Mexico, North Sea, and Trinidad deep water for fault picking. Shen et al. (Shen & Cao, 2020) developed a CNN-based model that recognizes ball seat event for the hydraulic fracturing operations. They used slurry rate, wellhead pressure, and their first and second derivative as input data to train a CNN model to predict ball seat events. Das et al. (Das & Mukerji, 2019) obtained petrophysical properties from prestack seismic data using CNNs. They used two approaches to train CNNs that map the correlation between prestack seismic data and petrophysical properties like porosity, volume of clay, and water saturation. In the first approach, they used cascaded CNNs with two separate networks. The first one predicts elastic properties (Vp, Vs, and RHOB) from prestack seismic data, and the second one predicts petrophysical properties from elastic properties. In the second approach, called end to end, CNN finds the correlation between petrophysical data and prestack seismic data directly. Chaki et al. (Chaki et al., 2020) made proxy models for flow simulators by utilizing deep and recurrent neural networks. They used Brugge reservoir model which includes 60,000 grid blocks with 20 production and 10 injection wells. This model has nine grid layers in vertical direction (stratigraphic grid layer). A permeability multiplier is assigned to each grid layer. They trained deep and recurrent neural networks considering permeability multiplier as the input and production/injection rates and cumulative production/injection as output. Their RNN models showed good quality, especially for predicting future rates that were seen by the model during training. Chaun et al. (Tian & Horne, 2017) used RNNs to analyze permanent downhole gauge (PDG) data for reservoir characterization. They considered synthetic and actual cases in which flow rate and time were inputs and pressure (with and without noise) was output for the NARX (nonlinear autoregressive exogenous) model. Bao et al. (Bao et al., 2020) investigated the performance of LSTM and Ensemble Kalman filter LSTM on building proxy model for the reservoirs. Their proxy model can predict production and water cut using previous production rate, water cut, injection rate, and bottom hole pressure as inputs. Srinath et al. (Madasu & Rangarajan, 2018) developed a real time LSTM-based model to predict surface pressure for multistage pumping data during hydraulic fracturing using flow rate and proppant concentration. They used data from 3 wells (Permian, Marcellus, and Niobrara regions) to train the model. For all three wells, model's prediction for surface pressure is within 5% of the actual values. Sun et al. (Sun et al., 2020) used both physics-based and data-driven models to predict screen out during hydraulic fracturing. Labeled data (screen out or not) at each time are considered as target and 14 features such as pressure, slurry rate, and proppant concentration as inputs for the LSTM model.

Frac treating pressure prediction using LSTM

In this section, the one section frac stage data will be used to build an LSTM model to use the last 60 seconds of slurry rate and proppant concentration data to predict treatment pressure for the next second. This is for showing the proof of concept. In this exercise, the first 60 s of data will be used to predict the next second (61st second). Thereafter, seconds 2 to 61 will be used to predict the next second (62nd second). This process is then continued. The figure below illustrates this concept. 60 s is an arbitrary window size that was chosen for this exercise, and other window sizes such as 30, 90, 120, etc., can be evaluated as well.

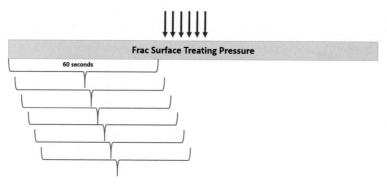

The first step is to import the main libraries followed by the one second frac data set that can be obtained from the portal listed below:

https://www.elsevier.com/books-and-journals/book-companion/9780128219294

```
import numpy as np
import pandas as pd
import matplotlib.pyplot as plt
from sklearn.preprocessing import MinMaxScaler
%matplotlib inline
data= pd.read_excel('Chapter6_Frac_Stage_Data.xlsx')
data.head()
Python output=Fig. 6.31
```

	Time	SLUR RATE	PROP CON	TR PRESS
0	1	49.4	0.0	8560
1	2	50.1	0.0	8537
2	3	50.1	0.0	8534
3	4	49.3	0.0	8617
4	5	49.2	0.0	8646

FIGURE 6.31 High frequency frac data head.

As illustrated, this data set has four main columns:

- Time in seconds
- Slurry rate denoted as "SLUR RATE" in barrels per min (bpm)
- Proppant concentration denoted as "PROP CON" in lbs per gallon (ppg)
- Treating pressure denoted as "TR PRESS" in psi

Next, let us plot the surface treating pressure data and slurry rate using the following lines of codes:

```
fig, ax1 = plt.subplots(figsize=(15,8))
plt.figure(figsize=(15,8))
ax2 = ax1.twinx()
ax1.plot(data['Time'], data['TR PRESS'], 'b')
ax2.plot(data['Time'], data['SLUR RATE'], 'g')
ax1.set_xlabel('Time (seconds)')
ax1.set_ylabel('Surface Treating Pressure (psi)', color='b')
ax2.set_ylabel('Slurry Rate (bpm))', color='g')
Python output=Fig. 6.32
```

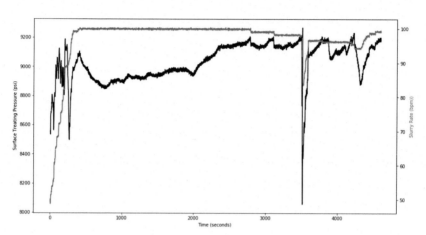

FIGURE 6.32 Surface treating pressure and slurry rate plot.

Next, let us divide the data into training and testing sets. The first 4500 s will be used for training, and the last 100 s will be used as testing in this example:

```
start_time = 1
end_time = 4500
#Define the data range used for training as follows:
filter = (data['Time'] > start_time) & (data['Time'] <= end_time)
data_training = data.loc[filter].copy()
data_training.tail()
Python output=Fig. 6.33
```

	Time	SLUR RATE	PROP CON	TR PRESS
4495	4496	98.7	1.0	9127
4496	4497	98.8	1.0	9125
4497	4498	98.5	1.0	9131
4498	4499	98.4	1.0	9136
4499	4500	98.6	1.0	9145

FIGURE 6.33 Tail of the training data.

Next, define the range for testing data as follows. Please note that this frac data set has 4600 rows in which the first 4500 rows are used for training and the last 100 rows are used for testing.

```
start_time2 = 4500
end_time2 = 4600
filter2 = (data['Time'] > start_time2) & (data['Time'] <= end_time2)
data_testing = data.loc[filter2].copy()
data_testing.tail()
Python output=Fig. 6.34
```

	Time	SLUR RATE	PROP CON	TR PRESS
4595	4596	99.1	1.0	9194
4596	4597	99.0	1.0	9201
4597	4598	99.0	1.0	9184
4598	4599	99.3	1.0	9178
4599	4600	99.0	1.0	9188

FIGURE 6.34 Tail of the testing data.

Next, drop the time column as follows:

```
training_data= data_training.drop(['Time'], axis=1)
```

Afterward, normalize the training data as follows:

```
scaler=MinMaxScaler()
training_data= scaler.fit_transform(training_data)
```

Next, first create an empty list for X_train and y_train as follows:

```
X_train= []
y_train= []
```

Afterward, write a for loop to iterate from 60 to the length of the "training_data." Next, within this for loop, take the empty list called "X_train" and append the training data from 0 to 60 (training_data[i-60:i]). Then, inside the same for loop, take the empty list called "y_train," and append the training data at the 61st row as "y_train." This allows the X_train to be the first 60 rows and y_train to be the 61st row.

```
for i in range(60,training_data.shape[0]):
    X_train.append(training_data[i-60:i])
    y_train.append(training_data[i,0])
```

Now, let us convert X_train and y_train into a numpy array prior to feeding it into an LSTM model. Typically, the LSTMs expect the data to be in a specific 3D array format. Therefore, let us convert the data as such:

```
X_train, y_train = np.array(X_train), np.array(y_train)
```

In addition, the shapes of X_train and y_train can be obtained as follows:

```
X_train.shape, y_train.shape
Python output=
((4439, 60, 3), (4439,))
```

As illustrated, X_train has a shape of 4439 (which is 4499 minus 60) and is a 3-dimensional data (3 input features). 4439 represents the "batch size," 60 represents the "time_span," and 3 represents "input_dimensions." Therefore, the input of an LSTM model is 3D array.

Since X_train and y_train have been defined, it is time to build the LSTM model. First, import the necessary libraries as follows:

- The first module from keras is called "Sequential" which is used for initializing the neural network.
- The second module from keras is called "Dense," and it is used for adding a connected neural network layer.
- "LSTM" as the name suggest is another module that is imported from keras, and it is for adding the LSTM layer.
- Finally, the "Dropout" module in keras is imported to ensure dropout layers are added to prevent overfitting.

```
from tensorflow.keras import Sequential
from tensorflow.keras.layers import Dense,LSTM, Dropout
```

In this exercise, the sequential methodology of building a DL is used. In this methodology, the layers can be added to the network very easily as illustrated below. First, make sure to make the seed number fixed (using a seed number of 100) as illustrated below. Then, place "Sequential" under the model name "Frac_LSTM." Feel free to change the model name to whatever is desired.

```
import tensorflow as tf
import random as python_random
def reset_seeds():
    np.random.seed(100)
    python_random.seed(100)
    tf.random.set_seed(100)
reset_seeds()
Frac_LSTM = Sequential()
```

Next, add the first LSTM layer with 200 units (units represent the dimensionality of the output space), "relu" activation function, and set "return_sequences" equal to "True." When this is set to true, it will return the last output in the output sequence. The output of an LSTM model could be either 2D array or 3D array depending on the "return_sequences" argument. Note that:

- If "return_sequences" is set to True, the output will be a 3D array (batch_size, time_steps, units).
- If "return_sequences" is set to False, the output will be a 2D array (batch_size, units).

Finally, set the "input_shape" to the shape of the training data set. "X_train.shape[1],3" will basically return "60,3." Afterward, add the "Dropout" layer of 0.3 which indicates 30% of the layers will be dropped. Add two more layers with the same configurations as the first layer as shown below. Note that after the first layer, there is no need to specify the "input_shape" for other layers. Therefore, "input_shape" argument for the second and third layers were excluded as illustrated below. Finally, add a "Dense" layer with "units = 1" which specifies the output of 1 unit.

```
Frac_LSTM.add(LSTM(units=200, activation='relu', return_sequences=
    True, input_shape=(X_train.shape[1],3)))
Frac_LSTM.add(Dropout(0.3))
Frac_LSTM.add(LSTM(units=200, activation='relu', return_sequences=
    True))
Frac_LSTM.add(Dropout(0.3))
Frac_LSTM.add(LSTM(units=200, activation='relu'))
Frac_LSTM.add(Dropout(0.3))
Frac_LSTM.add(Dense(units=1))
```

Next, obtain a summary of the Frac_LSTM model as follows:

```
Frac_LSTM.summary()
```
Python output=Fig. 6.35

```
Model: "sequential"

_____
Layer (type)                 Output Shape              Param #
=================================================================
lstm (LSTM)                  (None, 60, 200)           163200
_____
dropout (Dropout)            (None, 60, 200)           0
_____
lstm_1 (LSTM)                (None, 60, 200)           320800
_____
dropout_1 (Dropout)          (None, 60, 200)           0
_____
lstm_2 (LSTM)                (None, 200)               320800
_____
dropout_2 (Dropout)          (None, 200)               0
_____
dense (Dense)                (None, 1)                 201
=================================================================
Total params: 805,001
Trainable params: 805,001
Non-trainable params: 0
_____
```

FIGURE 6.35 Frac LSTM summary.

Next, compile the model using "adam" optimizer and "mean_-squared_error" as the loss function. This will use the mean of the squared errors as the loss function to minimize the error.

```
Frac_LSTM.compile(optimizer='adam', loss='mean_squared_error')
```

Next, fit the "Frac_LSTM" to the defined X_train and y_train, use 100 epochs or iterations, and use a batch size of 32. The higher the number of epochs, the more iterations it will run, and the longer time it will take the model to get generated. Please feel free to change the optimizer, number of epochs, batch size, dropout percentage, LSTM units, activation function, etc., to find the best hyperparameters for the model.

```
history=Frac_LSTM.fit(X_train,y_train, epochs=100, batch_size=32,
    shuffle=True)
```
Python output=Fig. 6.36

```
Epoch 90/100
4439/4439 [==============================] - 44s 10ms/sample - loss: 2.0579e-04
Epoch 91/100
4439/4439 [==============================] - 46s 10ms/sample - loss: 1.8662e-04
Epoch 92/100
4439/4439 [==============================] - 44s 10ms/sample - loss: 2.1588e-04
Epoch 93/100
4439/4439 [==============================] - 45s 10ms/sample - loss: 2.0377e-04
Epoch 94/100
4439/4439 [==============================] - 48s 11ms/sample - loss: 1.9440e-04
Epoch 95/100
4439/4439 [==============================] - 47s 11ms/sample - loss: 2.8725e-04
Epoch 96/100
4439/4439 [==============================] - 46s 10ms/sample - loss: 2.2165e-04
Epoch 97/100
4439/4439 [==============================] - 48s 11ms/sample - loss: 2.3116e-04
Epoch 98/100
4439/4439 [==============================] - 47s 11ms/sample - loss: 2.0920e-04
Epoch 99/100
4439/4439 [==============================] - 48s 11ms/sample - loss: 1.9204e-04
Epoch 100/100
4439/4439 [==============================] - 51s 11ms/sample - loss: 1.5980e-04
```
FIGURE 6.36 Last 10 epochs of the LSTM Model.

Now that a model has been trained, let us prepare a test data set. To predict the pressure of a test data set, it is important to obtain the last 60 s (rows) of the training data set. Therefore, define the last 60 s of the training data using the tail function as shown below:

```
past_60_secs= data_training.tail(60)
```

Next, append this last 60 s training data set to the testing data set as follows:

```
df=past_60_secs.append(data_testing,ignore_index=True)
df.head()
```
Python output=Fig. 6.37

	Time	SLUR RATE	PROP CON	TR PRESS
0	4441	98.6	1.0	9132
1	4442	98.7	1.0	9122
2	4443	98.6	1.0	9124
3	4444	98.8	1.0	9125
4	4445	98.8	1.0	9127

FIGURE 6.37 df head.

Next, drop the "Time" column from the df data frame as follows:

```
df=df.drop(['Time'], axis=1)
df.describe()
Python output=Fig. 6.38
```

	SLUR RATE	PROP CON	TR PRESS
count	160.000000	160.0	160.000000
mean	98.878125	1.0	9156.781250
std	0.307899	0.0	22.452934
min	98.200000	1.0	9118.000000
25%	98.600000	1.0	9137.000000
50%	98.900000	1.0	9158.500000
75%	99.100000	1.0	9176.000000
max	99.500000	1.0	9201.000000

FIGURE 6.38 df description.

As noted above, since the testing data (which were defined as the last 100 rows of the original imported excel file) were appended to the last 60 s of the training data, the total number of rows under "df" is 160. Next, normalize the test data as follows:

```
testing_inputs= scaler.transform(df)
```

Next, create two empty lists for X_test and y_test (as was previously performed for the training data) as follows:

```
X_test=[]
y_test=[]
```

Afterward, as was previously discussed, create a similar for loop and append to the created empty lists:

```
for i in range(60,testing_inputs.shape[0]):
    X_test.append(testing_inputs[i-60:i])
    y_test.append(testing_inputs[i,0])
```

Next, convert the X_test and y_test to a numpy array as follows:

```
X_test, y_test= np.array(X_test), np.array(y_test)
X_test.shape, y_test.shape
Python output=
((100, 60, 3), (100,))
```

Next, use the trained "Frac_LSTM" model to predict the "X_test," convert to a data frame, and call the column "Predicted TR PRESS."

```
y_pred=Frac_LSTM.predict(X_test)
y_pred=pd.DataFrame(y_pred,columns=['Predicted TR PRESS'])
y_pred.head()
Python output=Fig. 6.39
```

Predicted TR PRESS

0	0.964235
1	0.964230
2	0.964221
3	0.964215
4	0.964209

FIGURE 6.39 Predicted normalized treating pressure head.

After that, revert the predicted y_pred (predicted y values) and y_test (actual y values) to their original form (unnormalized form) to prepare for plotting as follows:

```
y_pred['Predicted TR PRESS']=y_pred['Predicted TR PRESS']*(data['TR
   PRESS'].max()-data['TR PRESS'].min())+(data['TR PRESS'].min())
y_test=pd.DataFrame(y_test,columns=['Actual TR PRESS'])
y_test['Actual TR PRESS']=y_test['Actual TR PRESS']*(data['TR
   PRESS'].max()-data['TR PRESS'].min())+(data['TR PRESS'].min())
y_test.head()
Python output=Fig. 6.40
```

Actual TR PRESS

0	9223.445736
1	9228.135659
2	9230.480620
3	9228.135659
4	9228.135659

FIGURE 6.40 Actual unnormalized treating pressure head.

Let us also obtain the unnormalized predicted treating pressure as follows:

```
y_pred['Predicted TR PRESS'].head()
Python output=Fig. 6.41
```

```
0     9224.724609
1     9224.717773
2     9224.707031
3     9224.700195
4     9224.692383
Name: Predicted TR PRESS, dtype: float32
```

FIGURE 6.41 Predicted unnormalized treating pressure head.

The next step is to visualize y_test and y_pred versus time to observe the match.

```
plt.figure(figsize=(14,5))
plt.plot(y_test, color='red', label='Actual TR PRESS')
plt.plot(y_pred['Predicted TR PRESS'], color='blue', label='
    Predicted TR PRESS')
plt.title('Frac Treating Pressure Actual Vs. Prediction')
plt.xlabel('Time (seconds)')
plt.ylabel('Surface Treating Pressure (psi)')
plt.legend()
Python output=Fig. 6.42
```

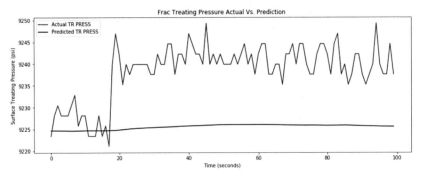

FIGURE 6.42 Actual vs. predicted treating pressure.

As illustrated in the figure above, the difference between actual and predicted pressures is roughly 20 psi; it is possible to, change window size, number of epochs, units, etc., to obtain a better match. Remember that this model was developed to predict the frac treating pressure at the next second for illustration purposes and proof of concept of the idea. The next step is to include more data and predict the frac treating pressure minutes into the future.

Nomenclature

e Error, difference between a nodes output and target
E Global error
F Activation function
g Gas specific gravity
h$_j$ Action potential at node j in hidden layer
I$_i$ Input signal at node i
j Error gradient in hidden layer
j Bias on node j
k Error gradient in output layer
n Number of incoming links to a node
P$_b$ Bubble point pressure (psi)
T Temperature('F)
w$_{ij}$ signal weight from node i to node j
α Learning Rate

References

Adeyemi, B., & Sulaimon, A. (2012). Predicting wax formation using artificial neural network. In *Nigeria annual international conference and exhibition, Lagos.*

Alakeely, A., & Horne, R. N. (August 1, 2020). *Simulating the behavior of reservoirs with convolutional and recurrent neural networks.* Society of Petroleum Engineers. https://doi.org/10.2118/201193-PA.

Ahmed, T. (2006). *Reservoir engineering handbook.* Burlington: Elsevier.

Amini, S., Mohaghegh, S. D., Gaskari, R., & Bromhal, G. (2012). Uncertainty analysis of a CO2 sequestration project using surrogate reservoir modeling technique. In *SPE western regional meeting, bakersfield.*

Arehart, R. (1990). Drill-bit diagnosis with neural networks. *SPE Computer Applications, 2*(04), 24–28.

Bao, A., Gildin, E., Huang, J., & Coutinho, E. J. (July 2020). Data-driven end-to-end production prediction of oil reservoirs by EnKF-enhanced recurrent neural networks. In *Paper presented at the SPE Latin American and Caribbean petroleum engineering conference, virtual.* https://doi.org/10.2118/199005-MS.

Belyadi, H., Belyadi, F., & Fathi, E. (2019). *Hydraulic fracturing in unconventional reservoirs.* Elsevier Inc.

Chaki, S., Zagayevskiy, Y., Shi, X., Wong, T., & Noor, Z. (January 2020). Machine learning for proxy modeling of dynamic reservoir systems: Deep neural network DNN and recurrent neural network RNN applications. In *Paper presented at the international petroleum technology conference, Dhahran, Kingdom of Saudi Arabia.* https://doi.org/10.2523/IPTC-20118-MS.

Crnkovic-Friis, L., & Erlandson, M. (September 28, 2015). *Geology driven EUR prediction using deep learning.* Society of Petroleum Engineers. https://doi.org/10.2118/174799-MS

Das, V., & Mukerji, T. (September 2019). Petrophysical properties prediction from pre-stack seismic data using convolutional neural networks. In *Paper presented at the SEG international exposition and annual meeting, San Antonio, Texas, USA.* https://doi.org/10.1190/segam2019-3215122.1.

Dramsch, J. S., & Lüthje, M. (November 30, 2018). *Deep-learning seismic facies on state-of-the-art CNN architectures*. Society of Exploration Geophysicists.

Gupta, K. D., Vallega, V., Maniar, H., Marza, P., Xie, H., Ito, K., & Abubakar, A. (January 1, 2019). *A deep-learning approach for borehole image interpretation*. Society of Petrophysicists and Well-Log Analysts.

https://www.forbes.com/sites/bernardmarr/2018/05/21/how-much-data-do-we-create-every-day-the-mind-blowing-stats-everyone-should-read/?sh=45826cef60ba.

Fausett, L. V. (1993). *Fundamentals of neural networks: Architectures, algorithms and applications*. Pearson.

Ford, D. A., & Kelly, M. C. (2001). Using neural networks to predict lithology from well logs. In *SEG annual meeting, San Antonio*.

Gartland, S., Owen, G., Cottis, R., & Turega, M. (1999). Neural network methods for the prediction of pitting potentials. In *CORROSION 99, San Antonio*.

Gharbi, R. B., & Elsharkawy, A. M. (1997). Universal neural network based model for estimating the PVT properties of crude oil systems. In *SPE Asia Pacific oil and gas conference and exhibition, Kuala Lumpur*.

Gunter, R., & Albert, T. (1992). Inversion of seismic waveforms using neural networks. In *1992 SEG annual meeting, New Orleans*.

Haghighat, S. A., Mohaghegh, S. D., Gholami, V., & Moreno, D. (2014). Production analysis of a niobrara field using intelligent top-down modeling. In *SPE western North American and rocky mountain joint meeting, Denver*.

Johnston, D. H. (1993). Seismic attribute calibration using neural networks. In *SEG annual meeting, Washington*.

Kalam, M., Al-Alawi, S., & Al-Mukheini, M. (1996). Assessment of formation damage using artificial neural networks. In *SPE formation damage control symposium, Lafayette*.

Kelleher, J. D. *Deep learning (the MIT press essential knowledge series)*. (pp. 177). MIT Press. Kindle Edition.

Kingma, D. P., & Ba, J. L. (2015). Adam: A method for stochastic optimization. In *International conference on learning representations*.

Kriesel, D. (2005). *A brief introduction to neural networks*. Bonn www.dkriesel.com.

Madasu, S., & Rangarajan, K. P. (November 2018). Deep recurrent neural network DRNN model for real-time multistage pumping data. In *Paper presented at the OTC arctic technology conference, Houston, Texas, USA*. https://doi.org/10.4043/29145-MS.

McCain, W. D., Jr. (January 1, 1999). *Properties of petroleum fluids* (2nd ed.). PennWell Corp, ISBN 9780878143351.

Mohaghegh, S. (2000). Virtual-intelligence applications in petroleum engineering: Part 1—artificial neural networks. *Journal of Petroleum Technology, 52*(09), 64−73.

Mohaghegh, S. D., Goddard, C., Popa, A., Ameri, S., & Bhuiyan, M. (2000). Reservoir characterization through synthetic logs. In *SPE eastern regional meeting, Morgantown*.

Mohaghegh, S., Grujic, O., Zargari, S., Kalantari-Dahaghi, A., & Bromhol, G. (2012). Top down, intelligent reservoir modeling of oil and gas producing shale reservoirs; case studies. *International Journal of Oil, Gas and Coal Technology, 5*(1).

Mohaghegh, S., McVey, D., Aminian, K., & Ameri, S. (1995). Predicting well stimulation results in a gas storage field in the absence of reservoir. *SPE Reservoir Engineering, 11*(04), 268−272.

Nakutnyy, P., Asghari, K., & Torn, A. (2008). Analysis of waterflooding through application of neural networks. In *Canadian international petroleum conference, Calgary*.

Omrani, S. P., Vecchia, A. L., Dobrovolschi, L., Van Baalen, T., Poort, J., Octaviano, R., Binn-Tahir, H., & Esteban Muñoz, E. (2019). *Deep Learning and Hybrid Approaches Applied to*

Production Forecasting. Paper presented at the Abu Dhabi International Petroleum Exhibition & Conference, Abu Dhabi, UAE, November 2019. https://doi.org/10.2118/197498-MS.

Raschka, S., & Mirjalili, V. *Python machine learning - second edition: Machine learning and deep learning with Python, scikit-learn, and TensorFlow.* (pp. 548). Packt Publishing. Kindle Edition.

Rashid, T. (2016). *Make Your Own Neural Network*, CreateSpace Independent Publishing Platform.

Schultz, R., & Chen, D. (2003). Dynamic neural network calibration of quartz transducers. In *SPE annual technical conference and exhibition, Denver.*

Siruvuri, C., Nagarakanti, S., & Samuel, R. (2006). Stuck pipe prediction and avoidance: A convolutional neural network approach. In *IADC/SPE drilling conference, Miami.*

Shen, Y., & Cao, D. (September 24, 2020). *Development of a ball seat event recognition algorithm with convolutional neural network for the real-time hydraulic fracturing analytics system.* Society of Petroleum Engineers. https://doi.org/10.2118/200003-MS.

Sun, J. J., Battula, A., Hruby, B., & Hossaini, P. (July 2020). Application of both physics-based and data-driven techniques for real-time screen-out prediction with high frequency data. In *Paper presented at the SPE/AAPG/SEG unconventional resources technology conference, virtual.* https://doi.org/10.15530/urtec-2020-3349.

Surguchev, L., & Li, L. (2000). IOR evaluation and applicability screening using artificial neural networks. In *SPE/DOE improved oil recovery symposium, Tulsa.*

Tang, H. (2008). Improved carbonate reservoir facies classification using artificial neural network method. In *Canadian international petroleum conference, Calgary.*

Tian, C., & Horne, R. N. (October 2017). Recurrent neural networks for permanent downhole gauge data analysis. In *Paper presented at the SPE annual technical conference and exhibition, San Antonio, Texas, USA.* https://doi.org/10.2118/187181-MS.

Thomas, A. L., & La Pointe, P. R. (1995). Conductive fracture identification using neural networks. In *The 35th U.S. Symposium on rock mechanics.* Reno: USRMS.

Vassallo, M., & Bernasconi, G. (2004). Bit bounce detection using neural networks. In *SEG annual meeting, Denver.*

Zhang, Q., Yusifov, A., Joy, C., Shi, Y., & Wu, X. (September 25, 2019). *FaultNet: A deep CNN model for 3D automated fault picking.* Society of Exploration Geophysicists.

Further reading

Osman, E., Abdel-Wahhab, O., & Al-Marhoun, M. (2001). Prediction of oil PVT properties using neural networks. In *SPE Middle East oil show, Bahrain.*

Publishing, AI. Python deep learning for beginners: Theory and practices step-by-step using TensorFlow 2.0 and keras (machine learning & data science for beginners) (pp. 140): AI Publishing LLC. Kindle Edition.

Chapter 7

Model evaluation

The main objective of machine learning is to build models that can find patterns in the data. These data-driven models are generally trained to deal with clustering or regression problems. The quality of a model can be assessed by their prediction accuracy when given new (unseen) data. In order to improve the quality or prediction performance of a model, first, it is necessary to define evaluation metrics and scoring systems. Then, these metrics in conjunction with an evaluation method can be used to pick the best strategies to improve model quality. Some model enhancement techniques include but are not limited to (i) hyperparameter optimization for a specific model, (ii) training different models, and, (iii) finally, changing the percentage of the train and test portions during data partitioning. In this chapter, evaluation metrics and techniques will be covered. Then, the best model selection methods are discussed. Finally, some model handling practices in Python such as model saving/loading are covered. All of these methods will be applied to oil and gas—related problems.

Evaluation metrics and scoring

There are different evaluation metrics that assess the performance of machine learning models. Binary classification, multinomial classification, and regression are the main learning techniques to be evaluated with specific metrics in this chapter. Some examples of the most widely used evaluation metrics for supervised classification problems are precision, accuracy, recall, F1 score, and confusion matrix. For unsupervised clustering problems, a silhouette score is recommended. Mean square error (MSE) and R^2 are the metrics that are mostly applied for regression models (e.g., neural networks). "Prediction of sand production," "rock typing," and "PVT estimation" are three examples that would be used for model evaluation in this chapter.

Binary classification: prediction of sand production

Sand production is one of the problems that many operators have been dealing with. Sand production can occur based on high flow rates, stress regime, rock geomechanical properties, and completion design. The prediction of sand production is very important since it helps production engineers to come up

Machine Learning Guide for Oil and Gas Using Python. https://doi.org/10.1016/B978-0-12-821929-4.00009-3

TABLE 7.1 Well attributes in Sand Production Data Set.

Symbol	Definition	Symbol	Definition
TVD	True vertical depth	BHFP	Bottom-hole flowing pressure
TT	Transmit time	DD	Drawdown pressure
COH	Cohesive strength of formation	EOVS	Effective overburden stress
Qg	Gas flow rate	SPF	Shots per foot
Qw	Water flow rate	Hperf	Perforation interval

with sand control strategies. Based on actual data for the wells that experienced sand production, it is possible to train a binary classifier model that predicts whether sand production occurs for a well. In the sand production prediction example, a data set that includes 29 wells from the Northern Adriatic basin (Moricca et al., 1994; Gharagheizi et al., 2017) was used. Sand production data set consists of attributes in Table 7.1.

The target variable in this example is sand production which is considered to be 1 if sand production occurred and 0 if sand was not produced from a well. Data set "Sand production" can be found in the following link: https://www.elsevier.com/books-and-journals/book-companion/9780128219294. Data loading and descriptive statistics are done by the following codes:

```
#data loading and descriptive statistics
import pandas as pd
dataset=pd.read_csv('Chapter7_Sand Production.CSV')
print(dataset.describe())
Python output=Fig. 7.1
```

The descriptive statistics of the data are shown in Fig. 7.1. Out of 29 wells, sand production was observed in 21 wells. Since data attributes have different

```
              No          TVD          TT          COH          Qg           Qw    \
count   29.000000    29.000000    29.000000    29.000000    29.000000    29.000000
mean    15.000000  2664.172414   109.931034    19.293103    70.751724   542.230966
std      8.514693  1027.947680    20.476347    11.056444    31.720460  1171.726103
min      1.000000   319.000000    85.000000     5.500000    23.000000     0.000000
25%      8.000000  1930.000000    98.000000    10.800000    48.000000    52.000000
50%     15.000000  2983.000000   100.000000    19.500000    69.800000    85.000000
75%     22.000000  3366.000000   120.000000    22.600000    93.400000   280.000000
max     29.000000  4548.000000   170.000000    53.200000   139.500000  5672.000000

              BHFP          DD         EOVS          SPF        Hperf   Sand Production Observed
count    29.000000   29.000000    29.000000    29.000000    29.000000                 29.000000
mean    185.113793   30.200000   435.586207     7.724138    10.965517                  0.724138
std      62.763364   31.623466   217.682849     7.457722     5.499888                  0.454859
min      67.000000    0.700000   111.000000     1.000000     3.500000                  0.000000
25%     140.400000    8.000000   242.000000     4.000000     6.000000                  0.000000
50%     184.600000   16.600000   492.000000     4.000000    11.000000                  1.000000
75%     222.100000   47.400000   601.000000     8.500000    15.500000                  1.000000
max     302.200000  124.400000   823.000000    33.000000    21.000000                  1.000000
```

FIGURE 7.1 Descriptive statistics of sand production data.

ranges, it is recommended to scale the data between 0 and 1 for the learning process (data set can also be standardized; the proper data transformation can be determined by the performance of the model which will be explained in this chapter). It is possible to perform feature ranking techniques mentioned in Chapter 6 like Spearman's rank correlation or random forest to see which attributes play an important role in sand production. In this section, the objective is to compare different evaluation metrics for a binary classification problem. "K-nearest neighbors" (KNN) algorithm would be used to train a model that predicts which well will produce sand. For the training process and more importantly model evaluation, data should be partitioned into train and test segments. Data scaling, partitioning, and training KNN model can be done with the following codes.

```
# Define input(x) and target(y)
x=dataset.iloc[:,1:11]
y=dataset.iloc[:,11].values
# Scale input data between 0 and 1
from sklearn.preprocessing import MinMaxScaler
sc=MinMaxScaler()
xnorm=pd.DataFrame(data=sc.fit_transform(x))
# Partition data into test and train
import numpy as np
seed=50
np.random.seed(seed)
from sklearn.model_selection import train_test_split
x_train, x_test, y_train, y_test=train_test_split(xnorm,
  y, test_size=0.3)
# Train binary classifier (k-neighbor) and predict on train and test data
  # set
from sklearn.neighbors import KNeighborsClassifier
KNC=KNeighborsClassifier(n_neighbors=3,leaf_size=3)
  KNC.fit(x_train,y_train)
y_train_predict=KNC.predict(x_train)
y_test_predict=KNC.predict(x_test)
```

The first evaluation metric is "accuracy," defined as the ratio of correct prediction to total prediction. The results of the trained KNN model prediction for the test data set are shown in Table 7.2. In the test data set(y_train), there are nine observations where six wells produced sand (1 values) and three wells experienced no sand production (0 values). When the model predicts value 1 or sand production correctly, it is called True Positive (TP). In this example, there are five TPs. When the model makes a correct prediction for 0 values or no sand production, it is called True Negative (TN). If the model makes a wrong prediction equal to 1 or sand production, it is defined as False Positive (FP) or type I error. Finally, if the model makes a wrong prediction equal to 0 or no sand production, it is defined as False Negative (FP) or type II error. In this example, there are six TPs, one TN, and two FPs. There was no FN in the model's prediction.

TABLE 7.2 Model Prediction for test data set.

y_test	y_test-predict	Prediction status
0	0	TN
1	1	TP
0	1	FP
1	1	TP
1	1	TP
1	1	TP
0	1	FP
1	1	TP
1	1	TP

Based on mentioned definitions, accuracy can be defined as:

$$accuracy = \frac{TP + TN}{TP + TN + FP + FN}$$

In this example, the accuracy is calculated to be $(6 + 1)/(6 + 1 + 2 + 0) = 0.78$. In Python, the accuracy of a classification problem can be calculated with the following code:

```
from sklearn.metrics import accuracy_score
accuracy_test=accuracy_score(y_test, y_test_predict)
print('Accuracy_test: %f' % accuracy_test)
Python output:
Accuracy_test: 0.777778
```

In binary classification, when the number of cases in each class is very imbalanced, accuracy may not be a good metric for the model's prediction performance evaluation (Albon, 2018). Some of the imbalance classes in the oil and gas industry can be wells that experience blowout or severe pipeline leakage/oil spills. For imbalanced data sets, it is better to use "Precision," defined as the ratio of correct or TPs to all the positives that were predicted by the model. It should be mentioned that class with fewer samples like the number of well blowouts is considered positive.

$$precision = \frac{TP}{TP + FP}$$

In sand production example, there are 6 TPs and 2 FPs. Therefore, precision is calculated as $(6)/(6 + 2) = 0.75$. In Python, the precision of a classification problem is obtained with the following codes:

```
from sklearn.metrics import precision_score
precision_test=precision_score(y_test, y_test_predict)
print('Precision_test: %f' % precision_test)
Python output:
Precision_test: 0.750000
```

The other evaluation metric for the classification problem is "Recall." It is defined as what proportion of all positive values are correctly predicted positives. All positive values include both TPs and FNs (FN is actually a positive value which is incorrectly predicted as negative). Recall identifies the ability of the model to predict positive values. Mathematically, Recall can be defined as:

$$recall = \frac{TP}{TP + FN}$$

For this example, there are 6 TPs and 0 FN. The recall is calculated as $(6)/(6 + 0) = 1$. In Python, recall of a classification problem is obtained with the following codes:

```
from sklearn.metrics import recall_score
recall_test=recall_score(y_test, y_test_predict)
print('Recall_test: %f' % recall_test)
Python output:
Recall_test: 1.000000
```

Selecting an appropriate metric is sometimes related to the objective of the classification problem. In some cases, the goal is to have high precision while in other cases, having high recall is a necessity. In order to have a balance between precision and recall, another metric, defined as "F1 score," is a harmonic mean of precision and recall. Considering that precision is the ratio of TPs to all the values that predicted positive and recall is the ratio of TPs to all positive values, F1 score can be defined as:

$$F1 = 2 \times \frac{Precision \times Recall}{Precision + Recall}$$

In this example, precision is 0.75 and recall is 1; therefore, F1 score is calculated as $2 \times (0.75 \times 1)/(0.75 + 1) = 0.857$. Python codes for calculating F1 score in sand production example is as follows:

```
from sklearn.metrics import f1_score
f1_test=f1_score(y_test, y_test_predict)
print('F1 score_train: %f' % f1_test)
Python Output=
F1 score_train: 0.857143
```

There are some other evaluation metrics for binary classifiers like
"Cohen_Kappa_score" or "Roc_auc_score" that are left for the reader.
However, it is also possible to have a comprehensive report for the most
important evaluation metrics for binary classification problems. In sklearn, by
importing classification_report, precision, recall, F1 score, and support for
each class would be summarized in a table. Support is the number of instances
in each class. For example, in the test data set for sand production problem, in
class 1 (wells with sand production), there are six wells. In class 0 (wells with
no sand production), there are three wells. Supports for classes 1 and 0 are six
and three, respectively. In the classification report, there are two averages.
"Macro" means the averaging process designates the same weight to each
class. "Weighted" calculates mean of scores, considering a weight for each
class proportional to the size of the data in that class. The classification report
for the test data set for sand production problem can be obtained with the
following codes. The output of the following codes is in Fig. 7.2.

```
from sklearn.metrics import classification_report
print(classification_report(y_test, y_test_predict))
Python output=Fig. 7.2
```

Multiclass classification: facies classification

Evaluation metrics for multiclass classification problems are almost the same
as binary classification. The other recommended metric for multiclass problems is the confusion matrix. In order to explain the confusion matrix, the
"facies classification" problem designed by "Dubois, Bohling, and Chakrabarti" (Dubois et al., 2007) is used in this section. The data can be found in the
following link: https://www.elsevier.com/books-and-journals/book-companion
/9780128219294.

In this problem, the objective is to estimate rock facies for a gas reservoir
in Kansas. Data were gathered from nine wells with different logs. The whole
data set includes nine rock classes or facies (model output), with seven log
attributes as input or predictor variables. There are 3232 samples in this
example. Facies classes and log attributes are shown in Tables 7.3 and 7.4.

In this problem, the goal is to train a classification model that predicts each
class for new (test) data. "Logistic regression" is used as a multiclass

	precision	recall	f1-score	support
0	1.00	0.33	0.50	3
1	0.75	1.00	0.86	6
accuracy			0.78	9
macro avg	0.88	0.67	0.68	9
weighted avg	0.83	0.78	0.74	9

FIGURE 7.2 Classification report for the test data set in sand production problem.

TABLE 7.3 Facies classes in the facies classification problem.

Class number	Facies
1	Nonmarine sandstone
2	Nonmarine coarse siltstone
3	Nonmarine fine siltstone
4	Marine siltstone and shale
5	Mudstone (limestone)
6	Wackestone (limestone)
7	Dolomite
8	Packstone-grainstone (limestone)
9	Phylloid-algal bafflestone (limestone)

TABLE 7.4 Log attributes in the facies classification problem.

Well log attributes (predictor)	
Gamma ray	GR
Resistivity	ILD_log10
Photoelectric effect	PE
Neutron-density porosity difference	DeltaPHI
Average neutron-density porosity	PHIND
Marine indicator	NM_M
Relative position	RELPOS

classifier. Other classifiers will be considered for this problem in Section Grid search and model selection. Similar to previous problems, it is necessary to scale all the input data between 0 and 1 since they are not in the same range. It is also necessary to partition the data into train and test. 25% of the data is considered for testing. Data import, scaling, partitioning, and training logistic regression model can be done with the following codes. The descriptive statistics of the data are shown in Fig. 7.3.

```
#Import the data and print descriptive statistics
import pandas as pd
import seaborn as sns
```

	Facies	Depth	GR	ILD_log10	DeltaPHI \
count	3232.000000	3232.000000	3232.000000	3232.000000	3232.000000
mean	4.422030	2875.824567	66.135769	0.642719	3.559642
std	2.504243	131.006274	30.854826	0.241845	5.228948
min	1.000000	2573.500000	13.250000	-0.025949	-21.832000
25%	2.000000	2791.000000	46.918750	0.492750	1.163750
50%	4.000000	2893.500000	65.721500	0.624437	3.500000
75%	6.000000	2980.000000	79.626250	0.812735	6.432500
max	9.000000	3122.500000	361.150000	1.480000	18.600000

	PHIND	PE	NM_M	RELPOS
count	3232.000000	3232.000000	3232.000000	3232.000000
mean	13.483213	3.725014	1.498453	0.520287
std	7.698980	0.896152	0.500075	0.286792
min	0.550000	0.200000	1.000000	0.010000
25%	8.346750	3.100000	1.000000	0.273000
50%	12.150000	3.551500	1.000000	0.526000
75%	16.453750	4.300000	2.000000	0.767250
max	84.400000	8.094000	2.000000	1.000000

FIGURE 7.3 Descriptive statistics for facies database.

```
dataset=pd.read_csv('Chapter7_Facies Data.CSV')
print(dataset.describe())
x=dataset.iloc[:,4:11]
y=dataset.iloc[:,0].values
#Scale the input data from 0 to 1
from sklearn.preprocessing import MinMaxScaler
sc=MinMaxScaler()
xnorm=pd.DataFrame(data=sc.fit_transform(x))
#Partition the data into train and test
from sklearn.model_selection import train_test_split
seed=50
x_train, x_test, y_train, y_test=train_test_split(xnorm, y,
    random_state=1)
#Import logistic regression and train the model
from sklearn.linear_model import LogisticRegression
LG=LogisticRegression(max_iter=200).fit(x_train,y_train)
y_predict=LG.predict(x_test)
```

"Confusion matrices" is one the most widely used metrics to evaluate the performance of multiclass classifiers. In the confusion matrix, sometimes shown as a heat map, each row represents an actual or target class. Each column shows the model's prediction for each class. Therefore, a value in each cell is a combination of prediction and actual value. Values on the diagonal of the matrix shows how many model predictions for each class is correct. For a good model, it is expected to see most of the non-zero values in the diagonal entries and zero values otherwise. If observation counts spread all over the matrix, it means the model does not perform properly. In Facies classification problem, the data set is partitioned into 75% for training and 25% for testing (this is the default for train_test_split). Therefore, the test data set includes 808 samples. The confusion matrix which compares true classes and predicted classes for the test data set, can be plotted in Python with the following codes. The confusion matrix is shown in Fig. 7.4.

```
import matplotlib.pyplot as plt
import numpy as np
from sklearn.metrics import confusion_matrix
matrix=confusion_matrix(y_test, y_predict)
cv=np.arange(1,10)
dataframe=pd.DataFrame(matrix,index=cv,columns=cv)
# Create heatmap
sns.heatmap(dataframe, annot=True, cbar=None, cmap="Blues")
plt.title("Confusion Matrix"), plt.tight_layout()
plt.ylabel("True Class"), plt.xlabel("Predicted Class")
plt.show()
Python output=Fig. 7.4
```

In this problem, there are nine facies classes. Therefore, the confusion matrix represents nine rows indicating true classes and nine columns representing models' prediction. Let's review the performance of the model for class 1 that is Nonmarine sandstone. In the test data set, 62 samples belong to class 1. Therefore, true classes are 62 items. By looking at all the values in row 1, it can be inferred that, out of 62 samples, the model predicts that 27 samples belong to class 1 (correct prediction), 34 samples belong to class 2, and one sample belongs to class 3. Values in column 1 depict that model predicted 38

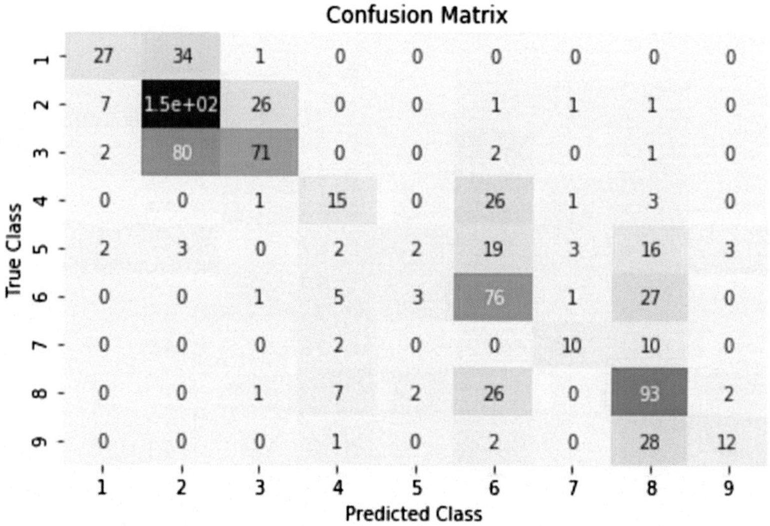

Confusion Matrix

	1	2	3	4	5	6	7	8	9
1	27	34	1	0	0	0	0	0	0
2	7	1.5e+02	26	0	0	1	1	1	0
3	2	80	71	0	0	2	0	1	0
4	0	0	1	15	0	26	1	3	0
5	2	3	0	2	2	19	3	16	3
6	0	0	1	5	3	76	1	27	0
7	0	0	0	2	0	0	10	10	0
8	0	0	1	7	2	26	0	93	2
9	0	0	0	1	0	2	0	28	12

True Class (vertical axis) / Predicted Class (horizontal axis)

FIGURE 7.4 Confusion matrix for facies classification problem.

samples as class 1 (sum of all the values in column 1). As mentioned earlier, 27 samples are correctly predicted to belong to class 1. However, seven samples that are predicted to belong to class 1 actually belong to class 2, two samples belong to class 3, and two samples belong to class 5. For each class, it is possible to evaluate the performance of the model by looking at the confusion matrix in detail.

As the number of classes increases, it may not be easy to understand the overall performance of the model with a confusion matrix. Therefore, a better option can be the use of a "classification_report". Precision, recall, and F1 score are the metrics associated with the classification report. For multiclass classification, precision for each class is the ratio of correctly predicted class to all the predicted classes. In this example, for class 1, 27 samples are predicted correctly, to belong to class 1 out of 38 predicted samples as class 1. Based on the mentioned values, the model's precision for predicting class 1 on the test data set is $27/38 = 0.71$. The recall is calculated as the ratio of correctly predicted class to all true class values. For class 1 in this example, 27 out of 62 samples are predicted correctly, so the recall is $27/62 = 0.44$. By having precision and recall, the F1 score can be calculated for each class. The classification report for facies classification problem can be calculated by the following codes.

```
from sklearn.metrics import classification_report
print(classification_report( y_test,y_predict))
Python output=Fig. 7.5
```

	precision	recall	f1-score	support
1	0.71	0.44	0.54	62
2	0.56	0.81	0.66	185
3	0.70	0.46	0.55	156
4	0.47	0.33	0.38	46
5	0.29	0.04	0.07	50
6	0.50	0.67	0.57	113
7	0.62	0.45	0.53	22
8	0.52	0.71	0.60	131
9	0.71	0.28	0.40	43
accuracy			0.56	808
macro avg	0.56	0.46	0.48	808
weighted avg	0.57	0.56	0.54	808

FIGURE 7.5 Classification report for facies logistic regression model.

Evaluation metrics for regression problems

There are two common evaluation metrics for regression models: MSE and Coefficient of determination, R^2. MSE or loss function measures the sum of squared errors for the regression model as defined in Eq. 6.10. The error was defined as the difference between the target output and the model's prediction. Squaring the error leads to dealing with positive numbers and penalizing higher error values. Higher the MSE means that the performance of the model gets worse. During neural network training, the objective was to minimize MSE or loss function for validation data set through weight optimization. MSEs for Shale Gas Wells problems were shown in Figs. 6.12 and 6.13. The other metric for regression problems is Coefficient of determination or R squared. It measures the amount of variance of target values to the model's prediction:

$$R^2 = 1 - \frac{\sum\limits_{i=1}^{n} \left(y_i - \widehat{y}_i \right)^2}{\sum\limits_{i=1}^{n} \left(y_i - \overline{y}_i \right)^2}$$

In this R^2 equation, yi is target value, $\widehat{y}_i$ is model's prediction, and $\overline{y}_i$ is the average of all the target values. R^2 values closer to 1 indicate better model performance. As shown in Chapter 6, R^2 in Python can be determined with "r2_score" method. R^2 for Shale Gas Well and PVT problems are shown in Figs. 6.14, 6.18−6.20.

Cross-validation

Multiple evaluation metrics were introduced to understand how well a machine learning model can perform on test data or the data that has never been exposed to the model. In order to select test data, the whole data set is partitioned to train and test in typically 70% and 30% ratios. The model is built by going through the learning process using the train data and then assessed by test data. The drawback of the mentioned procedure is, evaluation results can be biased by the samples that were picked in the test data set. Moreover, the model is deprived of some data (test) for the training process. In order to have a more generalized model elevation that covers the whole data set, "cross-validation" is recommended.

In "k-fold cross-validation," the most common version of cross-validation, the whole data set is partitioned into k "folds." K is typically between 5 and 10. In the first iteration of model training, the first fold indicates a test data set and the remaining (k-1) folds indicate a training data set. When the model is trained, the metric score would be calculated on the test data set which is fold one. Then, in the next iteration, the second fold indicates a test data set and the remaining (k-1) folds indicate a training data set. Next, the second model is built on the training data set (all the folds except the second one) and the score is determined on the second fold. This process continues until the kth iteration

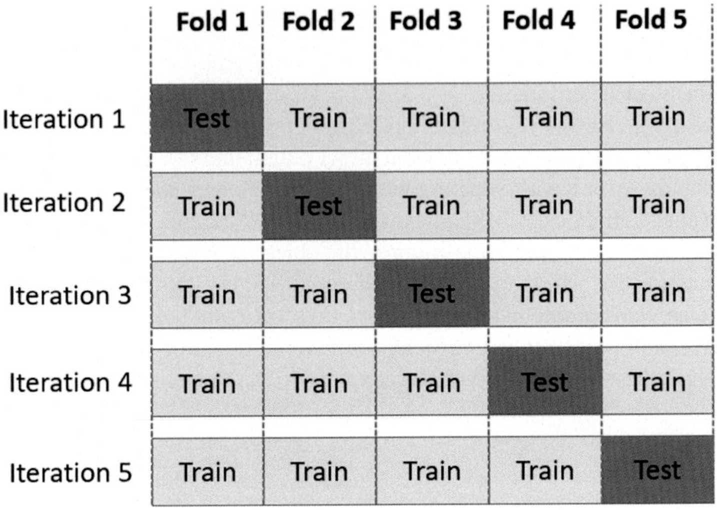

	Fold 1	Fold 2	Fold 3	Fold 4	Fold 5
Iteration 1	Test	Train	Train	Train	Train
Iteration 2	Train	Test	Train	Train	Train
Iteration 3	Train	Train	Test	Train	Train
Iteration 4	Train	Train	Train	Test	Train
Iteration 5	Train	Train	Train	Train	Test

FIGURE 7.6   Data partitioning for five-fold cross-validation.

is done. The evaluation score also would be determined for each iteration and reported at the end. An example of five-fold cross-validation is shown in Fig. 7.6.

Cross-validation for classification

In Python, cross-validation can be performed in the scikit-learn library by the "cross_val_score" function. For this function, it is required to specify at least five parameters: (i) the model for training and evaluation, (ii) input data set, (iii) output data set, (iv) number of folds (default is 3), and (v) metric score. Cross-validation would be applied to three machine learning examples covered in this chapter and Chapter 6. The Python code for sand production example is as follows (this is the continuation of the codes used to train the KNN for sand production problem at the beginning of this chapter):

```
from sklearn.model_selection import cross_val_score
scores=cross_val_score(KNC, xnorm, y,cv=5,scoring='accuracy')
print("Cross-validation scores: {}". format(scores))
print("Average cross-validation score: {}". format(scores.mean()))
Python output=
Cross-validation scores: [0.66666667 1. 1. 0.66666667 1.]
Average cross-validation score: 0.8666666666666666
```

In this example, KNN is used as the binary classifier model, "xnorm" and "y" are input and output data sets, the number of folds is 5, and the evaluation metric is "accuracy." Based on cross-validation results, fold accuracy scores range from 0.67 to 1 with an average of 0.87. Therefore, it is expected to see 0.86 accuracy as the overall performance of the model. However, the wide range in fold's accuracy score may imply that model is highly dependent on some specific folds that were used for training. This can be a result of the small size of the data set. Let's perform five-fold cross-validation for facies classification problem with the following Python code (this is the continuation of the codes used to train the logistic regression model for facies classification problem at the beginning of this chapter):

```
from sklearn.model_selection import cross_val_score
scores_kfold=cross_val_score(LG, xnorm, y,cv=5)
print("Kfold Cross-validation scores: {}". format(scores_kfold))
print("Average Kfold cross-validation score: {}". format
  (scores_kfold.mean()))
Python output=
Kfold Cross-validation scores: [0.52704791 0.53323029 0.48916409
  0.48297214 0.57120743]
Average Kfold cross-validation score: 0.5210339695953221
```

For facies classification problem, the accuracy scores range from 0.48 to 0.57 with an average value of 0.52. As compared to the sand production problem, the accuracy is lower. However, less variation is observed in score values due to the fact that more samples exist in facies classification problem.

Cross-validation for regression

In order to apply k-fold cross-validation to a regression model, the PVT estimation problem from Chapter 6 is used. In this problem, 249 PVT samples, including temperature, solution gas to oil ratio (RS), gas gravity, and Oil API were used to correlate them with bubble point pressure using neural networks. After training the neural networks, R^2 and MSE were calculated for test and train data sets. Now, five-fold cross-validation is added to that example to get MSE and R^2 for each fold using the following code in Python (make sure to use the following code as a continuation of PVT problem in Chapter 6).

```
import numpy as np
seed=50
np.random.seed(seed)
y_norm_Ravel=ynorm.values.ravel()
  from sklearn.model_selection import cross_val_score
scores_MSE=cross_val_score(clf, Xnorm,
y_norm_Ravel,cv=5,scoring='neg_mean_squared_error')
print("MSE_ Cross-validation scores: {}". format(scores_MSE))
print(" Average Kfold cross-validation MSE_score: {}". format
  (scores_MSE.mean()))
scores_R2=cross_val_score(clf, Xnorm, y_norm_Ravel,cv=5,scoring=
  'r2')
print(" R2_Cross-validation scores: {}". format(scores_R2))
print(" Average R2_Cross-validation scores: {}". format
  (scores_R2.mean()))
Python output=Fig. 7.7
```

```
MSE_ Cross-validation scores: [-0.0055198  -0.00155995 -0.01205835 -0.01779542 -0.00282173]
Average Kfold cross-validation MSE_score: -0.007951048760489427
R2_Cross-validation scores: [0.93905578 0.7947706  0.94509814 0.63662807 0.89653764]
Average R2_Cross-validation scores: 0.8424180443306877
```

FIGURE 7.7   Cross-validation report for PVT neural network model.

In PVT estimation problem's cross-validation, the R^2 ranges from 0.64 to 0.95 with an average value of 0.84 which is a reasonably good value for model performance.

Stratified K-fold cross-validation

In K-fold cross-validation, the whole data set is divided into K folds to train the model on K-1 folds and test the model on the remaining fold iteratively. The percentage of samples belonging to various target classes in each fold can be different. In order to honor the percentage of different classes in the whole data set in each fold, it is recommended to use "stratified" k-fold cross-validation (Muller & Guido, 2016). In sand production example, 72% of the observations (wells) experienced sand production and 28% didn't. Therefore, by using a stratified k-fold cross-validation, each fold includes 72% of sand produced wells and 28% of no-sand wells. It is worth mentioning that in *cross_val_score* when "cv" is an integer, stratified k-fold cross-validation would be the default. Stratified k-fold cross-validation for sand production problem can be done in Python with the following code:

```
from sklearn.model_selection import StratifiedKFold
skfold=StratifiedKFold(n_splits=5,shuffle=True, random_state=100)
scoresSK=cross_val_score(KNC, xnorm, y,cv=skfold,scoring=
   'accuracy')
print(" StratifiedKFold Cross-validation scores: {}". format
   (scoresSK))
print(" Average StratifiedKFold cross-validation score: {}". format
   (scoresSK.mean()))
Python output=
StratifiedKFold Cross-validation scores: [1.  0.83333333 0.83333333
   0.83333333 1. ]
Average StratifiedKFold cross-validation score: 0.9
```

Another cross-validation method is "Leave One Out." In this method, the number of folds is the same as the number of samples in the data set. The number of models that would be trained is the same as samples. If the number of samples is high, this method can be computationally exhaustive since it is required to have trained models equal to the number of data samples. For facies classification problem, leave one out cross-validation can be done with the following Python codes:

```
from sklearn.model_selection import LeaveOneOut
loo=LeaveOneOut()
scores_loo=cross_val_score(LG, xnorm, y, cv=loo)
print(" Number of cv iterations-Leave one Out: ", len(scores_loo))
print(" Mean accuracy Leave one Out: {}". format(scores_loo.mean()))
Python output=
Number of cv iterations-Leave one Out: 3232
Mean accuracy Leave one Out: 0.5590965346534653
```

The last cross-validation method in this chapter is Shuffle-Split. In this method, it is necessary to define the number of iterations or splits. Then, the percentage of the test and train should be specified. In each split, the data set would be partitioned into train and test based on a specified percentage. In k-fold cross_validation, onefold out of k is considered for the test. However, in Shuffle-Split cross-validation, there is control over the percentage of data for testing the model in each split. Shuffle-Split cross-validation for sand production example in Python is as follows:

```
from sklearn.model_selection import ShuffleSplit
sh_sp=ShuffleSplit(test_size =.25, train_size =.75, n_splits = 5,
    random_state=50)
scoresSP=cross_val_score(KNC, xnorm, y, cv=sh_sp)
print(" Cross-validation scores:{}". format(scoresSP))
print(" Mean accuracy: {}". format(scoresSP.mean()))
Python output=
Cross-validation scores: [0.75 1. 0.75 0.75 0.75]
Mean accuracy: 0.8
```

In this code, "ShuffleSplit" from sklearn is called. The number of iterations is 5. In each split, 75% of the data is used for training and 25% is used for testing. After performing cross-validation, the average accuracy is 0.8. Shuffle-Split cross-validation for facies classification example in Python is as follows:

```
from sklearn.model_selection import ShuffleSplit
sh_sp=ShuffleSplit(test_size =.25, train_size =.75, n_splits = 6,
    random_state=50)
scores_SP=cross_val_score(LG, xnorm, y, cv=sh_sp)
print(" Cross-validation scores:{}". format(scores_SP))
print(" Mean accuracy: {}". format(scores_SP.mean()))
Python output=
Cross-validation scores: [0.54455446 0.56064356 0.5779703
    0.54455446 0.55321782 0.57549505]
Mean accuracy: 0.5594059405940595
```

In facies classification example, the number of splits is 6. Data is partitioned into 75% and 25% training and testing consecutively. The average of cross-validation accuracy on test data sets for all the splits is 0.56. Shuffle-Split cross-validation for PVT estimation example in Python is as follows:

```
import numpy as np
seed=50
np.random.seed(seed)
from sklearn.model_selection import ShuffleSplit
from sklearn.model_selection import cross_val_score
sh_sp=ShuffleSplit(test_size =.3, train_size =.7, n_splits = 8,
  random_state=50)
scores=cross_val_score(clf, Xnorm, y_norm_Ravel, cv=sh_sp,
  scoring='r2')
print(" Cross-validation scores:{}". format(scores))
print(" Mean R2: {}". format(scores.mean()))
Python output=
Cross-validation scores: [0.89955581 0.94445178 0.77980894
  0.78714374 0.94654158 0.91353788 0.79064149 0.84369953]

Mean R2: 0.8631725957628656
```

In PVT estimation example, eight splits are considered with 70% of the data for neural network training and 30% for model testing. The evaluation metric in this example is R^2. Cross-validation R^2 scores for eight splits range from 0.78 to 0.95 with an average of 0.86.

Grid search and model selection

In previous sections, different model evaluation metrics in conjunction with cross_validation methods were introduced to assess the prediction performance of a model. The final goal for training a machine learning model is to find and train a model with the best performance. The model's performance is highly dependent on hyperparameter selection. In order to maximize the performance of a model, optimized combinations of hyperparameters must be picked. Moreover, for a specific problem like classification or regression, multiple models can be used. Therefore, to achieve the best prediction performance, it is necessary to select the best model with an optimized combination of hyperparameters. In this section, some techniques for getting the best combination of hyperparameters and selecting best models will be discussed.

Grid search for hyperparameter optimization

In order to optimize model performance and get the best set of hyperparameters, one should first generate a wide variety of combinations for hyperparameters. Then, by searching through combinations for hyperparameters, it is possible to find the combination that returns the highest model's performance. Finding the best set of hyperparameters by searching through all the combinations can be done by "grid search." To illustrate grid search, a for-loop would be used for generating search space for the sand production problem followed by finding the best set of hyperparameters.

In sand production example, KNN was used for binary classification. For this model, out of multiple hyperparameters, the number of neighbors, weight function, and leaf size are selected for a grid search. Different values and functions are assigned to hyperparameters. For each combination of hyperparameters, five-fold cross-validation was performed, and the average accuracy of the folds was calculated. At the end, there will be 24 different models with combinations of hyperparameters and their accuracy. The set that returns a model with the highest accuracy is the optimized hyperparameter. Initial grid search code in Python for the sand production problem is as follows:

```python
from sklearn.neighbors import KNeighborsClassifier
#Define k-nearest neighbor's combination of hyper-parameters
highest_score=0
i=0
for n in [2,3,4,5]:
    for w in['uniform','distance']:
        for l in[2,3,4]:
#Train a model with each combination of hyper-parameters and
   #calculate the accuracy
                KNCG=KNeighborsClassifier(n_neighbors=n,
                    weights=w, leaf_size=l)
                scores=cross_val_score(KNCG, xnorm, y,
                    cv=5,scoring='accuracy')
                scoreg=scores.mean()
                print("iteration:",i," score:",scoreg,
                    " NN:",n, " weight:",w," leaf size:",l)
                i+=1
#Find the hyper-parameters that returns a model with the highest
   #accuracy score
                if scoreg >= highest_score:
                    highest_score=scoreg
                    best_parameters={' n_neighbors':
                        n, 'weights': w,'leaf_size':l}
print(" Highest score: ", (highest_score))
print(" Best parameters: ", format(best_parameters))
```
Python output=Fig. 7.8

By looking at the results, it is observed that iterations 3, 4, and 5 return the highest scores. It means that combinations of 2 NN, "distance" weights, and leaf sizes equal to "2,3,4" are the best combinations that result in the highest score (accuracy) of 0.9. To combine grid search over hyperparameters with cross-validation "GridSearchCV" form sklearn library can be used. In GridSearchCV, the sets of all hyperparameters to be tuned are passed in as a dictionary and the model type is the other input parameter (which is the KNN model for sand production example). Afterward, the GridSearchCV method trains a model based on the best combination of hyperparameters given the input and output features. Grid search with cross-validation for sand

```
iteration: 0   score: 0.8933333333333333  NN: 2  weight: uniform  leaf size: 2
iteration: 1   score: 0.8933333333333333  NN: 2  weight: uniform  leaf size: 3
iteration: 2   score: 0.8933333333333333  NN: 2  weight: uniform  leaf size: 4
iteration: 3   score: 0.9  NN: 2  weight: distance  leaf size: 2
iteration: 4   score: 0.9  NN: 2  weight: distance  leaf size: 3
iteration: 5   score: 0.9  NN: 2  weight: distance  leaf size: 4
iteration: 6   score: 0.8666666666666666  NN: 3  weight: uniform  leaf size: 2
iteration: 7   score: 0.8666666666666666  NN: 3  weight: uniform  leaf size: 3
iteration: 8   score: 0.8666666666666666  NN: 3  weight: uniform  leaf size: 4
iteration: 9   score: 0.8666666666666666  NN: 3  weight: distance  leaf size: 2
iteration: 10  score: 0.8666666666666666  NN: 3  weight: distance  leaf size: 3
iteration: 11  score: 0.8666666666666666  NN: 3  weight: distance  leaf size: 4
iteration: 12  score: 0.7533333333333333  NN: 4  weight: uniform  leaf size: 2
iteration: 13  score: 0.7533333333333333  NN: 4  weight: uniform  leaf size: 3
iteration: 14  score: 0.7533333333333333  NN: 4  weight: uniform  leaf size: 4
iteration: 15  score: 0.8666666666666666  NN: 4  weight: distance  leaf size: 2
iteration: 16  score: 0.8666666666666666  NN: 4  weight: distance  leaf size: 3
iteration: 17  score: 0.8666666666666666  NN: 4  weight: distance  leaf size: 4
iteration: 18  score: 0.8  NN: 5  weight: uniform  leaf size: 2
iteration: 19  score: 0.8  NN: 5  weight: uniform  leaf size: 3
iteration: 20  score: 0.8  NN: 5  weight: uniform  leaf size: 4
iteration: 21  score: 0.8  NN: 5  weight: distance  leaf size: 2
iteration: 22  score: 0.8  NN: 5  weight: distance  leaf size: 3
iteration: 23  score: 0.8  NN: 5  weight: distance  leaf size: 4
Highest score:  0.9
Best parameters:  {' n_neighbors': 2, 'weights': 'distance', 'leaf_size': 4}
```

FIGURE 7.8   k-nearest neighbors model performance after 24 iterations.

production problem is done with Python with the following codes. Outputs from the best model are calculated using all the data set and compared with actual target values using the classification report.

```
#Grid Search with Cross-validation
from sklearn.model_selection import GridSearchCV
Neighbor=[2,3,4,5]
Weight=['uniform','distance']
Leaf=[2,3,4]
hyperparameters=dict(n_neighbors=Neighbor,
weights=Weight,leaf_size=Leaf)
KNN=KNeighborsClassifier()
gridsearch=GridSearchCV(KNN, hyperparameters, cv=5)
Best_Model=gridsearch.fit(xnorm, y)
print('Best n_neighbors:', Best_Model.best_estimator_.get_params
  ()['n_neighbors'])
print('Best weights:', Best_Model.best_estimator_.get_params()
  ['weights'])
print('Best leaf_size:', Best_Model.best_estimator_.get_params()
  ['leaf_size'])
B=Best_Model.predict(xnorm)
from sklearn.metrics import classification_report
  print(classification_report(y, B))
Python output=Fig. 7.9
```

```
Best n_neighbors: 2
Best weights: distance
Best leaf_size: 2
```

	precision	recall	f1-score	support
0	1.00	1.00	1.00	8
1	1.00	1.00	1.00	21
accuracy			1.00	29
macro avg	1.00	1.00	1.00	29
weighted avg	1.00	1.00	1.00	29

FIGURE 7.9 Grid search results for the sand production problem.

According to the grid search results, the best model is obtained using hyperparameters of the number of neighbors 2, weights distance, and leaf_size 2. Sometimes, when the number of hyperparameters' combinations is high, the grid search can be computationally exhaustive. In this case, it can be more efficient to search within a random set of hyperparameter combinations. In Python, "RandomizedSearchCV" can be used from sklearn library for randomized hyperparameter search. For this method (RandomizedSearchCV) similar to normal grid search, model type, hyperparameter space, and the number of cross-validation folds should be passed in as input parameters. Additionally, it is necessary to define the number of iterations which is equal to the number of randomly selected hyperparameter combinations. The default value for the number of iterations is 10. Python code for randomized hyperparameter search for sand production problem is as follows:

```
from sklearn.model_selection import RandomizedSearchCV
Neighbor=[2,3,4,5]
Weight=['uniform','distance']
Leaf=[2,3,4]
hyperparameters=dict(n_neighbors=Neighbor,
weights=Weight,leaf_size=Leaf)
KNN=KNeighborsClassifier()
gridsearch_randomized=RandomizedSearchCV(KNN,
hyperparameters,n_iter=12,random_state=1, cv=5)
Best_Model=gridsearch_randomized.fit(xnorm, y)
print('Best n_neighbors:', Best_Model.best_estimator_.get_params
   ()['n_neighbors'])
print('Best weights:', Best_Model.best_estimator_.get_params()
   ['weights'])
print('Best leaf_size:', Best_Model.best_estimator_.get_params()
   ['leaf_size'])
BRG=Best_Model.predict(xnorm)
from sklearn.metrics import classification_report
print(classification_report(y, BRG))
Python output=Fig. 7.10
```

```
Best n_neighbors: 2
Best weights: distance
Best leaf_size: 4
                precision    recall   f1-score    support

            0       1.00      1.00       1.00          8
            1       1.00      1.00       1.00         21

     accuracy                           1.00         29
    macro avg       1.00      1.00       1.00         29
 weighted avg       1.00      1.00       1.00         29
```

FIGURE 7.10 Random grid search results for the sand production problem.

With a randomized search, based on 12 iterations, the best model is achieved using hyperparameters with 2 number of neighbors, weights distance, and leaf_size of 4.

In the previous example, a grid search was applied for a binary classification problem. Next, we use a grid search to optimize a model for a multiclass classification problem, an example of facies classification. In facies classification example, "logistic regression" was used as the classification model. Two hyperparameters as "penalty" and "inverse of regularization strength" denoted as C would be used for the grid search. The penalty is considered to be l1 or l2. For hyperparameter C, 20 values from 1 to 10000 in log scale are selected. In the search process, the goal is to find the best combination of penalty and C providing the model with the highest accuracy. Grid search for facies classification problem can be done in Python with the following codes:

```
from sklearn.model_selection import GridSearchCV
penalty=['l1','l2']
C=np.logspace(0,4,20)
LRG=LogisticRegression(multi_class='auto',solver='liblinear',
  max_iter=200)
hyperparameters=dict(C=C,penalty=penalty)
gridsearch=GridSearchCV(LRG,hyperparameters, cv=5, verbose=0)
Best_Model=gridsearch.fit(xnorm, y)
print('Best penalty:', Best_Model.best_estimator_.get_params()
  ['penalty'])
print('Best C:', Best_Model.best_estimator_.get_params()['C'])
B=Best_Model.predict(x_test)
  from sklearn.metrics import classification_report
print(classification_report(y_test, B))
Python output=Fig. 7.11
```

Based on grid search results, the best model is achieved by having an l2 penalty and C equal to 48.3. The average precision of the optimized model (after grid search) is 0.58. It is also possible to use a randomized search if the number

```
Best penalty: l2
Best C: 48.32930238571752
```

	precision	recall	f1-score	support
1	0.60	0.58	0.59	62
2	0.59	0.68	0.63	185
3	0.66	0.56	0.60	156
4	0.56	0.48	0.52	46
5	0.33	0.08	0.13	50
6	0.53	0.67	0.59	113
7	0.44	0.50	0.47	22
8	0.58	0.66	0.62	131
9	0.79	0.63	0.70	43
accuracy			0.59	808
macro avg	0.57	0.54	0.54	808
weighted avg	0.58	0.59	0.58	808

FIGURE 7.11 Classification report for facies classification example using grid search.

of combinations is high. The last example for the hyperparameter grid search is the PVT problem which uses a neural network for regression. As mentioned in Chapter 6, "MLPRegressor" was used as a neural network training algorithm. For the PVT grid search optimization, the number of neurons in the hidden layer, activation function, penalty parameter (alpha), and initial learning rate were selected as hyperparameters to be tuned. The best model will be returned which is used when the best model is selected and the model's output is calculated for the whole data set. Five-fold cross-validation will be used to assess the performance of the best model using R^2 as the metric. Optimum combination of hyperparameters that are the result of the grid search optimization can be accessed by "best_estimator_". Python code for performing the grid search optimization for PVT estimation example is as follows:

```
from sklearn.model_selection import GridSearchCV
from sklearn.neural_network import MLPRegressor
import numpy as np
seed=50
np.random.seed(seed)
hyperparameters=[{'hidden_layer_sizes': [2,3,4,5,6,7],
  'activation':['relu','tanh'],'solver':['lbfgs'], 'alpha':
  [0.0001,0.001,0.01,0.1,1,10], 'batch_size':['auto'], 'learning_rate':
  ['constant'], 'learning_rate_init':[0.001,0.01,0.1,1], 'max_iter':
  [500]}]
MLPR=MLPRegressor()
gridsearch=GridSearchCV(MLPR, hyperparameters, cv=5, verbose=0)
Best_Model=gridsearch.fit(Xnorm, y_norm_Ravel)
print('hidden_layer_sizes:',
  Best_Model.best_estimator_.get_params()['hidden_layer_sizes'])
```

```
print('Best activation:',
   Best_Model.best_estimator_.get_params()['activation'])
print('Best alpha:', Best_Model.best_estimator_.get_params()['alpha'])
print('Best learning_rate_init:',
   Best_Model.best_estimator_.get_params()['learning_rate_init'])
from sklearn.model_selection import cross_val_score
y_norm_Ravel=ynorm.values.ravel()
scores_R2=cross_val_score(Best_Model.best_estimator_,Xnorm,
   y_norm_Ravel,cv=5,scoring='r2')
print(" R2_Cross-validation scores: {}". format(scores_R2))
print(" Average R2_Cross-validation scores: {}". format(scores_R2.
   mean()))
```

Python output:
hidden_layer_sizes: 7
Best activation: tanh
Best alpha: 0.0001
Best learning_rate_init: 1
R2_Cross-validation scores: [0.99381755 0.99472335 0.99585629
   0.9951116 0.99433003]
Average R2_Cross-validation scores: 0.9947677643426687

Based on grid search results, it is observed that the best neural network model was trained by 7 neurons in the hidden layer, the best activation function of tanh, alpha equal to 0.0001, and an initial learning rate of 1. By performing cross-validation over the whole PVT data set using the best model, average R^2 for five-fold is 0.99 which is very high. This grid search needed a long run time to be completed. As mentioned earlier, in order to perform a grid search in a shorter time, it is possible to perform a randomized grid search. Applying a randomized grid search for PVT estimation problem can be done with the following Python codes:

```
from sklearn.model_selection import RandomizedSearchCV
import numpy as np
seed=50
np.random.seed(seed)
hyperparametersR={'hidden_layer_sizes': [2,3,4,5,6,7],
   'activation': ['relu','tanh'],'solver':['lbfgs'], 'alpha':
   [0.0001,0.001,0.01,0.1,1,10], 'batch_size':['auto'], 'learning_rate':
   ['constant'], 'learning_rate_init':[0.001,0.01,0.1,1], 'max_iter':
   [200]}
MLPR=MLPRegressor()
from sklearn.neural_network import MLPRegressor
seed=50
np.random.seed(seed)
gridsearchR=RandomizedSearchCV(MLPR, hyperparametersR,
   random_state=1, cv=5, verbose=0,n_jobs=-1)
   Best_ModelR=gridsearchR.fit(Xnorm, y_norm_Ravel)
print('hidden_layer_sizes:',
   Best_ModelR.best_estimator_.get_params()
   ['hidden_layer_sizes'])
```

```
print('Best activation:',
  Best_ModelR.best_estimator_.get_params()['activation'])
print('Best alpha:', Best_ModelR.best_estimator_.get_params()['alpha'])
print('Best learning_rate_init:',
  Best_ModelR.best_estimator_.get_params()['learning_rate_init'])
from sklearn.model_selection import cross_val_score
scores_R2R=cross_val_score (Best_ModelR.best_estimator_, Xnorm,
  y_norm_Ravel,cv=5,scoring='r2')
print(" R2_Cross-validation scores: {}". format(scores_R2R))
print(" Average R2_Cross-validation scores: {}". format(scores_R2R.
  mean()))
```
Python output:
hidden_layer_sizes: 6
Best activation: tanh
Best alpha: 0.001
Best learning_rate_init: .01
R2_Cross-validation scores: [0.99458316 0.99552709 0.995924
  0.9933157 0.99023985]
Average R2_Cross-validation scores: 0.9939179607489855

With a randomized grid search, the best neural network model is trained by 6 neurons in the hidden layer, tanh activation function, alpha equal to 0.001, and initial learning rate of 0.01. These values are slightly different from what was found by the grid search as expected. The average R^2 for fivefold cross-validation is 0.99. The run time for the randomized grid search decreased significantly since it just went through 10 iterations (compared with 624 for GridSearchCV).

Model selection

In the previous section, the grid search was applied for just one model to find the hyperparameters that return the best performance model. For many pattern recognition, classification, clustering, and regression problems, it is possible to use different machine learning algorithms or models. The question here is which model and combination of hyperparameters would perform the best on a specific data set. In this case, the search space includes different learning algorithms (models) and hyperparameters.

Facies classification problem would be used to illustrate model selection more clearly. In the facies classification problem, logistic regression was used to classify rock types into nine different groups. In the previous section, the grid search resulted in the average precision for the best logistic regression model as 0.58. This time, in the search space, in addition to logistic regression, random forest classifier is also included. For random forest classifier, the number of estimators and classifier's maximum number of features is the hyperparameters in the search space. For logistic regression, "penalty and regularization strength" (C) is in the search space. The objective here is to discover which learning algorithm and combination of hyperparameters returns the best model. Python codes for model selection and hyperparameter optimization of facies classification problem is as follows:

```python
import numpy as np
from sklearn.linear_model import LogisticRegression
from sklearn.ensemble import RandomForestClassifier
from sklearn.model_selection import GridSearchCV
from sklearn.pipeline import Pipeline
np.random.seed(50)
pipe = Pipeline([("clf", LogisticRegression(solver='liblinear',
  max_iter=200))])

log_classifier = {"clf": [LogisticRegression(solver='liblinear',
  max_iter=200)],
            "clf__penalty": ['l1', 'l2'],
            "clf__C": np.arange(0.1, 100, 20)}

ranforest_classifier = {"clf": \
   [RandomForestClassifier(random_state=0)],
            "clf__n_estimators": np.arange(10, 300, 10),
            "clf__max_features": [1, 2, 3]}

grid = [ranforest_classifier,log_classifier]

gridsearch_models = GridSearchCV(pipe, grid, cv=5, verbose=0,
  n_jobs=-1)
Best_modelM = gridsearch_models.fit(x_train, y_train)

print(Best_modelM.best_estimator_.get_params()["clf"])
y_BM=Best_modelM.predict(x_test)
from sklearn.metrics import classification_report
print(classification_report(y_test, y_BM))
Python output:
RandomForestClassifier(max_features=2, n_estimators=80,
  random_state=0)
```

1	0.81	0.71	0.76	62
2	0.71	0.78	0.74	185
3	0.75	0.73	0.74	156
4	0.79	0.72	0.75	46
5	0.64	0.42	0.51	50
6	0.70	0.77	0.73	113
7	0.68	0.77	0.72	22
8	0.75	0.76	0.76	131
9	0.95	0.91	0.93	43
accuracy			0.74	808
macro avg	0.75	0.73	0.74	808
weighted avg	0.74	0.74	0.74	808

Classification report for facies classification example using model selection and grid search

From grid search results, it is observed that the random forest classifier out-performs logistic regression in the classification performance of facies classification problem. The best model was achieved by setting the number of estimators to 80 and a maximum feature of 2. The best model was used to predict different classes on test data set and compared with target values. Based on classification report results, the weighted average precision for nine classes using random forest classifier is 0.74 which is much better than logistic regression.

Partial dependence plots

When a machine learning model is trained, and its prediction performance is verified with evaluation techniques like cross-validation, it is possible to use the model for prediction or sensitivity analysis. It is sometimes required to understand what is the impact of one or several input attributes on the target value. One way is to create a matrix of all the input attributes and keep the value of each attribute equal to average or any other desired value. Then, vary the one specific attribute that is required for sensitivity analysis within the desired range. Finally, by using the matrix of input data (with one variable attribute) and predicting the model response, the dependency between target and input attribute can be identified.

In Python, the interaction between the target and an input variable can be visualized by "partial dependence plots (PDPs)." What PDP does, is to plot the average of the model's outcome over multiple combinations of input variables versus a specific input. In order to visualize the relationship between the target and two input variables, heat map, contour lines, or 3D plots can be used (scikit).

In Chapter 6, the neural network was used to correlate gas EUR to 13 different geology, drilling, and completion variables. In this chapter, PDPs are used to find the dependency between gas EUR and some important input variables. In Chapter 6, based on feature ranking analysis, it was determined that lateral length and proppant loading had the highest impact on the EUR. The partial dependence relations for pairs of some input variables and gas EUR can be plotted in Python by the following code:

```
import numpy as np
import pandas as pd
import matplotlib.pyplot as plt
from sklearn.pipeline import make_pipeline
from sklearn.neural_network import MLPRegressor
from sklearn.inspection import partial_dependence
from sklearn.inspection import plot_partial_dependence
from sklearn.preprocessing import MinMaxScaler
import seaborn as sns
dataset=pd.read_csv('Chapter7_Shale Gas Wells.csv')
X=dataset.iloc[:,0:13]
y=dataset.iloc[:,13].values
seed=15
np.random.seed(seed)
```

```
from sklearn.model_selection import train_test_split
X_train, X_test, y_train, y_test=train_test_split(X, y, test_size
  =0.25)
model=make_pipeline(MinMaxScaler(), MLPRegressor
  (hidden_layer_sizes=(25,25),
                        learning_rate_init=0.01,
                        early_stopping=True,max_iter=500))
model.fit(X_train, y_train)
print("Test R2 score: {:.2f}".format(model.score(X_test, y_test)))
features=['Stage Spacing','bbl/ft','Proppant Loading','Dip',
  'Thickness','Lateral Length','Injection Rate','Porosity',
  'Percentage of LG']
plot_partial_dependence(model,X, features,
  n_jobs=3, grid_resolution=20)
fig = plt.gcf()
fig.suptitle('Partial dependence of Gas EUR\n' 'for Stage Spacing,bbl/ft,
  Proppant Loading,Dip,Thickness,Lateral Length,Injection Rate,
  Porosity,Percentage of LG')
fig.subplots_adjust(hspace=0.3)
Python output=Fig. 7.12 Test R2 score: 0.79
```

Note: The three lines of code beginning with "fig.suptitle" should be in the same line in Jupyter.

The pipeline was used to connect the data preprocessing step (Min-MaxScaler) and the model (MLPRegressor) and assemble them together. The assembly eliminates the need to scale the data since MinMaxScaler is part of the process within the pipeline. As shown in Fig. 7.12, there is an increasing

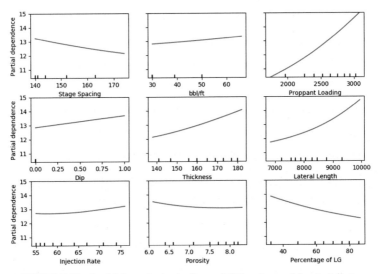

FIGURE 7.12 Partial dependence plot for gas EUR and several input attributes.

trend between gas EUR versus bbl/ft, proppant loading, Dip, thickness, and injection rate. As stage spacing and percentage of linear Gel decrease, gas EUR increases. Finally, porosity does not show a significant impact on gas EUR. It is also important to look at the interaction between two input attributes and the impact on the target output. The impact of the lateral length and proppant loading on gas EUR can be shown with a heat map using the following code:

```
features_LLPL=[('Lateral Length','Proppant Loading')]
plot_partial_dependence(model,X, features_LLPL, n_jobs=3,
  grid_resolution=20)
fig=plt.gcf()
fig.suptitle('Gas EUR vs. Lateral Length and Proppant Loading')
Python output=Fig. 7.13
```

As shown in Fig. 7.13, it is possible to find gas EUR at any combination of proppant loading and lateral length. Based on the heat map, it is observed that higher gas EURs are achieved by drilling longer wells and pumping more proppant. Besides heat maps, the impact of the interaction between proppant loading and lateral length on the gas EUR can be depicted by "3D interaction plots." The Python code for showing two attributes, interactions in three dimensions is as follows. In Fig. 7.14, 3D surface of the EUR versus different values of lateral length and proppant loading is depicted.

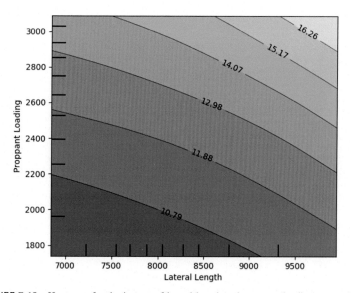

FIGURE 7.13 Heat map for the impact of lateral length and proppant loading on gas EUR.

```
from mpl_toolkits.mplot3d import Axes3D
fig=plt.figure()
features1=('Proppant Loading', 'Lateral Length')
pdp, axes=partial_dependence(model, X_train, features=features1,
  grid_resolution=30)
XX, YY=np.meshgrid(axes[0], axes[1])
Z=pdp[0].T
ax=Axes3D(fig)
surf=ax.plot_surface(XX, YY, Z, rstride=1, cstride=1, cmap=plt.cm.
  BuPu, edgecolor='k0')
ax.set_xlabel(features1[0])
ax.set_ylabel(features1[1])
ax.set_zlabel('Gas EUR')
ax.view_init(elev=30, azim=150)
plt.colorbar(surf)
plt.suptitle('Partial dependence of Gas EUR vs.\n' 'Lateral Length
  and Proppant Loading')
plt.subplots_adjust(top=1)
plt.show()
Python output=Fig. 7.14
```

Note: The two lines of code beginning with "plt.suptitle" should be in the same line in Jupyter.

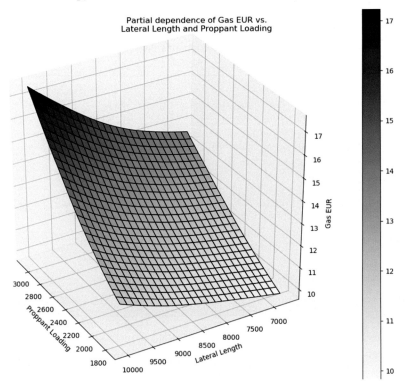

FIGURE 7.14 3D Plot showing the interaction between lateral length/proppant loading and gas EUR.

Size of training set

The reliability of a machine learning model depends on multiple factors like problem complexity (number of inputs, outputs, and interconnection), model selection, proper hyperparameters, and the amount of data available for training/testing. There are two important questions about the size of the required data for a reliable machine learning model. The first one is what is the minimum number of samples or observations that result in a high-performance model. The second question is whether adding new samples improves the performance of the model (generally, having more samples improves the model quality). For example, for the gas EUR prediction problem, one can ask: What is the minimum number of well data required for training a model with reliable prediction? Or what is the minimum number of PVT samples that are required to have a robust prediction for bubble point pressure? This can be a challenging question when the number of wells in the data set is important. For the green fields when the number of wells is low, it is very important to identify the minimum number of wells to make reliable models.

There are some statistical methods that can be used to specify the minimum number of samples in the data set using hypothesis testing and a confidence interval. However, in machine learning problems, the complexity of the model (number of hyperparameters or model's type) is very important. To evaluate the impact of a training sample size on the model performance, "Learning Curve" can be used (Albon, 2018). In order to use "learning_curve" in Python, one must specify the machine learning model, input and output (target) data, a range for training data samples, the number of folds for cross-validation, and finally score metric. After specifying these parameters, the Learning_curve algorithm starts with minimum sample size, splits it into folds, trains the model, and finally measures the score for train and test folds. Then, it increases the number of samples and repeats the mentioned procedure until it reaches a maximum number of samples in the range. At the end, the performance of the model (on both train and test data) versus the number of samples in the data can be visualized. As an example, Learning_curve is applied to the PVT problem. Python codes for using Learning Curve for PVT problem is as follows:

```
import numpy as np
import pandas as pd
from sklearn.model_selection import learning_curve
from sklearn.neural_network import MLPRegressor
import matplotlib.pyplot as plt
from sklearn.preprocessing import MinMaxScaler
dataset=pd.read_csv('Chapter7_PVT Data.csv')
X=dataset.iloc[:,0:4]
y=dataset.iloc[:,4].values
sc=MinMaxScaler()
Xnorm=pd.DataFrame(data=sc.fit_transform(X))
seed=50
```

```
np.random.seed(seed)
clfB = MLPRegressor(activation='relu',alpha=0.01,
    hidden_layer_sizes=7, learning_rate_init=0.01,
        max_iter=500, solver='lbfgs',random_state=20)
train_sample_sizes, train_MSEscores,test_MSEscores\
    =learning_curve(clfB,Xnorm,y,cv=5,

scoring='neg_mean_absolute_error',n_jobs=-1,
train_sizes=np.linspace(0.1,1.0,100),random_state=10)
train_MSEmean = np.mean(train_MSEscores, axis=1)
test_MSEmean = np.mean(test_MSEscores, axis=1)
plt.plot(train_sample_sizes, -train_MSEmean, color="k",
    label="Training MSE score")
plt.plot(train_sample_sizes, -test_MSEmean,'--', color="k",
    label="Cross-validation MSE score")
plt.title("Learning Curve for PVT Data Set")
plt.xlabel("Training Set Sample Size"), plt.ylabel("MSE Score"),
plt.legend(loc="best")
plt.tight_layout()
plt.show()
Python output=Fig. 7.15
```

As it is shown in Fig. 7.15, when the number of samples in the training data set increases the average MSE for training does not change significantly while the average MSE for the cross-validation decreases. When the number of samples in the training data set reaches 120, the average MSE for the cross-validation flattens to 105. It means that to have the highest MSE or the best

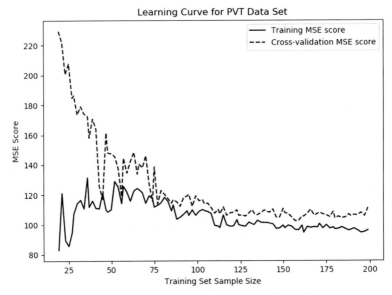

FIGURE 7.15 Learning curve for PVT estimation problem.

model (neural network) performance for PVT estimation problem, it is necessary to have at least 120 PVT samples. Adding more will not improve the model's performance significantly.

Save-load models

In this chapter, different methods to evaluate, validate, and optimize multiple machine learning algorithms were discussed. The final goal was to make the best performing model. When the best model is trained, it is necessary to save it as a file for further usage like a prediction on a new data set. When a scikit-learn model is built, it is possible to save it as a "pickle" file (which is a Python file format) or using "joblib library" as a ".joblib file" in the jupyter directory. Then, the saved model can be loaded to make predictions. For facies classification example, a random forest classifier is trained using the training data set by optimized hyperparameters from the grid search. Afterward, the model is saved. Finally, the saved model will be loaded in other Python files to make a prediction on a test data set. Saving a trained model for facies classification problem can be done in Python using the following codes:

```
import pandas as pd
from sklearn.pipeline import make_pipeline
from sklearn.preprocessing import MinMaxScaler
from sklearn.ensemble import RandomForestClassifier
dataset=pd.read_csv('Chapter7_Facies Data.CSV')
x=dataset.iloc[:,4:11]
y=dataset.iloc[:,0].values
  import numpy as np
seed=50
np.random.seed(seed)
from sklearn.model_selection import train_test_split
```

	precision	recall	f1-score	support
1	0.92	0.73	0.82	82
2	0.73	0.86	0.79	215
3	0.78	0.71	0.74	181
4	0.61	0.61	0.61	51
5	0.60	0.38	0.47	68
6	0.60	0.60	0.60	142
7	0.81	0.64	0.71	33
8	0.60	0.76	0.67	139
9	0.98	0.75	0.85	59
accuracy			0.71	970
macro avg	0.74	0.67	0.69	970
weighted avg	0.72	0.71	0.71	970

FIGURE 7.16   Random forest classification prediction of the test data set for facies problem using saved model.

```
x_train, x_test, y_train, y_test=train_test_split(x, y,
  test_size=0.3)
model = make_pipeline(MinMaxScaler(),RandomForestClassifier
  (max_features=2, n_estimators=80, random_state=0),)
rfmodel1=model.fit(x_train,y_train)
from joblib import dump
dump(rfmodel1,"rfmodel1.joblib")
```
Python output:
['rfmodel1.joblib']

```
import pandas as pd
from joblib import load
dataset=pd.read_csv('Chapter7_Facies Data.CSV')
x=dataset.iloc[:,4:11]
y=dataset.iloc[:,0].values
import numpy as np
seed=50
np.random.seed(seed)
from sklearn.model_selection import train_test_split
x_train, x_test, y_train, y_test=train_test_split(x, y,
  test_size=0.3)
Model=load('rfmodel1.joblib')
yModel=Model.predict(x_test)
from sklearn.metrics import classification_report
print(classification_report(y_test, yModel))
```
Python output: Fig 7.16

References

Albon, C. (April 10, 2018). *Machine learning with python cookbook: Practical solutions from preprocessing to deep learning* (1st ed.). O'Reilly Media. ISBN-10 : 9781491989388.

Dubois, M. K., Bohling, G. C., & Chakrabarti, S. (2007). Comparison of four approaches to a rock facies classification problem. *Computers & Geosciences, 33*(5), 599−617. https://doi.org/10.1016/j.cageo.2006.08.011.

Gharagheizi, F., Mohammadi, A. H., Arabloo, M., & Shokrollahi, A. (June 2017). Prediction of sand production onset in petroleum reservoirs using a reliable classification approach. *Petroleum, 3*(2), 280−285. https://doi.org/10.1016/j.petlm.2016.02.001.

https://scikit-learn.org/stable/auto_examples/inspection/plot_partial_dependence.html.

Moricca, G., Ripa, G., Sanfilippo, F., & Santarelli, F. J. (1994). *Basin scale rock mechanics: Field observations of sand production.* Delft, Netherlands: Rock Mechanics in Petroleum Engineering.

Muller, A. C., & Guido, S. (October 25, 2016). *Introduction to machine learning with python: A guide for data scientists* (1st ed.). O'Reilly Media. ISBN-10 : 1449369413.

Chapter 8

Fuzzy logic

In almost every complex problem, it is required to deal with uncertainty. The logic that has been used for most of the last 2000 years to address uncertainty in the problems comes from Aristotle's logic or Stoic logic. In this logic, binary faith resulted in one important rule that was either the fact or not the fact. This logic considered the world in a black and white or bivalent manner that, for example, says the sky is blue or not blue (Kosko, 1994). Gradually, from the 17th century, philosophers and scientists realized that some scientific statements cannot be completely true or false. They started to move from deterministic (black and white) statements to gray or fuzzy ones.

The Stoic logic was improved and formulated in an algebraic way by George Boole as Boolean Logic. His works also included the illustration of using probability theory to analyze large quantities of social data. Decades after Boole had published the *Laws of Thought*, in which he covered theories of logic and probabilities, Georg Cantor, a German mathematician, introduced Cantor set or "Set Theory." The works of Boole and Cantor were the foundation of probability theory, very powerful in dealing with uncertainties in problems. Probability theory can mostly be applied to the problems that are governed by randomness. For cases that partial unreliable information is available or there are imprecision and vagueness in language that defines the problem, probability theory is not very practical. In these cases, uncertainty and vagueness should be handled with a different technique (Rajasekaran & Pai, 2013).

From the beginning of the 20th century, many scientists like Charles Sanders Peirce, Jan Lukasiewicz, and Max Black started considering vagueness in the world of logic. They came up with many-valued logic (it assumes there are more than two truth values) and vague sets. In the 1960s, Lotfi A. Zadeh, professor at UC Berkley, presented his famous paper about Fuzzy Sets (Zadeh, 1965). In his work, he introduced membership functions to cover the range of every possible value. Then, he proposed fuzzy set operations to which is analogous to classical set operations. Zadeh's works and publications are the basis of fuzzy logic. Since his work, fuzzy logic has been developed considerably. Japanese engineers and scientists utilized the capabilities of fuzzy logic in a wide variety of systems and products with more than thousands of patents (Kosko, 1994). Fuzzy logic has been used extensively for system/process control, pattern recognition and classification, decision-making, optimization, and forecasting/prediction in a wide variety of industries and disciplines.

Machine Learning Guide for Oil and Gas Using Python. https://doi.org/10.1016/B978-0-12-821929-4.00003-2
Copyright © 2021 Elsevier Inc. All rights reserved.

There has been a lot of application for fuzzy logic in the oil and gas industry since four decades ago in various fields such as drilling, petrophysics, well stimulation design, system control, field development optimization, reservoir characterization, production allocation, etc. Mohaghegh was one of the pioneers in using fuzzy logic in the oil and gas industry. He used fuzzy logic in conjunction with neural networks and genetic algorithms to select the best candidate for well restimulation in Green River Basin (Mohaghegh, 2000). He also applied fuzzy logic for pattern recognition in unconventional resources and mapping natural fracture systems (Mohaghegh, 2017). Gholami et al. used fuzzy rock typing to upscale high-resolution geological models for flow simulation (Gholami & Mohaghegh, 2009). Xion and Holditch showed how to design optimal stimulation treatment and developed fuzzy evaluators for treatment type, barrier condition, injection method, and formation damage diagnostics (Xiong & Holditch, 1995). Rivera et al. used fuzzy logic to design a pressure control system for a fracturing facility, considering a wide range of fluid types and pumping conditions (Rivera, 1994). Aminzadeh et al. implemented fuzzy logic with neural networks to rank hydrocarbon reservoirs and identify high-grade prospects (Aminzadeh & Brouwer, 2006). Ahmed et al. utilized fuzzy logic to estimate the rate of penetration for a drilling bit using surface drilling parameters and drilling fluid properties (Ahmed et al., 2019). Fuzzy logic was also used to estimate rock strength by rock masses (Sari, 2016), production allocation in commingled production wells (Widarsono et al., 2005), and choke size control for production systems (Odedele & Ibrahim, 2014).

In this chapter, the classical set theory will be briefly discussed to prepare the reader for the fuzzy set theory. For the fuzzy sets, membership functions, fuzzy set operations, and approximate reasoning would be explained. Finally, fuzzy inference and fuzzy clustering will be discussed with oil and gas—related examples like choke control and rock-type classification.

Classical set theory

Classical sets are a collection of unique items usually denoted by curly brackets {}. RockType = {Sandstone, Limestone, Carbonate, Shale}. Rock-Type set includes different rock types.

As shown in Fig. 8.1, a Venn diagram can be used to visualize a set and shows its relationship with the other sets. Each circle represents a set; mutual members are shown by overlapping part of the circles. *Universal Set* denoted as U includes all the elements in the context which are being modeled using sets. Any subset of U can be considered as a set in the universe.

Membership of Elements of an object x is in a set A and is shown as $x \in A$. For example, Limestone $\in$ RockType means Limestone is a member of RockType set. Cardinality of sets is the number of items inside a set. Set cardinality is denoted as $|A|$ for a set A. For example, when cardinality of RockType is 4, $|RockType| = 4$. Null set is a set with no items inside of it. For example, A = {} shows a null set with cardinality of $|A| = 0$. Usually null sets are denoted as ϕ. Singleton set is a set

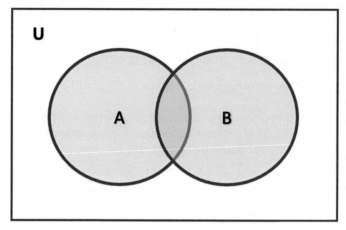

FIGURE 8.1 Venn diagram and union.

with cardinality of 1. For example, if A = {earth} is the set of all planets with life on them in solar system, |A| = 1 and is a singleton set. Subset and superset of a set is defined by choosing a number of items of a set. Choosing a number of items of set *A* defines a new set *B* that is a subset of A, and set A will be considered as superset of *B*. Assume a selected set of car companies denoted as *C* = {Toyota, Buick, Jeep, Mazda, Mercedes, BMW, Cadillac, Ford, Honda}, we can define a subset of *C* as Japanese car companies J = {Mazda, Toyota, Honda}. Subset relation is shown as J⊂C. J is a subset of *C* and *C* is considered as a superset of J which contains all the items in J.

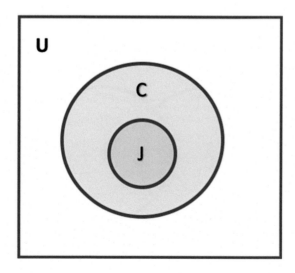

Subset and superset.

Set operations

Union of two sets includes all the elements inside both sets. For example,

$$A = \{a,\, b,\, c\}, \quad B = \{b,\, d\}, \quad A \cup B = \{a,\, b,\, c,\, d\}$$

Union can be shown using a Venn diagram as illustrated in the shaded area of Fig. 8.1.

Intersection of two sets includes the common items in both sets. Below please find one example of intersection between A and B and its corresponding Venn diagram in Fig. 8.2 where the shaded area is showing $A \cap B$.

$$A = \{a,\, b,\, c\}, \quad B = \{b,\, d\}, \quad A \cap B = \{b\}$$

Complement of a set A includes all elements in the universe U except the ones in set A. Complement of set A is shown as A^c. Assume a set of Japanese car companies J = {Mazda, Toyota, Honda}, the Jc is the set of all car companies except Mazda, Honda, and Toyota. The J^c which is the complement of set J can be seen in Fig. 8.3 as the shaded area.

Note that the intersection of a set and its complement is a null set, and the union of a set and its complement is the universe $J \cap J^c$ and $J \cup J^c = U$.

Difference of set A with respect to set B is shown as A-B which includes all elements of A that are not an element of B. $A - B = \{a,c\}$ and $B - A = \{d\}$. The Venn diagram of $A - B$ can be seen in Fig. 8.4.

Set properties

Law of Commutativity: Union and intersection operations are commutative, meaning the order of sets does not change the result.

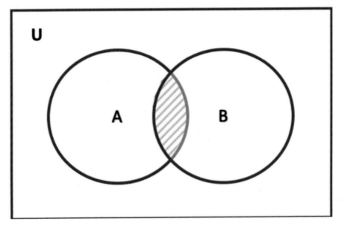

FIGURE 8.2 Intersection.

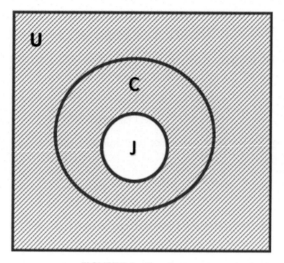

FIGURE 8.3   Complement.

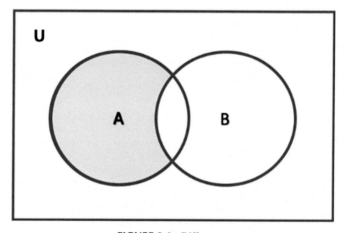

FIGURE 8.4   Difference.

$$A \cup B = B \cup A$$
$$A \cap B = B \cap A$$

(8.1)

Law of Associativity: Union and intersection operations performed on three sets follow associativity such that the order of applying the operation does not matter, note that parenthesis has the highest precedence so the operation in the parenthesis is done first:

$$A\cup(B\cup C)=(A\cup B)\cup C$$
$$A\cap(B\cap C)=(A\cap B)\cap C \tag{8.2}$$

Law of Distributivity: When applying union and intersection on three sets, the first operation can be distributed as follows:

$$A\cup(B\cap C)=(A\cup B)\cap(A\cup C)$$
$$A\cap(B\cup C)=(A\cap B)\cup(A\cap C) \tag{8.3}$$

Idempotent Law: This law states that union or intersection of a set with itself will be the set.

$$A\cup A=A$$
$$A\cap A=A \tag{8.4}$$

Identity Law: Considering a null set and a universe set U, the following holds for operations on set A, ϕ, E.

$$A\cup\Phi=A$$
$$A\cup U=U$$
$$A\cap\Phi=\Phi$$
$$A\cap U=A \tag{8.5}$$

Absorption Law: The union of a set A with the intersection of the same set A with some other set B is A. The same holds for the intersection of a set A with the union of set A with some other set B.

$$A\cup(A\cap B)=A$$
$$A\cap(A\cup B)=A \tag{8.6}$$

Involution Law: Applying the complement on a set two times will result in the set itself.

$$(A^c)^c=A \tag{8.7}$$

Transitivity Law: Transitivity law holds for the subset relationship, if a set A is a subset of set B which is a subset of C, then set A will be a subset of C.

$$if\ A\subset B\ and\ B\subset C\ then\ A\subset C \tag{8.8}$$

De Morgan Law: The complement of union of two sets can be calculated as the intersection of the complement of those sets. Also, the complement of intersection of two sets can be calculated as the union of the complement of those sets.

$$(A\cup B)^c=A^c\cap B^c$$
$$(A\cap B)^c=A^c\cup B^c \tag{8.9}$$

Fuzzy set

Definition

In the classical set theory, it was explained that there are two groups in the universe of discourse. One group considers all the members of the universe, and the other group includes nonmembers. Imagine a universe that includes reservoir rocks as shale and sandstone. Therefore, in this universe, any rock should be either shale or sandstone to be a member. Any other rock that contains, for example, 80% sandstone and 20% limestone is not considered a member. In the classical set, membership values to the universe are crisp and considered 0 for nonmember and 1 for the member. However, in the fuzzy set, it is possible to assume partial membership and assign varying degrees of membership to a set. It should be noted that fuzzy sets are mainly used to deal with uncertainty and precision. The uncertainty is somehow related to linguistic variables such as old, young, tall, short. Fuzzy sets can be used to quantify and analyze linguistic variables. For example, consider the age and linguistic variables old and young. Let's assume people who are below 45 years as young and above 45 years as old. In this example, the universe of discourse includes young and old people. Depending on the age, a person is either old or not-old which represents a membership of 0 or 1 (Fig. 8.5).

This kind of crisp cut is not always defendable, as one might argue, the threshold should be 50 instead of 45. Since the concept of becoming an elderly happens during time rather than at one specific age, so the old variable can be viewed as a fuzzy concept and showed as a curve rather than a line.

As illustrated in Fig. 8.6, a curve is plotted for two fuzzy sets young and old over the universe of age defined as [0,100]. There are different ways to plot this curve. Here, it is assumed that membership to the old set starts around age

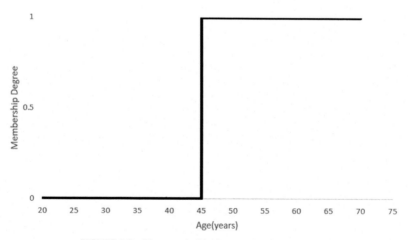

FIGURE 8.5 Young and old crisp membership functions.

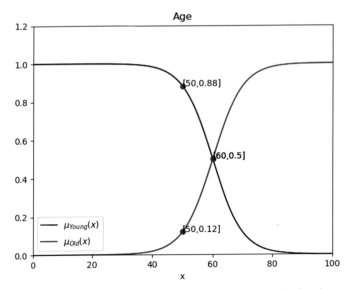

FIGURE 8.6 Young and old age fuzzy set and their membership functions.

40 and gradually increases until 80 years old. This allows for more logical decisions (policy) making for different ages, specifically ones in the range 40−80 which have some membership to the young group as well as to the old group. For example, for age 50 the membership to the young group is 0.88 and to the old group is 0.12, while it is 0.5 for the old and young group for age 60. The fuzzy set concept allows for a smoother and more effective control in many systems including temperature control, automatic transmission in cars, handwriting recognition, and oil and gas problems such as investment decisions, candidate selection for restimulation, and choke management.

Mathematical function

Remember that in classical set theory the membership for element in the universe X is 0 or 1, meaning considering a set A with an item $x \in X$ is either in the set (1) or not in the set (0). However, for fuzzy sets, there is a range of membership defined for each element in the universe X, with range [0,1] inclusive. The function that formulates the membership value for a fuzzy set is called a *membership function* and is shown for a fuzzy set A and universe X as follows:

$$\mu_A(x), \quad \text{for } x \in X. \tag{8.10}$$

In order to define a fuzzy set A in universe X given a membership function $\mu_A(x)$, a set of ordered pairs is defined as follows:

$$A = \{(x, \mu_A(x)) | x \in X\} \tag{8.11}$$

An example of fuzzy set is shown in Fig. 8.6 where the membership to young and old fuzzy sets for age 50 is $\mu_{Young}(50) = 0.88$, $\mu_{Old}(50) = 0.12$.

Support of a fuzzy set is the set of elements in the universe X with a nonzero membership function, $\mu_A(x) > 0$, $x \in X$. The **core** of a fuzzy set A is the set of elements with $\mu_A(x) = 1$. **Boundary** of fuzzy set A is a portion of the universe with nonzero membership which is not equal to 1. A crossover point of a fuzzy set A is the set with $\mu_A(x) = 0.5$. Core, support, and boundary are shown in Fig. 8.7.

Membership functions type

A variety of membership functions can be used for fuzzy sets. These membership functions are one dimensional for each set; however, when introducing more complex fuzzy systems, higher-order membership functions can be derived from multiple membership functions. Based on the behavior of the input variables, the following one-dimensional forms of membership functions can be used: (i) Triangular, (ii) Trapezoidal, (iii) Gaussian, and (iv) Sigmoidal (Himanshu & Lone, 2019). Some membership functions have linear form, e.g., Triangular and Trapezoidal which have some limitations because of their linear forms. However, linear membership functions are simpler for illustration and initial design of a fuzzy system. Smooth membership functions, e.g., Gaussian and Sigmoidal, are more complicated to design but more effective

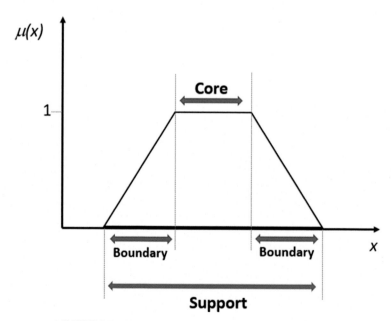

FIGURE 8.7 Representation of the core, support, and boundary.

for automated control. Also, note that most membership functions can be defined as a one-sided function to cover the whole range of universe values. Triangular membership functions are characterized with three parameters [a,b,c] in Eq. (8.12) and an example is illustrated in Fig. 8.8.

$$\mu_{Triangular}(x) = \max\left(\min\left(\frac{x-a}{b-a}, \frac{c-x}{c-b}\right), 0\right) \tag{8.12}$$

Trapezoidal membership functions can be characterized with four parameters [a,b,c,d] and those are specified as follows.

$$\mu_{Trapezoid}(x) = \max\left(\min\left(\frac{x-a}{b-a}, 1, \frac{d-x}{d-c}\right), 0\right) \tag{8.13}$$

where $\frac{x-a}{b-a}$ represents the ascending edge and $\frac{d-x}{d-c}$ is the descending edge with negative slope. A sample of trapezoidal membership function can be seen in Fig. 8.9.

Trapezoidal membership function will be primarily used in this chapter. Note that a one-sided trapezoidal membership function can be specified by having either $a = b$ or $c = d$.

The next membership function is Gaussian function which follows a Gaussian distribution and is characterized with two parameters (m, σ). The function form can be seen below and an example is illustrated in Fig. 8.10.

$$\mu_{Gaussian}(x) = exp\left(-\frac{1}{2}\left(\frac{x-m}{\sigma}\right)\right) \tag{8.14}$$

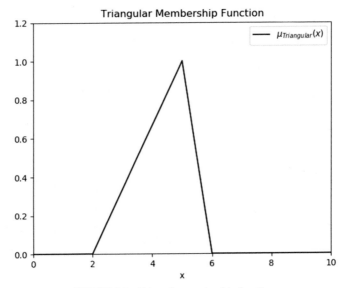

FIGURE 8.8 Triangular membership function.

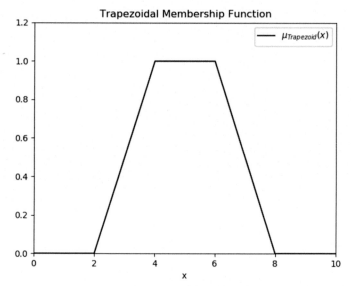

FIGURE 8.9   Trapezoidal membership function.

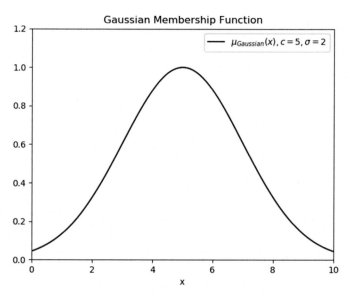

FIGURE 8.10   Gaussian membership function.

Next membership function is sigmoidal, characterized with two parameters [a,c] and the form of the function is as follows:

$$\mu_{Sigmoidal} = \frac{1}{1 + exp(-a(x - c))} \tag{8.15}$$

An example of sigmoidal membership function is used for Young and Old fuzzy sets example in Fig. 8.6.

Fuzzy set operations

Subset of a fuzzy set A is a subset of fuzzy set B, $(A \subset B)$ if the following holds true:

$$\forall x \in X, \ \mu_A(x) \leq \mu_B(x) \tag{8.16}$$

Fig. 8.11 illustrates this concept.

Complement of a fuzzy set A is a fuzzy set Ac with its membership function defined as follows:

$$\mu_{A^c}(x) = 1 - \mu_A(x) \tag{8.17}$$

Union of two fuzzy sets A and B is a fuzzy set U with its membership function as maximum of the two sets A and B membership function values defined as follows:

$$\mu_C(x) = max\{\mu_A(x), \ \mu_B(x)\} \tag{8.18}$$

A sample union of fuzzy sets A and B can be seen in Fig. 8.12.

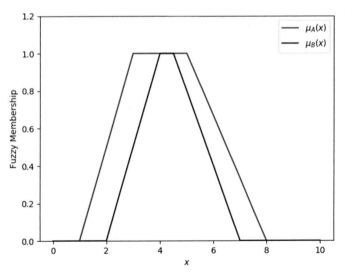

FIGURE 8.11 Subset.

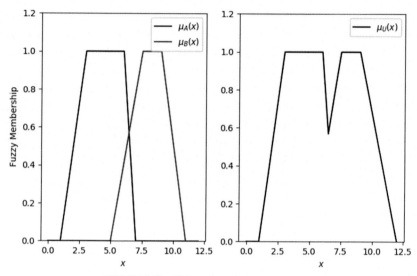

FIGURE 8.12   Union of two fuzzy sets A and B.

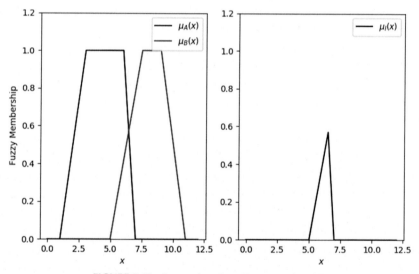

FIGURE 8.13   Intersection of two fuzzy sets A and B.

Intersection of two sets A and B is a fuzzy set I with its membership function as a minimum of the two sets A and B defined as follows:

$$\mu_C(x) = \min\{\mu_A(x), \ \mu_B(x)\} \tag{8.19}$$

A sample intersection of two sets A and B can be seen in Fig. 8.13.

Fuzzy inference system

A fuzzy inference system is a principled control system based on the fuzzy theory. The main components of this system are (i) Fuzzification, (ii) Fuzzy Rules Definition, (iii) Fuzzy Inference, and (iv) Defuzzification of the decision (Himanshu & Lone, 2019). All these steps are illustrated in a temperature control system and also a choke size control example in the next section. These four steps help us to define a logical framework based on the linguistic values defined and characterized using fuzzy sets. This system is similar to how humans use reasoning statements to solve some problems. This kind of control deems powerful in many engineering problems including oil and gas operations. There are three methods to perform a fuzzy inference system as Mamdani, Takagi-Sugeno, and Tsukamoto (Himanshu & Lone, 2019). In this book, the Mamdani Inference is used for two examples. The general workflow of the fuzzy inference system is shown in Fig. 8.14.

Input fuzzification

For (i) fuzzification, each input variable (I_i) and output variable (O_i) is considered over a meaningful universe (range of values). Also, a set of linguistic valuations for that input variable are used. For example, for temperature, a range [30, 120] can be defined and two linguistic variables "Cold" and "Hot" can be used to characterize the state of the temperature in our system. As part of fuzzification step, a membership function has to be defined for all the variables in that universe ($\mu_i(x)$, $x \in I_i$). Note that membership function can be of different forms, e.g., triangular, trapezoid, Gaussian. For example, membership functions $\mu_{Cold}(x)$ and $\mu_{Hot}(x)$ for linguistic values "Cold" and "Hot" can be considered as shown in Fig. 8.15 with the following membership function definitions:

$$\mu_{Cold}(x) = \max\left(\min\left(1, \frac{80-x}{80-60}\right), 0\right) \tag{8.20}$$

$$\mu_{Hot}(x) = \max\left(\min\left(\frac{x-60}{80-60}, 1\right), 0\right) \tag{8.21}$$

Consider a temperature control system where the fan speed is controlled given the input temperature and humidity level. In order to perform

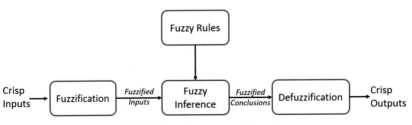

FIGURE 8.14 Fuzzy inference system.

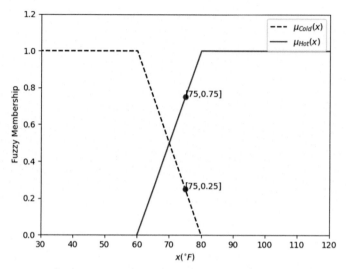

FIGURE 8.15 Temperature fuzzy sets cold and hot.

(i) fuzzification, the fuzzy sets are defined for the variables of this system (both inputs and output variables). The membership functions can be seen in Fig. 8.16. The temperature is described using "Cold" and "Hot" linguistic values using a trapezoidal membership function as explained above. The humidity is described using "Low" and "High" values with Trapezoidal membership function. The Fan Speed is defined using three membership functions corresponding to Low, Medium, and High Fan Speeds, using Trapezoidal, Triangular, and Trapezoidal membership functions. The universe for each membership function can be observed in Fig. 8.16 as x-axis limits.

Next step is the Fuzzification of each crisp input value Ii, where degrees of membership (values between 0 and 1) are assigned to different fuzzy sets defined for that input variable Ii. For example, for a temperature control system assuming temperature value of 75°F, the membership values to fuzzy sets "Cold" and "Hot" can be obtained by plugging in 75 into $\mu_{Cold}(x)$ and $\mu_{Hot}(x)$ which results in 0.25 and 0.75, respectively. Now these membership values can be used in fuzzy rules defined in next step (ii) Fuzzy rules definition, to derive implications and consequently decisions.

Fuzzy rules

Another important component of fuzzy sets are (ii) fuzzy rules definition, how they should be interpreted, and how decisions can be inferred from them. Fuzzy rules are defined as follows:

$$If \ x \ is \ C \rightarrow y \ is \ D. \tag{8.22}$$

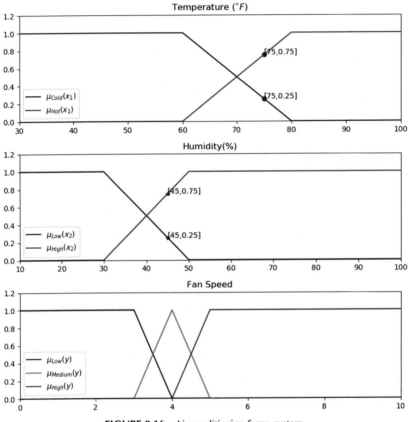

FIGURE 8.16 Air conditioning fuzzy system.

which is interpreted as if input x is in fuzzy set C (Premise) then the decision to make is in fuzzy set D (Conclusion). In this rule, the sets C and D are fuzzy sets defined to characterize linguistic variables for input and output, respectively. The first expression is called the premise and the second expression is conclusion of the rule. These rules are used to derive conclusions (control systems) based on crisp inputs fed into the fuzzy system.

For a system with N input variables $I_1, \cdots, I_N$ with each input variable having $C_1, \cdots, C_N$ different linguistic values, the total number of rules that can be derived is $2^{2N-1} \prod_{i=1}^{N} C_i$ where 2^{2N-1} accounts for different combination of rules (AND, OR) as well as using the fuzzy set or its complement (NOT). Note that AND operation in logic is defined as Intersection, OR operation is defined as Union, and NOT is defined as Complement which were defined in the previous sections. For example, in the temperature control system defined earlier with two input variables, Temperature and Humidity. The output or control action is Fan Speed. There are two linguistic values Cold and Hot for the Temperature and Low and High for Humidity and three linguistic values Low, Medium, and High for Fan Speed. For this system, four rules can be derived as shown in Table 8.1. Note that another set of 12 rules can be defined if the complement of the input values (NOT) are considered resulting in a total of 16 rules using AND combination of different fuzzy sets. Additionally, if the OR combination of input values' fuzzy sets are used, 16 more rules can be derived resulting in a total of $2^{2 \times 2 - 1}(2 \times 2) = 32$ possible rules.

For this example the following set of rules are considered for the temperature control system:

1) If Temperature is Cold AND Humidity is Low $\rightarrow$ Fan Speed Low,
2) if Temperature is Cold AND Humidity is NOT Low $\rightarrow$ Fan Speed Medium,
3) if Temperature is Hot OR Humidity is High $\rightarrow$ Fan Speed Medium,
4) if Temperature is Hot and Humidity is High $\rightarrow$ Fan Speed High.

Note that usually an aggregation of rules is used for decision-making and these rules are defined by domain experts based on different values of inputs and corresponding control action.

Inference

In order to draw conclusions from a set of rules defined for a system ((iii) inference step), the strength of the premise of each rule, referred to as "Firing Strength" of the premise, is calculated given a set of input values. The firing strength of a rule shown as $\mu_{premise(i)}(x_1, x_2)$ quantifies the strength of the rule premise given a set of crisp input values x_1, x_2. Note that the premise can have different combination of input variables, e.g., AND, OR, NOT. Calculating the firing strength of each rule is different based on fuzzy set operations used in that rule (see Fuzzy set operations section). For example, considering input values: Temperature denoted as $x_1 = 75°F$ and Humidity denoted as $x_2 = 43\%$,

TABLE 8.1 Temperature humidity rule space.

Fan speed		Humidity	
	AND	Low	High
Temperature	Cold	Low	Med
	High	Med	High

the calculated firing strength of these four rules defined for our air conditioning system is as follows:

1) If Temperature is Cold AND Humidity is Low →Fan Speed Low

$$\mu_{premise(1)}(75, 43) = \min(\mu_{Cold}(75), \ \mu_{Low}(43)) = \min(0.25, 0.35) = 0.25$$

2) If Temperature is Cold AND Humidity is NOT Low → Fan Speed Medium

$$\mu_{premise(2)}(75, 43) = \min(\mu_{Cold}(75), 1 - \mu_{Low}(43)) = \min(0.25, 0.65) = 0.25$$

3) If Temperature is Hot OR Humidity is High → Fan Speed Medium

$$\mu_{premise(3)}(75, 43) = \max(\mu_{Hot}(75), \ \mu_{High}(43)) = \max(0.75, 0.65) = 0.75$$

4) If Temperature is Hot and Humidity is High → Fan Speed High

$$\mu_{premise(4)}(75, 43) = \min(\mu_{Hot}(75), \ \mu_{High}(43)) = \min(0.75, 0.65) = 0.65$$

Note that the firing strength for any other combination of rules can be calculated similarly. Also, in this example, all rules have nonzero firing strength values resulting in all rules being active. However, if some rules have firing strength equal to 0, those rules are considered inactive and will not be considered for decision-making. Next, it is shown how the active rules can be used for deriving fuzzy implication for a set of rules.

The implied fuzzy set for each rule's decision (D_i) given input values (e.g., (x_1,x_2)) is calculated as intersection of the calculated firing signal $\mu_{premise(i)}(x_1, x_2)$ and membership function for the same rule's decision $(\mu_{D_i}(y))$ as follows:

$$\mu_{D_i}(y) = \min\left(\mu_{premise(i)}(x_1, x_2), \ \mu_{D_i}(y)\right) \tag{8.23}$$

These implied fuzzy sets can be calculated for decision values (fan Speed) for all four rules: $D_1 = $ Low, $D_2 = $ Medium, $D_3 = $ Medium, and, $D_4 = $ High as follows:

$$\mu_{D_1}(y) = \min\left(\mu_{premise(1)}(75, 43), \ \mu_{Low}(y)\right) = \min(0.25, \mu_{Low}(y)) = 0.25$$

$$\mu_{D_2}(y) = \min\left(\mu_{premise(2)}(75, 43), \mu_{Medium}(y)\right) = \min(0.25, \mu_{Medium}(y)) = 0.25$$

$$\mu_{D_3}(y) = \min\left(\mu_{premise(3)}(75, 43), \mu_{Medium}(y)\right) = \min(0.75, \mu_{Medium}(y)) = 0.75$$

$$\mu_{D_4}(y) = \min\left(\mu_{premise(4)}(75, 43), \ \mu_{High}(y)\right) = \min(0.65, \mu_{High}(y)) = 0.65$$

You can see $\mu_{Low}(y)$, $\mu_{Medium}(y)$, $\mu_{High}(y)$ in Fig. 8.16. The implied fuzzy sets $\mu_{D_1}(y)$, $\mu_{D_2}(y)$, $\mu_{D_3}(y)$, $\mu_{D4}(y)$ are plotted as illustrated in Fig. 8.17.

Next, all of these implied fuzzy sets have to be aggregated to one fuzzy set to make the final decision. For aggregating the n sets implied from n rules, an intersection of all the implied sets have to be used to derive the final decision fuzzy set as follows:

$$\mu_{Aggr}(y) = \max_{i=1}^{n} \mu_{D_i}(y) \tag{8.24}$$

The aggregated fuzzy set for temperature control example is illustrated in Fig. 8.18.

Defuzzification

Finally, in the step (iv) defuzzification, the aggregated fuzzy set has to be interpreted as a crisp output (control action) for the fuzzy inference system. There are different approaches for interpreting the aggregate fuzzy set. Some of these approaches include (i) Center of gravity (centroid), (ii) Maxima method (Himanshu & Lone, 2019). Note that these are some sample methods of defuzzification and other methods can be designed to serve the needs of the system. These methods will be explained here and illustrated in the following example.

(i) Center of Gravity or centroid defuzzification method uses the area under the implied aggregated fuzzy set, using the following formula:

$$y = \frac{\int y\mu_{Aggr}(y)dy}{\int \mu_{Aggr}(y)dy} \tag{8.25}$$

where $\mu_{Aggr}(y)$ is the aggregated implied fuzzy set and the integral part in the denominator $\int \mu_{Aggr}(y)$ shows the area below the $\mu_{Aggr}(y)$. The centroid can be calculated for nonsmooth membership functions by

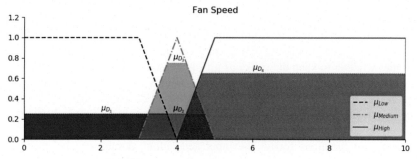

FIGURE 8.17 Implied fuzzy sets for four rules.

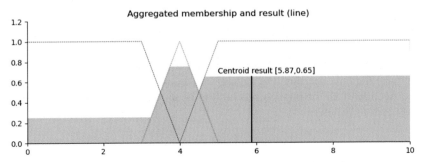

FIGURE 8.18   The aggregated decision fuzzy set and the result.

dividing the area into N familiar shapes and calculating the area below each part denoted as Ai as well as the centroid of each region denoted as yi and averaging using the following equation:

$$y = \frac{\sum_{i}^{N} y_i A_i}{\sum_{i}^{N} A_i} \tag{8.26}$$

(ii) Maxima defuzzification method is based on finding the maximum membership obtained from the implied fuzzy set ($\mu_{Aggr}(y)$) among the universe of decision values. There are different ways of choosing the maxima, e.g., First of Maxima (FOM), Last of Maxima (LOF) which, respectively, chooses the minimum and maximum decision values y that gives the maximum membership value.

Fuzzy inference example: choke adjustment

This example is based on a work by Odedele et al. (Odedele & Ibrahim, 2014) which uses fuzzy logic inference model to control choke size and consequently well production by considering a range of values for tubing head pressure, gas–liquid ratio (GLR), and production rate. The controller in this example receives tubing head pressure, GLR, and production rate as input variables and decides whether to keep the choke as it is, increase the choke size, or decrease the choke size (output). The output of this system is a recommendation of how much (%) the choke size should be increased or decreased. To model this problem using fuzzy inference system, the steps shown in the previous section should be taken as follows: (i)Fuzzification: It is necessary to generate fuzzy membership functions for all the input and output variables to be used to fuzzify the crisp input values. (ii) Fuzzy Rules Definition: define the fuzzy rules in order to make the appropriate decision, based on different combinations of the input parameters. (iii) The inference step aggregates the rules output, and finally, (iv) defuzzification would be performed to get a crisp output to make a decision accordingly. Most of the fuzzy logic operation can be done in Python with "scikit-fuzzy" of "skfuzzy" toolbox (pythonhosted). In order to use this toolbox in Python, it is necessary to install it with the "pip install scikit-fuzzy" command. Let's start solving this problem by defining all the variables and generating trapezoidal membership functions in Python with the following codes:

```python
import numpy as np
import skfuzzy as fuzz
import matplotlib.pyplot as plt
# Define universe variables
# Tubing Head pressure (x_THP) values range from 70 to 540 psi
# Gas Liquid ratio (x_GLR) values range from 0 to 27 Mscf/bbl
# Oil production rate (x_q) values range from 60 to 649 bbl/day
# Choke Adjustment values (x_choke) range from -15 to 30 percentage
# change
x_THP=np.arange(70, 541, 0.01)
x_GLR=np.arange(0, 28, 0.01)
x_q=np.arange(60, 650, 0.01)
x_choke=np.arange(-15, 30, 0.01)


# Create trapezoidal fuzzy membership functions for low, medium, and
# high linguistic variables.
THP_low=fuzz.trapmf(x_THP, [70, 70, 80, 120])
THP_med=fuzz.trapmf(x_THP, [80, 120, 230, 300])
THP_hi=fuzz.trapmf(x_THP, [230, 300, 540, 540])
GLR_low=fuzz.trapmf(x_GLR, [0, 0, 2, 4])
GLR_med=fuzz.trapmf(x_GLR, [2, 4, 7, 10])
GLR_hi=fuzz.trapmf(x_GLR, [7, 10, 27, 27])
q_low=fuzz.trapmf(x_q,[60, 60, 90, 160])
q_med=fuzz.trapmf(x_q, [90, 160, 260, 330])
q_hi=fuzz.trapmf(x_q, [260, 330, 650, 650])
choke_increase=fuzz.trapmf(x_choke,[-15, -15, -12,-9])
choke_fixed=fuzz.trapmf(x_choke, [-12, -9, 8,13])
choke_decrease=fuzz.trapmf(x_choke, [8,13, 30,30])


# Plot value ranges and the trapezoidal membership functions
fig, (ax0, ax1, ax2,ax3)=plt.subplots(nrows=4, figsize=(8, 9))
ax0.plot(x_THP, THP_low, 'b', linewidth=1.5, label='Low')
ax0.plot(x_THP, THP_med, 'g', linewidth=1.5, label='Medium')
ax0.plot(x_THP, THP_hi, 'r', linewidth=1.5, label='High')
ax0.set_title('Tubing Head Pressure')
ax0.legend()
ax1.plot(x_GLR, GLR_low, 'b', linewidth=1.5, label='Low')
ax1.plot(x_GLR, GLR_med, 'g', linewidth=1.5, label='Medium')
ax1.plot(x_GLR, GLR_hi, 'r', linewidth=1.5, label='High')
```

```
ax1.set_title('GLR')
ax1.legend()
ax2.plot(x_q, q_low, 'b', linewidth=1.5, label='Low')
ax2.plot(x_q, q_med, 'g', linewidth=1.5, label='Medium')
ax2.plot(x_q, q_hi, 'r', linewidth=1.5, label='High')
ax2.set_title('Flow Rate')
ax2.legend()
ax3.plot(x_choke, choke_increase, 'b', linewidth=1.5, label='Low')
ax3.plot(x_choke, choke_fixed, 'g', linewidth=1.5, label='Medium')
ax3.plot(x_choke, choke_decrease, 'r', linewidth=1.5, label='High')
ax3.set_title('Choke Change( %)')
ax3.legend()
# Not Showing top and right axes
for ax in (ax0, ax1, ax2,ax3):
    ax.spines['top'].set_visible(False)
    ax.spines['right'].set_visible(False)
    ax.get_xaxis().tick_bottom()
    ax.get_yaxis().tick_left()
plt.tight_layout()
```
Python output=Fig. 8.19

As it is illustrated in Fig. 8.19, input variables are divided into three classes as low, medium, and high with trapezoidal membership functions within variable ranges shown in the Python code using np.arange. To create fuzzy membership functions, it is necessary to import "skfuzzy" and use ".trapmf" as a trapezoidal fuzzy membership function. For each class (low, medium, and high), it is necessary to provide four points as edges of the trapezoid to generate membership functions. As an example, when "THP_med = fuzz.trapmf(x_THP, [80, 120, 230,300])" is used to define medium class for tubing head pressure, medium membership to THP values below 80 psi is zero. As THP starts increasing from 80 to 120 psi, medium membership starts increasing from 0 to 1 linearly. Then, medium membership remains 1 as THP increases from 120 to 230 psi. Finally, medium membership starts decreasing from 1 to 0 as THP increases from 230 to 300 psi. For the percentage of the choke size, three classes are considered to reduce the choke (RC), maintain the choke (No Change-NC), and increase the choke (IC). It is worth mentioning that in the "No Change" scenario, it is also possible to adjust the choke with medium level changes. For this controller, it is possible to increase the choke size by at most 15% and decrease the choke size by at most 30%. Negative values mean that choke should increase and vice versa. To keep the well's operational condition in the safe region, it is decided not to increase the choke more than 15%.The next step is to define fuzzy rules to make a controlled decision, based on input variables. For example, a fuzzy rule can be considered as if tubing head pressure is low, GLR is low, and the production rate is low then increase the choke size. The other rule is if tubing head pressure is high, GLR is low, and the production rate is high then reduce the choke size. Since there are three input variables and each variable is divided into three classes, total possible number of rules is $2^{2 \times 3 - 1} \times 3^3 = 864$, considering both OR and AND

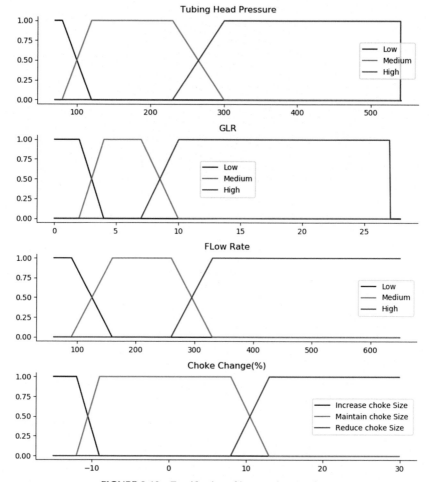

FIGURE 8.19 Fuzzification of input and output data.

combination and NOT of each predicate. In this example, 10 rules were considered as listed in Table 8.2.

Let's consider a set of input values where THP is 265 psi, GLR is 20 mscf/bbl, and the production rate is 300 bbl/day. In order to fuzzify these crisp values according to the defined fuzzy inference system above, "interp_membership" command should be used. For example, in "THP_mem_med = fuzz.interp_membership(x_THP, THP_med, 265)" command, membership of 265 psi to medium class is found to be 0.5 (Fig. 8.19). Python code to find membership values for each input parameter at the mentioned values (265 psi, 20 mscf/day, and 300 bbl/day) are as follows. Membership values to classes low, medium, and high for the mentioned values are shown in Table 8.3.

TABLE 8.2 Fuzzy rules for choke control.

Rules	1	2	3	4	5	6	7	8	9	10
THP(psi)	Low	Low	Low	Medium	Medium	Medium	High	High	High	High
GLR (mscf/bbl)	Low	Medium	High	Low	Medium	High	Low	Medium	High	Low
Production rate(bbl/day)	Low	Low	Low	Medium	Medium	Medium	High	High	High	Low
Choke size control	Increase choke	Increase choke	Increase choke	No change	No change	No change	Reduce choke	Reduce choke	Reduce choke	Increase choke

TABLE 8.3 Fuzzy membership values for THP 265 psi, GLR 20 mscf/bbl, and q 300 bbl/day.

265 psi membership to low THP	0
265 psi membership to medium THP	0.5
265 psi membership to high THP	0.5
20 mscf/bbl membership to low GLR	0
20 mscf/bbl membership to medium GLR	0
20 mscf/bbl membership to high GLR	1
300 bbl/day membership to low q	0
300 bbl/day membership to medium q	0.43
300 bbl/day membership to high q	0.57

```
# Activating of fuzzy membership functions for THP 265 psi, GLR
  # 20 mscf/day and q 300 bbl/day
THP_mem_lo=fuzz.interp_membership(x_THP, THP_low, 265)
THP_mem_med=fuzz.interp_membership(x_THP, THP_med, 265)
THP_mem_hi=fuzz.interp_membership(x_THP, THP_hi, 265)
GLR_mem_lo=fuzz.interp_membership(x_GLR, GLR_low, 20)
GLR_mem_med=fuzz.interp_membership(x_GLR, GLR_med, 20)
GLR_mem_hi=fuzz.interp_membership(x_GLR, GLR_hi, 20)
q_mem_lo=fuzz.interp_membership(x_q, q_low, 300)
q_mem_med=fuzz.interp_membership(x_q,q_med, 300)
q_mem_hi=fuzz.interp_membership(x_q, q_hi, 300)
```

Based on the fuzzy rules that are shown in Table 8.2 and memberships of input parameters to each class as shown in Table 8.3, and (iii) inference step explained in previous section, the firing strength of all rules can be calculated. Only rules 6 and 9 will have nonzero firing strength and hence are active. Let's see how the firing strength of rule 6 can be computed. Rule 6 states that if THP is medium AND GLR is high AND q is medium THEN maintain the choke size or no change in the choke (NC). Since AND operator is applied, minimum value among the three input memberships should be picked to calculate the firing strength of the rule 6. The minimum membership value is 0.43. This value is assigned to maintain the choke size class to calculate firing strength for rule 6 as shown in Fig. 8.20.

The same procedure should be performed to compute the firing strength of rule 9. When two firing strengths are calculated, it is necessary to perform fuzzy aggregation using the "max" operator to find an overall fuzzy set. These

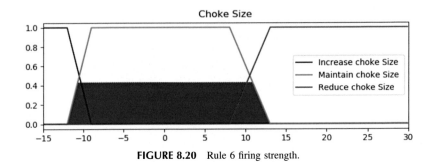

FIGURE 8.20 Rule 6 firing strength.

steps were explained in details in previous section. Python codes for computing firing strengths and fuzzy aggregation are as follows:

```
# Firing strength calculation for Rule 6
# If THP is medium AND GLR is high AND q is medium THEN Choke Medium
active_rule6=np.amin([THP_mem_med, GLR_mem_hi, q_mem_med])
choke_activation_6=np.fmin(active_rule6, choke_fixed)
# Firing strength calculation for Rule 9
# If THP is high AND GLR is high AND q is high THEN Reduce high Choke
active_rule9=np.amin([THP_mem_hi, GLR_mem_hi, q_mem_hi])
choke_activation_9=np.fmin(active_rule9, choke_decrease)
choke0=np.zeros_like(x_choke)
#Plot Membership functions for rules
fig, ax0=plt.subplots(figsize=(8, 2))
ax0.fill_between(x_choke, choke0, choke_activation_6, facecolor='b',
alpha=0.7)
ax0.plot(x_choke, choke_activation_6, 'b', linewidth=0.5,
  linestyle='--')
ax0.fill_between(x_choke, choke0, choke_activation_9,
  facecolor ='r', alpha=0.7)
ax0.plot(x_choke, choke_activation_9, 'r', linewidth=0.5,
  linestyle='--')
ax0.set_title('Output membership activity')
ax0.plot(x_choke, choke_increase, 'b', linewidth=1.5,
  label='Increase choke size')
ax0.plot(x_choke, choke_fixed, 'g', linewidth=1.5,
  label='Maintain choke size')
ax0.plot(x_choke, choke_decrease, 'r', linewidth=1.5,
  label='Reduce choke size')
ax0.set_title('Choke Size')
ax0.legend()
# Not Showing top and right axes
for ax in (ax0,):
    ax.spines['top'].set_visible(False)
    ax.spines['right'].set_visible(False)
    ax.get_xaxis().tick_bottom()
    ax.get_yaxis().tick_left()
plt.tight_layout()
```

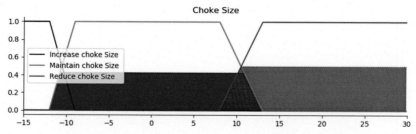

FIGURE 8.21 Aggregation of rule 6 and rule 9 using max operator.

```
# Aggregate two output membership functions or firing strengths
# together
aggregated=np.fmax(choke_activation_6, choke_activation_9)
```
Python output=Fig. 8.21

The last step of this problem is to calculate the choke control decision in step (iv) defuzzification which helps to come up with a crisp value for the percentage of choke change (decision variable). This can be done using "defuzz" command and specifying the aggregated firing strengths and defuzzification algorithm. In this example, "center of gravity" is used for defuzzification (centroid keyword is used in defuzz command for the center of gravity defuzzification). Defuzzification and the results in Python can be found with the following codes:

```
# compute defuzzified result using center of gravity
choke=fuzz.defuzz(x_choke, aggregated, 'centroid')
choke_activation=fuzz.interp_membership(x_choke, aggregated, choke)
# Plot the results
fig, ax0=plt.subplots(figsize=(8, 3))
ax0.plot(x_choke, choke_increase, 'b', linewidth=0.5, linestyle='--')
ax0.plot(x_choke, choke_fixed, 'g', linewidth=0.5, linestyle='--')
ax0.plot(x_choke, choke_decrease, 'r', linewidth=0.5, linestyle='--')
ax0.fill_between(x_choke, choke0, aggregated, facecolor='Orange',
   alpha=0.7)
ax0.plot([choke, choke], [0, choke_activation], 'k', linewidth=1.5,
   alpha=0.9)
ax0.set_title('Aggregated membership and result (line)')
# Not Showing top and right axes
for ax in (ax0,):
    ax.spines['top'].set_visible(False)
    ax.spines['right'].set_visible(False)
    ax.get_xaxis().tick_bottom()
    ax.get_yaxis().tick_left()
plt.tight_layout()
print (choke)
```
Python output=Fig. 8.22

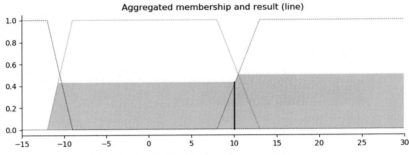

FIGURE 8.22 Defuzzification of aggregated fuzzy sets.

Based on Fuzzy control system decision, it is necessary to decrease the choke by 10.1% considering input variables for Tubing head pressure (265 psi), GLR (20mscf/bbl), and oil flow rate (300 bbl/day). This problem was solved step by step to show how fuzzy inference systems work. This problem can also be handled with "skfuzzy control" package which works for fuzzy system design. Choke control with control package can be solved with the following Python codes:

```
import numpy as np
import skfuzzy
from skfuzzy import control
# Define input variable for a fuzzy control system
THP=control.Antecedent(np.arange(70, 541, 1), 'THP')
GLR=control.Antecedent(np.arange(0, 28, 1), 'GLR')
q=control.Antecedent(np.arange(60, 650, 1), 'q')
choke=control.Consequent(np.arange(-15,30,0.01), 'choke')
# Create trapezoidal fuzzy membership functions
THP['low']=skfuzzy.trapmf(THP.universe, [70, 70, 80, 120])
THP['med']=skfuzzy.trapmf(THP.universe, [80, 120, 230, 300])
THP['hi']=skfuzzy.trapmf(THP.universe, [230, 300, 540, 540])
GLR['low']=skfuzzy.trapmf(GLR.universe, [0, 0, 2, 4])
GLR['med']=skfuzzy.trapmf(GLR.universe, [2, 4, 7, 10])
GLR['hi']=skfuzzy.trapmf(GLR.universe, [7, 10, 27, 27])
q['low']=skfuzzy.trapmf(q.universe,[60, 60, 90, 160])
q['med']=skfuzzy.trapmf(q.universe, [90, 160, 260, 330])
q['hi']=skfuzzy.trapmf(q.universe, [260, 330, 650, 650])
choke['IC']=skfuzzy.trapmf(choke.universe,[-15, -15, -12,-9])
choke['NC']=skfuzzy.trapmf(choke.universe, [-12, -9, 8, 13])
choke['RC']=skfuzzy.trapmf(choke.universe, [8,13, 30, 30])
# Plot Choke Membership Function
choke.view()
Python output=Fig. 8.23
```

"Skfuzzy" and control API from skfuzzy should be imported to solve this problem. "control.Antecedent" was used to define input variables (THP, GLR, and q) with their ranges for the fuzzy inference system. "Control.Consequent" was used to define the ranges for the fuzzy control decision which is choke adjustment (%). Then, trapezoidal fuzzy membership functions were assigned to each input variable (skfuzzy.trapmf). The plot of each membership function can be created with ".view" function. Choke membership functions and fuzzy classes are shown in Fig. 8.23. The next step is to define the fuzzy rules and use them in the control system. In order to have 3D visualization and show all the ranges in the input data, seven more rules are added to the rules that were defined earlier in Table 8.2. This can be done with the following Python codes:

```
# Defining Rule in a fuzzy control system to connect inputs to choke
# control decision
rule1=control.Rule(THP['low'] & GLR['low'] & q['low'], choke['IC'])
rule2=control.Rule(THP['low'] & GLR['med'] & q['low'], choke['IC'])
rule3=control.Rule(THP['low'] & GLR['hi'] & q['low'], choke['IC'])
rule4=control.Rule(THP['med'] & GLR['low'] & q['med'], choke['NC'])
rule5=control.Rule(THP['med'] & GLR['med'] & q['med'], choke['NC'])
rule6=control.Rule(THP['med'] & GLR['hi'] & q['med'], choke['NC'])
rule7=control.Rule(THP['hi'] & GLR['low'] & q['hi'], choke['RC'])
rule8=control.Rule(THP['hi'] & GLR['med'] & q['hi'], choke['RC'])
rule9=control.Rule(THP['hi'] & GLR['hi'] & q['hi'], choke['RC'])
rule10=control.Rule(THP['hi'] & GLR['low'] & q['low'], choke['IC'])
#Additional rules for 3D plot
rule11=control.Rule(THP['low'] & GLR['hi'] & q['hi'], choke['NC'])
rule12=control.Rule(THP['med'] & GLR['med'] & q['low'], choke['IC'])
rule13=control.Rule(THP['hi'] & GLR['med'] & q['low'], choke['NC'])
rule14=control.Rule(THP['hi'] & GLR['med'] & q['med'], choke['NC'])
rule15=control.Rule(THP['med'] & GLR['med'] & q['hi'], choke['RC'])
rule16=control.Rule(THP['low'] & GLR['med'] & q['med'], choke['IC'])
rule17=control.Rule(THP['low'] & GLR['med'] & q['hi'], choke['NC'])
# Define base classes for Fuzzy control system with 10 rules
choke_control=control.ControlSystem([rule1, rule2, rule3, rule4,
    rule5, rule6, rule7, rule8, rule9, rule10,rule11,rule12,rule13,rule14,
    rule15,rule16,rule17])
#Result computing from control system
choking=control.ControlSystemSimulation(choke_control)
# Provide inputs to the ControlSystem using input
choking.input['THP']=265
choking.input['GLR']=20
choking.input['q']=300
#compute the result of fuzzy system
choking.compute()
print (choking.output['choke'])
#plot the rules and final decision
choke.view(sim=choking)
Python output=Fig. 8.24
```

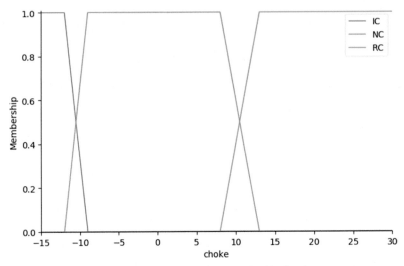

FIGURE 8.23 Choke trapezoid membership function.

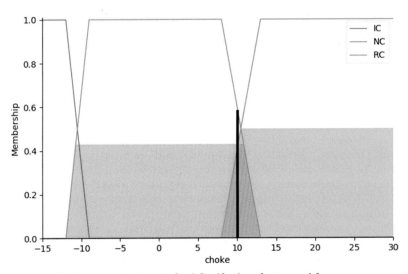

FIGURE 8.24 Final results for defuzzification of aggregated fuzzy sets.

Each rule was defined with control.Rule. Then the control system was created by using "control.ControlSystem" function considering all 10 rules. To simulate all the rules in a system and apply the controller, "control.Control SystemSimulation" was used. Finally, input parameters were identified using ".input" and the choke control decision was determined by the ".compute" function. The defuzzified output for the choke size is 10.1 which is interpreted as decreasing the choke by 10.1%.

In order to see the decision values for different input values, next a 3D plot of tubing head pressure (THP), flow rate (q), and a fixed GLR of 5 mscf/bbl can be coded as follows:

```
n=50
THP_highres=np.linspace(70, 540, n+1)
q_highres=np.linspace(60, 640, n+1)
x, y=np.meshgrid(THP_highres, q_highres)
z=np.zeros_like(x)
# Loop through the input values to calculate the corresponding control
# values
for i in range(n+1):
    for j in range(n+1):
        choking.input['THP']=x[i, j]
        choking.input['GLR']=5
        choking.input['q']=y[i, j]
        choking.compute()
        z[i, j]=choking.output['choke']
# Plot the 3D plot
import matplotlib.pyplot as plt
from mpl_toolkits.mplot3d import Axes3D
fig=plt.figure(figsize=(8, 10))
ax=fig.add_subplot(111, projection='3d')
surf=ax.plot_surface(x, y, z, cmap='YlGnBu', linewidth=0.5)
fig.colorbar(surf, ax=ax, shrink=0.3,aspect=5)
ax.set_xlabel('THP (psi)')
ax.set_ylabel('q (bbl/d)')
ax.set_zlabel('Choke Adjustment (-%)')
Python output=Fig. 8.25
```

The 3D plot in Fig. 8.25 shows what control actions for changing the choke should be taken when flow rate and tubing head pressure are changing within their ranges at constant GLR. For the conditions that flow rate is lower than 300 bbl/day and tubing head pressure is lower than 250 psi, the choke should be increased at most by 10%. For the conditions that flow rate is higher than 300 bbl/day and tubing head pressure is higher than 250 psi choke size should be decreased by at most 20%.

Fuzzy C-means clustering

In the clustering problems which are in the category of unsupervised learning, the objective is to divide a data set to classes such that the samples in each class are similar to each other as compared with samples in the other classes. In Chapter 4, different clustering techniques were discussed. In this section, the implication of fuzzy logic in the unsupervised clustering problems is explained. In order to see how fuzzy logic is used in clustering algorithms, let's have a quick review of K-means clustering from Chapter 4.

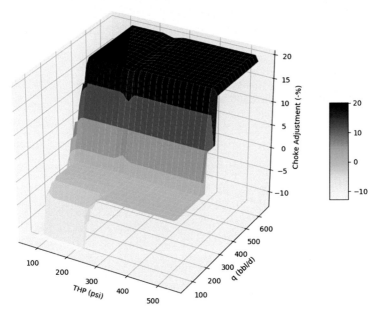

FIGURE 8.25 3D choke adjustment decision values for THP and q ranges.

In K-means clustering, first, the number of clusters is set and then random centroids are assigned to each cluster. Afterward, the distances (i.e., euclidean) between the data points and centroids are calculated. Data points are assigned to each class based on having a minimum distance to class centroids. New centroids for each class are calculated and then data points are reassigned to each class based on a newly calculated distance to each class centroid. These steps (calculating new centroids, calculating distances, and reassigning data to each class) are repeated until centroids remain the same as previous step (converged). In this technique, each data point belongs to just one class based on having a minimum distance to the class' centroid. In fuzzy clustering, which is also called "fuzzy c-means" clustering, it is possible that each data point belongs to many classes by calculating membership of each data point to the class centers.

In order to understand how fuzzy c-means clustering works, let's assume $X = \{x_1, x_2, \cdots, x_n\}$ is the data set space. Each data point has m attributes so x_i can be defined as a vector of attributes $x_i = (x_{i1}, x_{i2}, \cdots, x_{im})$. The space is m-dimensional. Therefore, x_{ij} can be considered as jth attribute for the ith data point. Imagine there are c clusters in the data set so the set of clusters is defined as $C = \{C_1, C_2, \cdots, C_c\}$. For each cluster in the C space, a centroid v is determined which belongs to V space defined as $V = \{v_1, v_2, \cdots, v_c\}$. Since the data set is m-dimensional, each centroid has m attributes as $v_i = (v_{i1}, v_{i2}, \cdots, v_{im})$. For each data point, it is possible to determine its degree of membership (u_{ik}) to each class using Eq. (8.28). Since there are n data points and c clusters, a fuzzy partition matrix is formed as $U = (u_{ik})_{(n \times c)}$ where u_{ik} is the degree of membership of the ith data point to k^{th} cluster and $\sum_{k=1}^{c} u_{ik} = 1$. The final goal in the fuzzy c-means clustering algorithm is to minimize an objective function $J_h(U, V)$ which is the quadratic sum of the weighted distances of each data point to each class centroid (Ren et al., 2016):

$$J_h(U, V) = \sum_{k=1}^{c} \sum_{i=1}^{n} u_{ik}^f d_{ik}^2 \qquad (8.27)$$

In Eq. (8.27), d_{ik} is the euclidean distance between *ith* data set and *kth* class centroid, and f is class fuzziness index. As f increases, it is expected to see fuzzier memberships. Generally, f should be in the range of 1.5−2.5 and is typically assumed to be 2. Fuzzy c-means clustering is an iterative process that starts from randomized initialization of membership degrees and updating centroids and membership degrees until minimization criteria for objective function are met. In each iteration, membership values (uik) and centroids (v_k) are calculated and updated using Eqs. (8.28) and (8.29):

$$u_{ik} = \frac{1}{\sum_{j=1}^{c} \left(\frac{d_{ik}}{d_{jk}} \right)^{2/(f-1)}} \qquad (8.28)$$

$$v_k = \frac{\sum_{i=1}^{n} u_{ik}^f x_i}{\sum_{i=1}^{n} u_{ik}^f} \qquad (8.29)$$

The algorithm for fuzzy c-means clustering works by iterative process through the following steps:

1 Initializing values for class number (c), fuzziness index (f), maximum number of iterations (I_{max}), and a threshold for error (e).
2 Initializing fuzzy membership matrix $U^{(0)}$ randomly.
3 Determining class centroids ($V^{(t)}$) by Eq. (8.29) at step t.
4 Calculating the objective function (J_h^t) by Eq. (8.27). If $J_h^t - J_h^{t-1} < e$ and t is less than maximum number of iterations (Imax) then stop the algorithm; otherwise repeat steps 2, 3, and 4.

Fuzzy c-means clustering is a simple algorithm with low computation cost and fast run time. However, similar to other clustering techniques, it may suffer from the fact that initialization of the fuzzy membership matrix, initialization for the number of clusters, existence of noise and outliers may affect the final results.

In the following example, fuzzy c-means clustering would be used to determine multiple clusters for a porosity_permeability data set. This synthetic data set includes 120 samples with porosity and permeability values from core data. The objective is to define different rock types based on samples with two attributes. Generally, there are many attributes for samples in rock typing clustering problems. When the number of attributes is more than 2, it is not easy to visualize different classes versus multiple attributes. In order to differentiate and visualize all the classes versus multiple attributes, samples with two attributes are used in this example. It is also worth mentioning that typically, there is a logarithmic trend in permeability versus porosity. Therefore, permeability values should be used in logarithmic scales. The data set should also be scaled to the range of 0−1 due to the fact that porosity and permeability values are in different ranges. Classification of rock types using porosity-permeability data using fuzzy c-means clustering can be done with the following Python codes:

https://www.elsevier.com/books-and-journals/book-companion/9780128219294

```python
import pandas as pd
import numpy as np
import matplotlib.pyplot as plt
import skfuzzy
import math
#Specify colors for different classes
  colors=['b', 'grey', 'g', 'r', 'c', 'm', 'y', 'k', 'lime', 'purple']
#import Dataset and change permeability to log-scale
dataset = pd.read_csv(
'Chapter8_Fuzzy_Clustering_Porosity_Permeability.CSV')
ds_log=pd.DataFrame.copy(dataset)
ds_log['Permeability']=ds_log['Permeability'].apply(math.log10)
# Scale the data from 0 to 1
from sklearn.preprocessing import MinMaxScaler
scaler=MinMaxScaler()
scaler.fit(ds_log)
ds_log_scaled=scaler.transform(ds_log)
#Transpose Scaled data for Fuzzy Cluster Algorithm
ds_log_scaled=ds_log_scaled.T
#Plot permeability vs porosity
plt.figure()
plt.plot(dataset['Porosity'],dataset['Permeability'],'ro')
plt.xlabel('Porosity(%)')
plt.ylabel('Permeability(md)')
#Plot permeability vs porosity with scaled data
plt.figure()
plt.plot(ds_log_scaled[0,:],ds_log_scaled[1,:],'ro')
plt.xlabel('Porosity')
plt.ylabel('Permeability')
```
Python output=Fig. 8.26

In Fig. 8.26, permeability versus porosity values are shown before and after logarithmic scaling of permeability and scaling the data between 0 and 1. In order to run fuzzy c-means clustering algorithm, it is required to call "skfuzzy.cluster.cmeans" from skfuzzy library. Several for loops would be used to vary number of clusters from 2 to 9 by changing the number of centroids, training the algorithm, and visualizing the results. The following Python code performs fuzzy c-means clustering and visualizations:

```
# Defining loops for Fuzzy C-means clustering and visualization with 8
#plots
import numpy as np
seed=50
np.random.seed(seed)
fig1, axes1=plt.subplots(2, 4, figsize=(12, 8))
fig1.suptitle('Fuzzy c-means clustering for Log scaled data')
fpcs=[ ]
n=2
for ax in axes1.reshape(-1):
    cntr,u,u0,d,jm,p,fpc=skfuzzy.cluster.cmeans(ds_log_scaled,
    n, 1.5, error=0.001, maxiter=500,init=None)
```

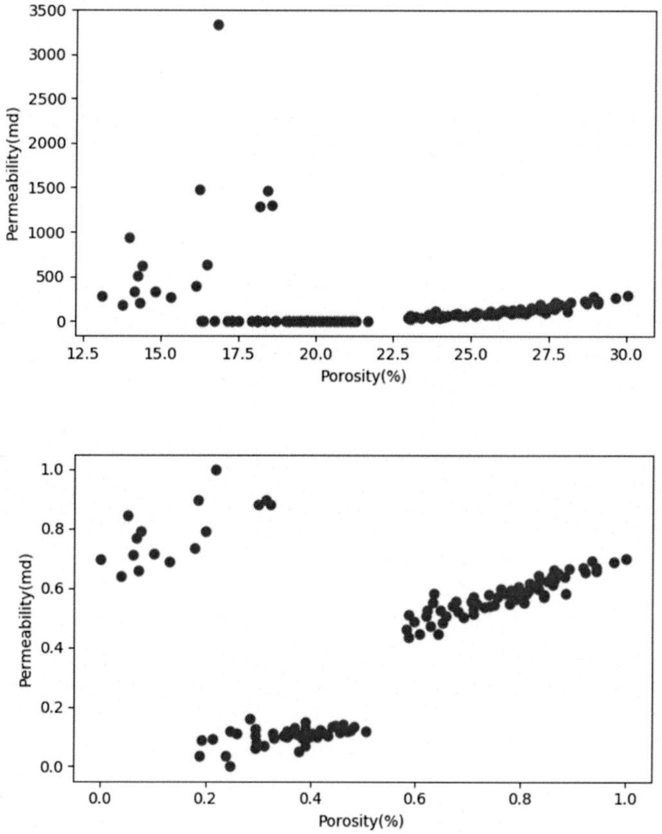

FIGURE 8.26 Permeability versus porosity values (bottom plot shows scaled data).

```
# Plotting defined classes, for each data point in the data set
cluster_membership=np.argmax(u, axis=0)
for i in range(n):
    ax.plot(ds_log_scaled[0,:][cluster_membership==i],
            ds_log_scaled[1,:][cluster_membership==i], '.',
            color=colors[i])
    # Mark the centroid for each class
    for x in cntr:
        ax.plot(x[0], x[1], 'r*')
    ax.set_title('Centers={0}; FPC={1:.2f}'.format(n,
    fpc))
    # Fuzzy partition coefficient storing
    fpcs.append(fpc)
    n = n + 1
```
Python output=Fig. 8.27

In this example, the number of classes (n) varied from 2 to 9. For each value of the class number, skfuzzy.cluster.cmean was used to classify the "ds_log_scaled" data set with class fuzziness index of 1.5, error threshold (as stopping criteria) of 0.001, and a maximum number of iterations of 500. If the error which is the difference between objective functions at two consecutive iterations gets below 0.001 or the number of iterations gets to 500, the algorithm will stop. When the algorithm stops, at each number of classes, it returns several fuzzy c-means clustering parameters. "Cntr" is a 2d array that includes the coordinates of the centroid for each class. "u" is the finalized fuzzy partition matrix (2d array) that shows the membership of each data point to each cluster. "u0" is the initial random guess for the fuzzy partition matrix. "d" is a matrix (2d array) that includes a final euclidean distance of each data point to the class centroids. "p" is the number of iterations. Finally, "fcp," a very important parameter, is the fuzzy partition coefficient that ranges from 0 to 1 and represents a validity index to evaluate the performance of the fuzzy clustering algorithm (Bezdek, 2013). The higher FCP values indicate better clustering quality. FCP is calculated by the following equation:

$$FCP = \frac{1}{N} \sum_{k=1}^{N} \sum_{i=1}^{c} u_{ik}^2 \qquad (8.30)$$

In Fig. 8.27, porosity-permeability data set is clustered from two classes to 9 classes. Each class is shown with different colors, and centroids are shown in red stars. For each number of classes, FCP is shown on top of the plot. In these examples, the best number of classes to group rocks based on porosity and permeability is 3 which also is represented with the highest FCP of 0.99. In order to find the optimum number of classes, it is also possible to plot the fuzzy partition coefficient versus the number of classes and find the maximum value. This can be done with the following Python codes:

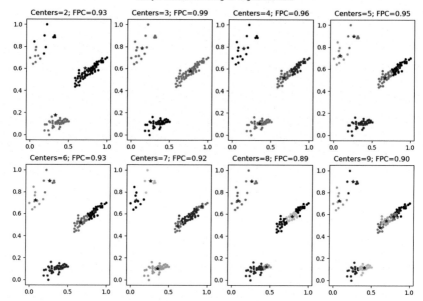

FIGURE 8.27   Fuzzy c-means clustering for log-scaled data varying number of classes from 2 to 9.

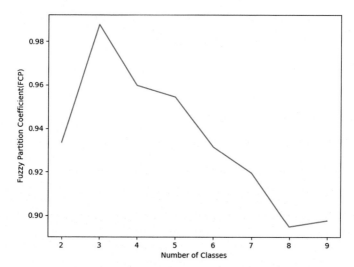

FIGURE 8.28   Fuzzy partition coefficient versus the number of classes.

```
#Plot fuzzy partition coefficient vs number of classes
plt.plot(np.arange(2,10), fpcs)
plt.xlabel("Number of Classes")
plt.ylabel("Fuzzy Partition Coefficient(FCP)")
```
Python output=Fig. 8.28

As illustrated in Fig. 8.28, when the number of classes increases from 2 to 3, fuzzy partition coefficient also reaches the maximum value of 0.99. When the number of classes increases to values higher than 3, the fuzzy partition coefficient starts decreasing. This means that based on this particular porosity/permeability data set, the optimum number of classes or rock types is 3.

References

Ahmed, S.,A., Elkatatny, S., Ali, A. Z., Mahmoud, M., & Abdulraheem, A. (March 22, 2019). Rate of penetration prediction in shale formation using fuzzy logic. In *International petroleum technology conference*. https://doi.org/10.2523/IPTC-19548-MS.

Aminzadeh, F., & Brouwer, F. (January 1, 2006). *Integrating neural networks and fuzzy logic for improved reservoir property prediction and prospect ranking*. Society of Exploration Geophysicists.

Bezdek, J. C. (2013). *Pattern recognition with fuzzy objective function algorithms*. United States: Springer US.

Gholami, V., & Mohaghegh, S. D. (January 1, 2009). *Intelligent upscaling of static and dynamic reservoir properties*. Society of Petroleum Engineers. https://doi.org/10.2118/124477-MS.

Himanshu, S., & Lone, Y. A. (November 30, 2019). *Deep neuro-fuzzy systems with Python: With case studies and applications from the industry* (1st ed.). Apress, 978-1-4842-5360.

Kosko, Bart (June 1, 1994). *Fuzzy thinking: The new science of fuzzy logic, Hyperion*. Reprint Edition. ISBN-10: 078688021X.

Mohaghegh, S. (November 1, 2000). *Virtual-intelligence applications in petroleum engineering: Part 3—fuzzy logic*. Society of Petroleum Engineers. https://doi.org/10.2118/62415-JP.

Mohaghegh, S. D. (August 7, 2017). Mapping the natural fracture network in Utica shale using artificial intelligence (AI). In *Unconventional resources technology conference*. https://doi.org/10.15530/URTEC-2017-2669739.

Odedele, T., & Ibrahim, H. D. (July 2–4, 2014). Oil well performance diagnosis system using fuzzy logic inference models. In *Proceedings of the world congress on engineering 2014* (Vol. I). London, U.K: WCE 2014.

https://pythonhosted.org/scikit-fuzzy/.

Rajasekaran, S., & Pai, G. A. Vijayalakshmi (June 16, 2013). *Neural networks, fuzzy logic and genetic algorithms: Synthesis and applications*. PHI Learning Private Limited, ISBN 978-81-203-2186-1.

Ren, M., Liu, P., Wang, Z., & Yi, J. (2016). A self-adaptive fuzzy c-means algorithm for determining the optimal number of clusters. *Computational Intelligence and Neuroscience, 2016*, 12. https://doi.org/10.1155/2016/2647389.

Rivera, V. P. (January 1, 1994). *Fuzzy logic controls pressure in fracturing fluid characterization facility*. Society of Petroleum Engineers. https://doi.org/10.2118/28239-MS.

Sari, M. (January 1, 2016). Estimating strength of rock masses using fuzzy inference system. In *International society for rock mechanics and rock engineering*.

Widarsono, B., Atmoko, H., Yuwono, I. P., Saptono, F., Tunggal, & Ridwan. (January 1, 2005). *Application of fuzzy logic for determining production allocation in commingle production wells*. Society of Petroleum Engineers. https://doi.org/10.2118/93275-MS.

Xiong, H., & Holditch, S. A. (January 1, 1995). *An investigation into the application of fuzzy logic to well stimulation treatment design*. Society of Petroleum Engineers. https://doi.org/10.2118/27672-PA.

Zadeh, Lotfi A. (January 6, 1965). Fuzzy sets. *Information and Control, 8*(3), 338–353.

Chapter 9

Evolutionary optimization

Optimization is the final and the most important step for most oil and gas−related problems. Optimization can be defined as selecting a set of attributes to deliver the best outcome of a process. Simply stated, optimization chooses the input values (within a reasonable range) to either maximize or minimize an objective function. Some famous optimization examples in the oil and gas industry are maximizing recovery factor, maximizing the net present value (NPV) of a project, minimizing capital expenditure known as CAPEX, minimizing water production, maximizing hydrocarbon production rates, minimizing environmental footprints, minimizing model's error (history matching), etc. These maximizing/minimizing processes can be done by optimizing some problems such as well spacing, completions design, injection rates, and tubing size selection. Note that some of the machine learning techniques covered in previous chapters are actually optimization problems. In clustering problems, the goal is to find the optimum number of classes to differentiate heterogeneity in the data. In the neural networks, the goal is to optimize weights to minimize the model's error (the difference between target values and prediction). All of these examples show the importance of the optimization process.

Optimization algorithms are mainly classified into two types as "deterministic" and "stochastic." Deterministic optimization deals with problems that can be modeled with robust mathematical formulations as objective functions. By using some linear/nonlinear programming, it would be possible to change the independent variables and study their impact on objective function to move towards the optimum point (minima/maxima). If the derivative optimization algorithm (i.e., gradient descent explained in Chapter 6) is used, the optimum solution is deterministic. Since there is randomness neither in the outcome of the objective function nor in the optimization algorithm, the final optimization result will not change if the optimization process is repeated many times. However, if random variables exist either in the objective functions (i.e., Monte Carlo simulations) or in the optimization algorithms, the optimization process is considered stochastic. Some examples of stochastic optimization methods are simulated annealing, cross-entropy method, random search, swarm algorithms, and evolutionary algorithms.

Machine Learning Guide for Oil and Gas Using Python. https://doi.org/10.1016/B978-0-12-821929-4.00005-6
419

In this chapter, two stochastic optimization techniques such as genetic algorithm and particle swarm would be discussed. A brief theory for each method is provided with oil and gas—related examples in Python.

Genetic algorithm

In the middle of the 20th century, computer scientists started learning from nature, especially natural evolution, to deal with challenging optimization problems that required extensive search among highly populated alternatives. Genetic algorithm, introduced by Rechenberg and developed by John Holland in the 1960—70s (Rajasekaran & Pai, 2013), is mainly based on Darwin's evolution theory of "survival of the fittest." The survival of the fittest is linked to the idea that the most successful organisms in surviving are the ones that best adjust to their environment. In a population of organisms, parents produce offspring. The fittest offspring survive and then become parents and this process continues through selection, crossovers, mutation, and inheritance. In 1990s, John Koza introduced "genetic programming" which utilizes genetic algorithm to find the best computer programs for solving a task (Mohaghegh, 2000). Genetic algorithm mimics the process of natural selection. Therefore, it is necessary to have a population of solutions/organisms which are referred to as chromosomes in biology. Each chromosome consists of genes which can be binary values, numbers, symbols, or characters. Next, it is required to evaluate the fitness of each solution/chromosome by the objective function. Chromosomes with higher fitness values go through a mating process with crossovers and mutations of genes and produce new chromosomes (offspring). New chromosomes with higher fitness have a higher probability to be selected for the mating process and produce a new generation. The fittest chromosome, the solution of the optimization, will be found after several iterations (generations).

The genetic algorithm is a popular optimizer with advantages like no need for any derivative calculations. It also can deal with both continuous and discrete functions and work with large search spaces with a lot of parameters. However, since it is a stochastic process, the solution might be uncertain and unrepeatable (tutorialspoint).

There have been several applications of genetic algorithms in the oil and gas industry. The first application of the genetic algorithm in the oil and gas industry was developed by Goldberg et al. for his Ph.D. work supervised by John Holland (Goldberg, 1983). In his work, a learning classifier system (LCS) was proposed to optimize gas pipeline performance by receiving some input variables like inlet/outlet pressures and flow rates, time of day, time of year, pressure rate change, and temperature. LCS decision was to select an optimal flow rate or detect gas leakage. Mohaghegh et al. showed the usage of hybrid neurogenetic approach to optimize hydraulic fracture design for refrac operation for the Clinton Sand in Northeast Ohio (Mohaghegh, 2000). He used the neural networks to predict wells' post fracture deliverability from some important completion attributes. Then, he utilized the neural network as a fitness function for a genetic algorithm to find the best combination of completion parameters for getting the highest post fracture deliverability. Jefferys used a genetic algorithm for minimizing the weight of a column for an offshore structure by searching through different structure designs with various axial loads and radial pressures (Jefferys, 1993). Martinez et al. showed how to implement genetic algorithm to optimize oil production for the wells operating with gas lift. The objective of this work was to assign the optimum amount of gas to each individual well in the field considering the limited gas supply (Martinez et al., 1994). Sarich used genetic algorithm to optimize investment decision-making. In this work, the goal was to select the optimum number of wells to be drilled out of 30 potential wells with capital budget constrain of $30,000,000 and to maximize the NPV (Sarich, 2001). Emrick et al. used a genetic algorithm to optimize well placement with some nonlinear constraints. They used numerical simulator as fitness evaluator to find the optimized number, location, and trajectory of producer/injectors that results in maximized NPV (Emerick et al., 2009). Morales et al. applied a genetic algorithm in gas condensate reservoir for well placement optimization using a heterogeneous reservoir simulation model. They came up with the best well coordinates (x,y,z) and direction that maximized cumulative gas production (Morales et al., 2010). Another application of a genetic algorithm in the oil and gas industry is related to Chaari et al. work that used integrated neural network and genetic algorithm to model two phase-pressure drops in pipes (Chaari et al., 2020). They used a genetic algorithm to find the optimum number of input parameters to train a neural network to predict pressure drops in the pipes with the highest accuracy (least amount of error). The genetic algorithm is also used in many other applications for the oil and gas industry like petrophysics (Fang et al., 1992), seismic inversion (Salamanca et al., 2017), steam injection optimization (Bybee, 2005), and reservoir characterization (Romero et al., 2000).

Genetic algorithm workflow

In order to understand the workflow of a genetic algorithm, it is necessary to review some of the related terminologies. "Population" is a space of all the solutions for the problem to be optimized. "Chromosome" is actually a solution to the problem. "Gene" is the building block of each chromosome. In the computer binary language, 0 and 1 are assigned to the values of the gene as "Allele." In the real-world population space, solutions are not in the binary format. For computational easiness, it is preferred to convert solution space into binary system space which is called "Genotype" through the "Encoding" process. "Fitness" evaluator is generally a mathematical function or objective function that should be maximized or minimized through the optimization process (fitness evaluator can be different from objective function in some cases). The fitness evaluator determines how good a solution is in the population space. Schematic of "Genotype" and its components are shown in Fig. 9.1:

The genetics algorithm consists of the following steps which are performed repeatedly until a criterion is reached: (i) "Population Initialization," (ii) "Fitness Calculation," (iii) "Parent Selection," (iv) "Crossover Operation," and (v) "Mutation." A schematic of these steps can be seen in Fig. 9.2. Next, these steps will be explained and the implementation in Python will be illustrated with a maximization example. The following codes consist of initial variable settings for this example.

```
#Variables Initialization
import random
import numpy as np
import matplotlib.pyplot as plt
from mpl_toolkits.mplot3d import axes3d
seed=11
random.seed(seed)
up_limit, low_limit=30, -30
```

The first step in the genetic algorithm is (i) "Population Initialization." The space of the initial solutions, characterized by the number of solutions and the size of chromosome, should be selected in a way so it covers diversified combinations of solutions. A large population size needs high computational power while a small size decreases mating and offspring quality. The size of the initial population can be a tuning parameter to improve the performance of genetic algorithm. Generally, random initialization is mostly used to generate

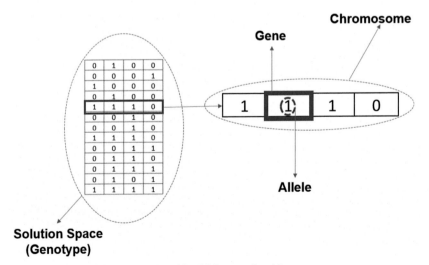

FIGURE 9.1 Schematic of "Genotype" and its components.

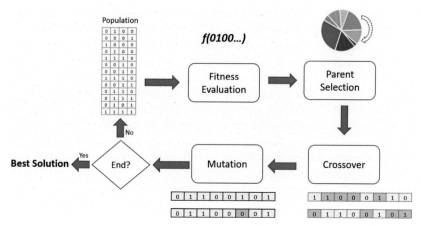

FIGURE 9.2 Genetic algorithm workflow.

initial population; the Python code for this method can be seen as follows. However, in some cases, heuristic initialization is used to seed a randomly initialized population with a few good solutions (based on pre-knowledge about the problem) to improve the convergence of the algorithm towards the optimal solution.

```
#Population Initialization
# Define the Method to Generate a population by choosing samples
#uniformly at random from the range specified
def gen_sample_population(size, x_limits, y_limits):
    x_low, x_up=x_limits
    y_low, y_up=y_limits
    population=np.zeros((size,2))
    for i in range(size):
        population[i,0]=random.uniform(x_low,
            x_up)
        population[i,1]=random.uniform(y_low,
            y_up)
    return population
# Producing a population
population=gen_sample_population(size=50,
                x_limits=(low_limit, up_limit),
                y_limits=(low_limit, up_limit))
# Plot the individual pairs in a sample generation
fig, ax0=plt.subplots(1,1)
ax0.set_title('A Sample Generation')
ax0.set_xlabel('X')
ax0.set_ylabel('Y')
ax0.plot(population[:,0],population[:,1],'ro')
Python output=Fig. 9.3
```

In this code, a method is defined named gen_sample_population that generates a sample population of size = 50 by drawing random numbers from a uniform distribution (random.uniform) within a defined range for X and Y passed as tuples x_limits and y_limits, respectively. In order to have repeatable results the seed for the illustrated randomly drawn population sample should be fixed with random.seed(seed) method which sets the seed to 11. A plot of this sample population is shown in Fig. 9.3.

Note that the population environment or model can be (a) steady state (incremental) or (b) generational. In a steady-state model, a few number of offsprings replace the same number of parents. In generational populations, the number of offsprings is equal to the population size. In other words, offsprings replace all the previous generations or parents.

The next step of the genetic algorithm, after the initialization of the population, is (ii) "Fitness Calculation," where the quality of each solution is calculated by "fitness function." This function receives the solution as an input and returns an output that represents how fit or appropriate the solution is for the optimization problem. By using fitness function, it is possible to compare the quality of all the solutions in the population and determine which ones are the fittest. In many cases, the fitness function can be the same as the objective function. A simple fitness function with one global maxima with form $f(x,y)=5-(x^2+y^2)$ is defined in following codes and the function is plotted.

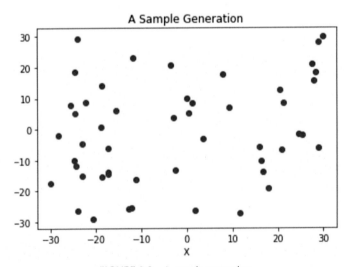

FIGURE 9.3   A sample generation.

```
#Fitness Function
# defining the fitness function which is f(x,y)=5 -(x^2 + y^2)
def function(point):
    return (5-(point[0] ** 2+point[1] ** 2))
#plot the fitness function
from mpl_toolkits.mplot3d import Axes3D
n=50
# linspace returns n equally spaced values over the interval low_limit
#to up_limit
x_vals=np.linspace(low_limit, up_limit, n+1)
y_vals=np.linspace(low_limit, up_limit, n+1)
# meshgrid returns the coordinate matrices by combining all the
#combinations of x_vals and y_vals and creates an n*n x and y spaces
x, y=np.meshgrid(x_vals, y_vals)
z=np.zeros_like(x)
# two nested loops to iterate over all 50 * 50 space for x and y values
for i in range(n+1):
    for j in range(n+1):
        z[i,j]=function([x[i,j], y[i,j]])
#Plot the 3D plot
fig=plt.figure()
#adding 3d projection for the plot
ax0=fig.add_subplot(111,projection='3d')
surf=ax0.plot_surface(x, y, z, rstride=1, cstride=1, cmap='viridis',
linewidth=0, antialiased=False)
#r'' will allow for adding latex code in the title, $$ indicates a latex
#formula
ax0.set_title(r'$5-(x^2+y^2)$')
ax0.set_xlabel('X')
ax0.set_ylabel('Y')
plt.tight_layout()
```
Python output=Fig. 9.4

In the code shown above, the fitness function is defined and is plotted as a 3D plot as illustrated in Fig. 9.4.

If the fitness function is very complex and its computation is computationally expensive, some approximation is required to simplify the fitness function. If the objective is to maximize f(x), fitness function can be the same as f(x). However, for minimization problems, fitness function should be transformed to -1×f(x). Positive and nonzero values are more favorable for genetic algorithm. Therefore, fitness functions should be transformed to meet these conditions (tutorialspoint).

The next step is (iii) "Parent Selection." When the fitness of all the solutions is determined, parent selection process chooses which solutions should be paired as parents to generate the offsprings for the next iteration. In the parent selection and reproduction process, it is very important to keep a

$$5 - (x^2 + y^2)$$

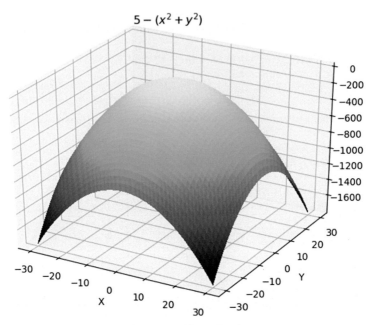

FIGURE 9.4 Fitness function.

balance between coupling the fittest solutions and maintaining the diversity (covering the whole space of solutions) in the selection to prevent premature convergence and getting stuck in local optima. There are various parent selection methods such as Roulette wheel, Boltzmann, Tournament, Rank, Steady-state, and Random selection.

Roulette wheel selection is considered a Fitness Proportionate Selection method that is widely used in genetic algorithms. In this method, a probability is assigned to each solution proportional to its fitness value. Solutions with higher probability have a better chance to be selected for mating and reproduction. To understand Roulette wheel selection, let's assume a circular wheel that is divided into n portions equal to the number of solutions in the population. Solutions, fitness values, and Roulette wheel (pie chart) are shown in Table 9.1 and Fig. 9.5. Higher fitness values result in a larger portion sizes.

In order to select a solution as a parent, the wheel rotates. At the end of the rotation, the portion of the wheel which is in front of the selection point is chosen as the first parent. The chance of election is proportional to the size of each portion or the fitness value. To select the next parent, the first selected parent could be removed from the population. The wheel rotates again, and at the next stop, the one in front of the selection point is selected. It is possible to have two selection points. Therefore, two parents can be selected at each wheel spin. The detail of the Roulette wheel selection method will be explained in an example.

TABLE 9.1 Chromosomes and fitness values for roulette wheel selection.

Solution	Fitness
S1	12
S2	2
S3	4.5
S4	3
S5	8
S6	5
S7	1
S8	7

```
#Parent Selection: Roulette
# select fit individuals as parents
def roul_choice(sorted_population):
    func_vals=np.array([function(x) for x in sorted_population])
    min_fitness=func_vals[0]
# Make all fitness values positive so probabilities can be calculated
#next
    if min_fitness < 0:
        func_vals += -min_fitness
    fitness_sum=sum(func_vals)
# Draw a random variable from uniform distribution
    rand=random.uniform(0, 1)
    acc=0
```

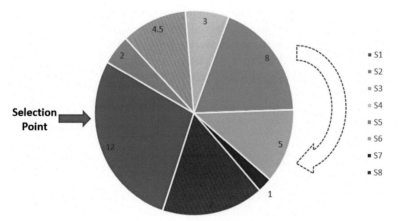

FIGURE 9.5   Roulette wheel selection schematic.

```
    for i in range(len(sorted_population)):
        fitness=func_vals[i]
# Probability of each individual being selected is calculated
        prob=fitness/fitness_sum
        acc += prob
        if rand <= acc:
            return sorted_population[i]
```

In the codes shown above, the method roul_choice where the parents are selected at random can be seen, with their chance of selection proportional to their fitness value. In other words, the individuals with more fitness have a higher probability of being selected as parents. This logic is implemented by dividing the interval (0,1) (our roulette) into pieces (slices of the pie), the size of each slice is equal to the probability of that slice $\frac{fitness}{fitness\_sum}$. For example, if there are four individuals x1,$\cdots$,x4 with fitness values of 3, 6, 9, and 12, the probability of each being selected is 0.1, 0.2, 0.3, and 0.4, respectively. Therefore, if the random draw falls into intervals [0,0.1], (0.1,0.3], (0.3, 0.6], (0.6, 1], the individuals x1,x2,x3,x4 will be chosen, respectively.

In the "Tournament" selection method, k random solutions from the population are selected. The one out of many solutions with the highest fitness is considered a parent. By repeating the same procedure, the next parents can be selected. There are some other parent selection methods like "Rank" selection, "Random" selection, and Boltzman selection.

(iv) "Crossover Operation" is the next step after the parents are selected. When the parents are selected and coupled, their chromosomes should be combined to generate offspring. The combination of the genetic information of the chromosomes is done with crossover operation. In "One Point" or "Single Site" crossover operation, a break location in the chromosomes is selected randomly and divides the chromosomes into two segments (tutorialspoint). The gene information from the second segments is swapped between chromosomes to generate offspring. This process is shown in Fig. 9.6. In "Two-point" crossover operation, instead of one break location, two break locations are selected randomly. Then gene information between two break locations is exchanged between parents to generate offspring as shown in Fig. 9.7. In "Uniform" crossover operation, a random value between 0 and 1 is assigned to each gene in the chromosome. If the random number is equal or greater than 0.5 for a gene, that gene is swapped as shown in Fig. 9.8.

Below is a simple crossover which takes an average of parents values to create children (for the real numbers, geometric mean or arithmetic mean for the cross over can be used):

```
#Crossover
def crossover(point_1, point_2):
    return [(point_1[0] + point_2[0])/2,(point_1[1] + point_2[1])/2]
```

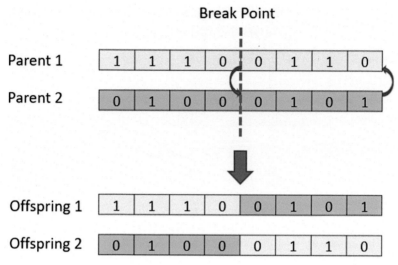

FIGURE 9.6   One point crossover.

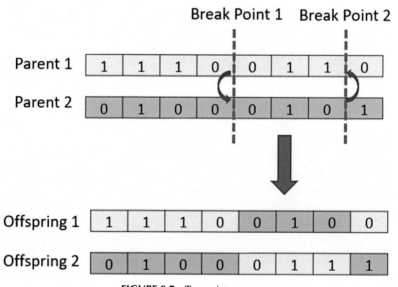

FIGURE 9.7   Two-point crossover.

There are other crossover operations like "matrix," "arithmetic recombination," and "Davis Order." One important characteristic of crossover is the crossover rate/probability, defined as the ratio of offspring chromosomes that

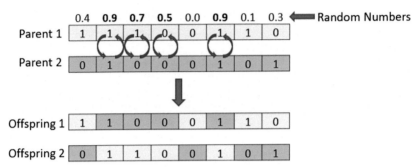

FIGURE 9.8   Uniform crossover.

are generated by crossover operation to the initial parent population. For example, if out of 100 parent chromosomes, 70 offspring chromosomes are made by crossover, then the crossover rate is 0.7. The crossover rate ranges from 0 to 1.

(v) "Mutation" happens after the offspring chromosomes are generated by crossover operation, where some of the genes are tweaked to add diversity to the new generation. This process is called Mutation. The most common mutation operator is "Bit Flip." In the "Bit Flip" mutation, random bit/bits are selected and then the **allele** (value) of the bit/bits is flipped as shown in Fig. 9.9. The mutation rate can be used to understand what percentage of total bits went through mutation. The mutation rate is generally less than the crossover rate.

```
#Mutation
def mutate(point):
    mute_x=point[0] + random.uniform(-0.5, 0.5)
    mute_y=point[1] + random.uniform(-0.5, 0.5)
# Guarantee to be kept inside boundaries
    mute_x=min(max(mute_x, low_limit), up_limit)
    mute_y=min(max(mute_y, low_limit), up_limit)
    return [mute_x,mute_y]
```

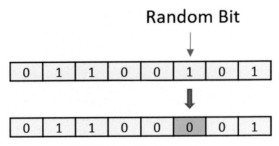

FIGURE 9.9   Bit flip mutation.

In this code, mutation is done by adding a small random value to the individual. It is necessary to make sure the mutated values fall into the specified range. All these steps (ii)−(v) are wrapped in the following method called make_next_generation.

```
#Make Next Generation: steps (ii)-(v)
# (ii) Fitness Function used for (iii) Parents selection and then (iv)
#Crossover and (v) Mutation used to Generate next generation
def offspring_generation(curr_population):
    next_generation=np.zeros(np.shape(curr_population))
# Sort items in the previous_population array based on their function
#value
    sorted_population=sorted(curr_population,key=function)
    population_size=len(curr_population)
    for i in range(population_size):
        # (iii) Parent Selection
        first_parent=roul_choice(sorted_population)
        second_parent=roul_choice(sorted_population)
        # Crossover
        offspring=crossover(first_parent, second_parent)
        # Mutation
        offspring=mutate(offspring)
        next_generation[i,:]=offspring
    return next_generation
```

In these lines of codes, all the steps for (iii) parent selection using (ii) fitness function and also performing (iv) crossover and (v) mutation for creating the new generation are illustrated.

When the new generation of the solutions is created, the process is repeated by going through steps (ii) fitness evaluation, (iii) parent selection, mating, (iv) crossover, and (v) mutation. The mentioned loop continues until the termination condition for the genetic algorithm is achieved. The termination conditions can be one of the followings: (a) when population has not improved (considering a threshold) in the past N number of iterations, (b) when reaching to specified number of iterations, (c) obtaining a defined value for fitness function. Genetic algorithm workflow is shown in following codes:

```
#Genetics Algorithm Workflow
from matplotlib.patches import ConnectionPatch
#Perform the Genetics Algorithm to generate generations and find the
#best individuals in each generation
generations=100
best_val=float('-inf')
best_pair=np.zeros((2))

prev_best_pair=np.zeros_like(best_pair)
```

```
best_population=-1
# Threshold for improvement in order to continue the search
threshold=10e-5
stop=False
coordsA="data"
coordsB="data"
fig, ax0=plt.subplots(1,1)
ax0.set_xlabel('X')
ax0.set_ylabel('Y')
ax0.set_title(r'$5-(x^2+y^2)$')
ax0.grid(True)
# Mark the optimum Individuals [0,0] as a star
ax0.plot([0], [0], 'g*', markersize=15, linewidth=0.15,
  label='Optimum Individual')
ax0.legend(loc='lower left')
# (i) Population Initialization
population=gen_sample_population(size=50, x_limits=(low_limit,
  up_limit), y_limits=(low_limit, up_limit))
# (ii) Fitness Function Calculation
func_vals=np.array([function(x) for x in population])
gen_max_val=max(func_vals)
best_population=1
# Update the best value obtained so far
best_val=gen_max_val
best_pair=population[np.argmax(func_vals),:]
print('Generation:' ,1, ' Best individual: ',best_pair)
# Loop through the number of generations
for i in range(1, generations):
    print("Iteration number: ", i + 1)
# (ii) Fitness Function used for (iii) Parents selection and then (iv)
  #Crossover and (v) Mutation used to Generate next generation
    population=offspring_generation(population)
    func_vals=np.array([function(x) for x in population])
    gen_max_val=max(func_vals)
# check if this generation maximum value (gen_max_val) is better than
#our best value (best_val) found thus far.
    if (gen_max_val > best_val):
        diff=abs(gen_max_val - best_val)
# If the improvement amount (diff) is less than threshold set
#stop=True => Break the loop, otherwise continue the search
        stop=True if diff < threshold else False
        best_population=i + 1
# Update the best value obtained so far
        best_val=gen_max_val
        prev_best_pair=np.copy(best_pair)
```

```
        best_pair=population[np.argmax(func_vals),:]
        print('Generation:' ,i + 1, ' individual: ',best_pair,
          'Improves the optimal value by amount ', diff)
# If it is after iteration 1 plot the arrow to show the evolution to the
#new best_pair
        if(i > 1):
# plots the best_pair as points with 'o' marker on the plot
            ax0.plot([best_pair[0]],[best_pair[1]], 'o')
# Mark each best_pair point with its iteration number (i + 1)
            ax0.text(best_pair[0] + 0.001 ,
            best_pair[1]+0.015 ,'{}'.format(i + 1))
# Create the arrow from prev_best_point to best_point to show the
#evolution to the new individual (best_pair)
            con=ConnectionPatch(prev_best_pair,
            best_pair, coordsA=coordsA, coordsB=coordsB,
            arrowstyle="-|>", shrinkA=2, shrinkB=2,
            mutation_scale=15, fc="w")
            ax0.add_artist(con)
        if (stop):
            break
# Print the final best individuals
print('Final Result is from iteration: ',best_population,
  ' for coordinates: ', best_pair , ' best value: ', best_val)
```
Python output=Fig. 9.10

In this code, all the steps (i)–(v) are shown and steps (ii)–(v) are repeated until the improvement achieved is less than the threshold defined as 10e-5 or the maximum number of iteration generations = 100 is reached. The result of this workflow can be observed in Fig. 9.10. As observed, the genetic algorithm result is approaching the optimum values [0,0] marked as a star, over multiple iterations. The generations 3, 4, 12, 16, 18, 19, 20, 22, 25, 27, 31, 62 improve the best value. The best value is from iteration 62 with pair [0.0122, -0.0087] and value 4.9998. The algorithm is terminated when generation 62 resulted in improvement of 6.8536e-05 which is less than the threshold defined as 10e-05. This is interpreted as the solution close enough to the optimum value, and the algorithm can be terminated.

Genetic algorithm example: EUR optimization

In this example, the objective is to use a genetic algorithm to find optimum drilling and completions parameters that maximize a well's estimated ultimate recovery (EUR). In order to do that, it is necessary to have a model that predicts EUR of a well with respect to various geological, drilling, and completion parameters. This model would be the fitness function that evaluates the effectiveness of drilling and completions jobs. Shale Gas Wells

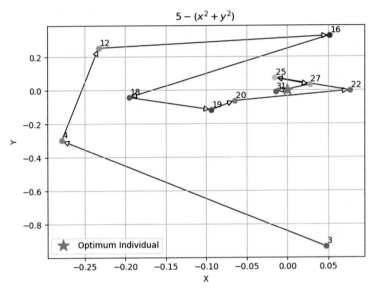

FIGURE 9.10 Genetics algorithm evolution.

example that was covered in Chapters 6 and 7 is used for optimization. The genetic algorithm is implemented to find the optimum lateral length and proppant amount that leads to the highest EUR for a well in a specific location. Similar to previous examples, the first step is to load the Shale Gas Well's data, define input and output, normalize the data to 0 to 1 range, split the data into train and test, and finally train a neural network. The mentioned steps can be done in Python with the following codes:

```
import numpy as np
import pandas as pd
import matplotlib.pyplot as plt
from sklearn.pipeline import make_pipeline
from sklearn.neural_network import MLPRegressor
from mpl_toolkits.mplot3d import Axes3D
from sklearn.preprocessing import MinMaxScaler
#Import the data set
dataset=pd.read_csv('Chapter9_Shale Gas Wells.csv')
X=dataset.iloc[:,0:13]
y=dataset.iloc[:,13].values
# Fix the seed number for splitting the data into training and testing
    seed=15
np.random.seed(seed)
from sklearn.model_selection import train_test_split
X_train, X_test, y_train, y_test=train_test_split(X, y,
    test_size=0.25)
```

```
#Specifying the neural network topology with MLPRegressor including 2
  #hidden layers
global modelNN
modelNN=make_pipeline(MinMaxScaler(), MLPRegressor(
  hidden_layer_sizes=(25,25),learning_rate_init=0.01,
  early_stopping=True,max_iter=500))
#Training the model

modelNN.fit(X_train, y_train)
print("Test R2 score: {:.2f}".format(modelNN.score(X_test,
  y_test)))
```

Python Output =
Test R2 score: 0.79

Pipeline was used to normalize the data and then feed it to the neural network. There are two hidden layers with 25 neurons per layer, with the initial learning rate of 0.01. The R^2 for the test data set is 0.79. The next step is to import the genetic algorithm optimizer and call it to find the maximum values for the well's EUR with two variables: lateral length and proppant loading. "Genetic algorithm" library should be installed by "pip install genetic algorithm" command in Anaconda prompt (pypi). The genetic algorithm generally finds the minimum of a function. Therefore, the neural network function should be multiplied by -1 to return the maximum value. Afterward, it is necessary to specify the range of variables as the solution space. A 2D array, named "varbound," is used to define the minimum and maximum of each variable. In the Shale Gas Wells example, there are 13 attributes. From these 13 attributes, lateral length and proppant loading should vary from 4500 to 11,500 ft and 1100 to 3200 #/ft, respectively. The rest of the variables are equal to their average values in the main data set (X). The Dip would be zero, meaning the well is down-dip. In reality, it is possible to see where the well is located and use the corresponding geological parameters such as porosity, formation thickness, and water saturation. Next, a genetic algorithm can be used to optimize design variables such as drilling and completions. The Python codes for making the neural network output negative and defining the variables are as follows:

```
from geneticalgorithm import geneticalgorithm as ga
def f(X):
# Reshape the column vector X to a row vector X_Row
    X_Row=X.reshape(1,-1)
# Return negative value for since GA minimizes the objective function
    return -1*modelNN.predict(X_Row)
#Lateral Length (4,500 ft to 11,500 ft) and Proppant loading (1100#/ft
  #to 3200#/ft). The rest of the attributes should be equal to average.
#Dip attribute should be zero.
varbound=np.array([[0.0,0.0]]*13)
avg_vals=X.mean(axis=0)
```

```
varbound[:,0]=avg_vals; varbound[:,1]=avg_vals
varbound[5,0] =4500
varbound[5,1]=11500
varbound[3,0]=0
varbound[3,1]=0
varbound[12,0]=1100
varbound[12,1]=3200
```

The next step is to define genetic algorithm parameters in algorithm_param. There is no maximum number of iterations for this example; however, the algorithm stops when, after 100 iterations, no improvement is achieved. The population size for the solutions or chromosomes is 100. Mutation probability defining what percentage of bits experience mutation is 0.1 (10%). Elit_ratio, a small portion of the fittest chromosomes that would be carried to the next generation without any change is 0.1. The crossover probability (or rate) that defines what percentage of offspring chromosomes are generated by crossover operator is 0.5 (50%). Parent_portion which indicates what percentage of the new population includes the parent chromosomes is 0.2. It means in the new iteration, 20% of the population are parents and 80% are the offspring chromosomes. The crossover type is uniform, meaning random numbers are assigned to each bit, and the ones with random values higher than 0.5 are switched. Next, it is required to identify a genetic algorithm model with the function "f" (it is actually the negative of neural network), the dimension of the solution (13 in this example), boundaries of the solutions (variable_boundaries), and finally, algorithm parameters. After specifying all the model requirements, it is possible to run the genetic algorithm with model.run command. Convergence report and solution of the genetic algorithm can be accessed with "model.report" and "model.output_dict," respectively. Running the genetic algorithm and viewing the results can be done in Python with the following codes:

```
algorithm_param={'max_num_iteration': None, 'population_size':100,
    'mutation_probability':0.1,'elit_ratio': 0.1,
    'crossover_probability': 0.5,'parents_portion': 0.2,
    'crossover_type':'uniform','max_iteration_without_improv':100}
model=ga(function=f,dimension=13,variable_type='real',
    variable_boundaries=varbound,
        algorithm_parameters=algorithm_param)
model.run()
convergence=model.report
solution=model.output_dict
Python output=Fig. 9.11
```

The output of the genetic algorithm is shown in Fig. 9.11. The convergence report is actually a plot that shows the objective function values versus the number of iterations. After 100 iterations from iteration number 70, no improvement was observed in the fitness function. The solution of the optimization is in an array including all the attributes. Maximum gas EUR was calculated to be 20.8Bcf with the optimized lateral length and proppant loading of 11498 ft and 3199.9 lb/ft (attribute 6th and 13th in the array). In order to visualize the space of solutions, a surface of calculated EUR values versus combinations of lateral length and proppant loading values is plotted with the following Python codes:

```
n=100
m=50
# linspace returns n and m equally spaced values over the interval
    #passed to it
x_vals=np.linspace(4500,11500,n+1)
y_vals=np.linspace(1100,3200,m+1)
# meshgrid returns the coordinate matrices by combining all the
    #combinations of x_vals and y_vals and creates an n*m x and y spaces
x,y=np.meshgrid(x_vals, y_vals)
z=np.zeros_like(x)
item=np.array([0.0]*13)
item [:]=avg_vals
for i in range(m+1):
    for j in range(n+1):


The best solution found:
[1.47640316e+02 3.51343874e+01 8.20158103e+02 0.00000000e+00
1.62365613e+02 1.14983729e+04 6.30790514e+01 7.33754941e+00
7.01049012e+03 1.92134387e+01 6.48454545e+01 9.30256917e-01
3.19921888e+03]

Objective function:
-20.758964946179123
```

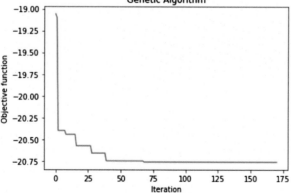

FIGURE 9.11 Genetic algorithm results for lateral length and proppant loading optimization.

```
            item[3]=0
            item[5]=x[i,j]
            item[12]=y[i,j]
            z[i,j]=-1*f(item)
fig=plt.figure()
#adding 3d projection for the plot
ax0=plt.axes(projection='3d')
surf=ax0.plot_surface(x, y, z, rstride=1, cstride=1,
  cmap='viridis', linewidth=0, antialiased=False)
ax0.set_xlabel('Lateral Length(ft)')
ax0.set_ylabel('Proppant Loading(#/ft)')
ax0.set_title('EUR(Bcf)')
plt.tight_layout()
Python output=Fig. 9.12
```

As illustrated in Fig. 9.12, the maximum value of EUR is observed at the largest values of lateral length and proppant loading which is almost similar to the genetic algorithm optimized results.

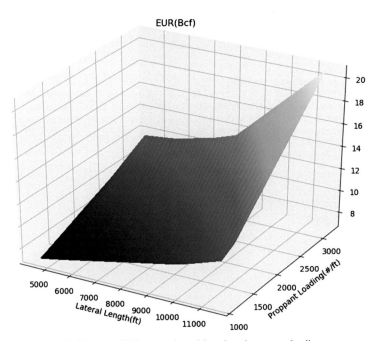

FIGURE 9.12 EUR versus lateral length and proppant loading.

Particle swarm optimization

In the previous section, the genetic algorithm, which is a nature-based search optimization, was explained. The genetic algorithm actually honors the theory that Darwin formulated for evolution according to natural selection. In this section "particle swarm optimization" (PSO) which is another nature-based optimization algorithm would be covered.

Swarm is considered a population of moving particles (in nature it can be bees, birds, or fish) that try to make a cluster while each particle seems to move in a disorganized manner. In nature, swarms showed some sort of sociocognitive behavior across the particles, helping them solve problems like finding food or escaping from predators. The cultural behavior of the swarm particles can be categorized as evaluating, comparing, and imitating (Eberhart et al., 2001). At first, particles need to evaluate the environment to differentiate attracting or repelling features in order to make an appropriate decision. Afterward, individuals (particles) compare themselves with others and find which ones are in better positions. Finally, the individuals imitate the ones that are in superior positions. The mentioned principle helps particles adapt to the environment, deal with challenges, and solve difficult problems. Scientists and engineers tried to use the same principles of particle swarm intelligence behavior to solve complex optimization problems.

Kennedy and Eberhart introduced particle swarm optimization in 1995 at a neural networks conference and then published the *Swarm Intelligence* book in 2001 (Eberhart et al., 2001). Similar to the genetic algorithm, particle swarm optimization has the capacity to optimize the problems with none-differentiable objective function (the main requirement for gradient-based optimizer). PSO is applied in various optimization problems like neural network training, fuzzy control, pattern recognition, robotics/nanorobotics, and even movie especial effects.

There are several applications of PSO in the oil and gas industry. Mohamed et al. used PSO for reservoir history matching (Mohamed et al., 2011). The objective of history matching of a synthetic reservoir simulation model was to estimate fault characteristics such as throw thickness, good and poor permeability factors. They used oil and water rates as two objective functions for history matching. Isebor et al. applied multiple derivative-free optimizers such as particle swarm to come up with the best development plan (Isebor et al., 2013). They used two-dimensional reservoir models, including channels with water injection and oil-producing wells. The development plan was optimized by deciding the number, location, and control strategy of the wells. Al-Nemer et al. (Al-Nemer et al., 2015) evaluated the performance of four stochastic optimization algorithms like particle swarm for generating automated type-curve matching for pressure transient analysis (well-test). A dual-porosity reservoir model (proposed by Warren and Root) was used to simulate pressure data with added noise that would be considered real pressure data. Next,

PSO was applied to estimate fracture permeability, matrix porosity, skin factor, the radius of investigation, storativity ratio, and interporosity flow parameter. Jesmani et al. applied PSO to find the best well placement for two different case studies (Jesmani et al., 2015). In one case study, they used a synthetic reservoir model with geological data taken from the Norne field in the Norwegian Sea. They used PSO to maximize NPV by finding the best combination of wells' distance, lateral length, orientation, and keeping the wells in a specific part of the reservoir. Ottah et al. utilized PSO to history match a reservoir and find aquifer properties with material balance technique (Ottah et al., 2015). Self et al. applied particle swarm to reduce drilling costs by finding optimum operational parameters (Self et al., 2016). In this work, they considered the rate of penetration (ROP) as objective function and minimized cost per foot by adjusting downhole weight on bit, revolutions per minute, pull depth, and bit combinations (In chapter 5, a supervised ML model to estimate ROP using various input drilling features was build and discussed). Al-Mudhafar et al. showed the usage of particle swarm in optimizing the CO_2-EOR process for unconventional oil reservoirs (Al-Mudhafar et al., 2017). They used a black oil reservoir model with five injectors/producers to forecast 30 years of performance of a shale oil reservoir that went through CO_2 flooding. They considered eight optimization parameters like injection/production/soaking duration, minimum bottom hole pressure, and maximum oil/injection rates. Khan et al. implemented PSO to maximize oil recovery from a naturally fractured carbonate reservoir including smart wells (Khan et al., 2018). The objective of this study was to determine the optimum setting of the inflow control valve (choke setting) to maximize oil recovery. Yan et al. applied PSO for multilayer geosteering (Yan et al., 2018). They wanted to use logging while drilling (LWD) measurements and use particle swarm to estimate some parameters like layer boundaries and layer resistivity. There have been other applications of PSO in the oil and gas industry like geophysical data inversion (Fernandez-Martinez et al., 2008), location of the critical failure surface for rock slope stability analysis (Javadzadeh & Javadzadeh, 2008), preselection of reservoir models (Le Ravalec-Dupin et al., 2010), and production optimization (Zhao et al., 2011).

Particle swarm optimization theory

In order to understand the basics of particle swarm theory, it is useful to look at a famous example of birds' swarm decision-making. When flying birds decide to land, they need to solve a complex problem and find the optimized landing point with the maximum chance of finding food and minimum risk of predators. The fact that swarm of birds can communicate and share information as social behavior enables them to find the optimum solution. The most important factor in the social behavior of birds' swarm is that each bird (particle) balances its own knowledge with swarm knowledge. Birds start searching for the solution (let's assume food) and then they communicate the solution that they find. Next, they follow the bird that has the best solution which means is closest to the food.

The goal for particle swarm is to find the optimum solution for an objective function denoted as $f(X)$ where X is a vector of n variables, $X = [x_1, x_2, x_3, ..., x_n]$. In the PSO, X is also considered as position vector and $f(X)$ is fitness or objective function. With fitness function, it would be possible to evaluate the goodness of each position and compare it with other positions. At each time or iteration step t, a swarm of P particles (each particle is denoted with index i) includes position and velocity vector for particle i as $X_i^t = (x_{i1}, x_{i2}, x_{i3}, ..., x_{in})^t$ and $V_i^t = (v_{i1}, v_{i2}, v_{i3}, ..., v_{in})^t$. The objective for all the particles in the swarm is moving toward the optimum after each iteration until they get to the global optimum. For the ith particle , velocity vector at iteration $t+1$ *denoted as* V_{ij}^{t+1} determines the new position X_{ij}^{t+1} of the jth variable of that particle with following equation (Almeida & Leite, 2019):

$$X_{ij}^{t+1} = X_{ij}^t + V_{ij}^{t+1} \tag{9.1}$$

where i varies from 1 to P and j varies from 1 to n. To find the new position of each particle, it is necessary to calculate V_{ij}^{t+1} with the following equation:

$$V_{ij}^{t+1} = wV_{ij}^t + c_1 r_1^t \left(pbest_{ij} - X_{ij}^t \right) + c_2 r_2^t \left(gbest_{ij} - X_{ij}^t \right) \tag{9.2}$$

The velocity of each particle at iteration t+1 is influenced by three components as inertia $\left(wV_{ij}^t \right)$, individual cognitive component $c_1 r_1^t \left(pbest_{ij} - X_{ij}^t \right)$, and social component $c_2 r_2^t \left(gbest_{ij} - X_{ij}^t \right)$. In inertia term, w, which is called inertia weight constant, controls the balance between global search (exploration) and local search (exploitation). When w is high, the particle is mainly following the previous direction, while lower values for w lets the particle to move into different regions. The individual cognitive component calculates the difference between the current position (X^t_{ij}) of the particle and particle's best previous position $(pbest_{ij})$ and multiplies it by a weight factor c_1 and a random number r_1^t. This component determines the importance of the particle's own experience regarding finding the optima in the search space. The random parameter r_1^t prevents the particle from becoming stuck in local optimum. The social component considers the fact that particles can communicate with each other and share the information about their previous experiences. After particles' communication, it is possible to find the best value of the fitness function obtained by a particle in the swarm as global best point or $gbest_{ij}$. Therefore, the difference between a particle's location (X^t_{ij}) and the global best point $(gbest_{ij})$ is the driving force that directs the particle towards global optima (Almeida & Leite, 2019). This driving force is multiplied by social learning factor (c_2) and random variable of r_2^t. Fig. 9.13 shows how new position of particle i is calculated.

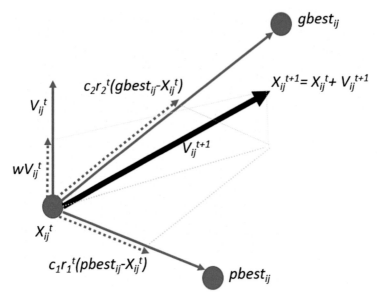

FIGURE 9.13 The new position and velocity vector for particle i in the swarm.

PSO can be performed with the following steps:

(i) Initialize particles' positions (X^0_{ij}), velocities (V^0_{ij}), and best position ($pbest_{ij}$)
(ii) Evaluate fitness values for each particle based on their initial positions and then calculate global best position
(iii) Update particles' best previous position ($pbest_{ij}$) and global best position ($gbest_{ij}$)
(iv) Update particles' velocities and positions
(v) If termination condition is satisfied then end the process otherwise go back to step (iii).

To illustrate the steps of PSO, consider the objective function $-50\cos(x)\cos(y)+x^2+y^2$ which has four local minimums [-3, -3], [-3, 3], [3, -3], [3, 3] with a value of -31, and a global minima at [0, 0] with a value of -50. The 3D plot of this function is shown in Fig. 9.14.

Fig. 9.14 can be plotted using the following lines of Python codes:

```
#3D plot
import numpy as np
import math
import random
import matplotlib.pyplot as plt
#function that models the problem
def function(position):
```

$$-50\cos(x)\cos(y) + x^2 + y^2$$

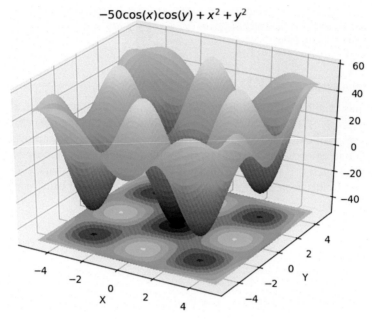

FIGURE 9.14 Swarm optimization example.

```
    return -50*math.cos(position[0])*math.cos(position[1]) +\
    position[0]**2 + position[1]**2
#plot the fitness function
from mpl_toolkits.mplot3d import Axes3D
n=100
low_limit=-5
up_limit=5
# Creates n equally spaced values between low_limit and up_limit
x_vals=np.linspace(low_limit,up_limit,n+1)
y_vals=np.linspace(low_limit,up_limit,n+1)
# Create n*n matrices x and y with all combinations of x_vals and y_vals
x,y=np.meshgrid(x_vals, y_vals)
z=np.zeros_like(x)
for i in range(n+1):
    for j in range(n+1):
        z[i,j]=function([x[i,j],y[i,j]])
#Plot the 3D plot
fig=plt.figure()
#adding 3d projection to the plot
ax0=plt.axes(projection='3d')
surf=ax0.plot_surface(x, y, z, rstride=1, cstride=1,
  cmap='viridis', linewidth=0, antialiased=False)
# Plot the contour plot of z values
```

```
ax0.contourf(x, y, z, zdir='z', offset=-50, cmap='viridis')
ax0.set_title(r'$-50\cos(x)\cos(y) + x^2 + y ^2$')
ax0.set_xlabel('X')
ax0.set_ylabel('Y')
plt.tight_layout()
Python output=Fig. 9.14
```

In this example, all five steps of PSO problems are illustrated. First, all the parameters w, c1, c2, and other variables are initialized. The stopping criterion is also set to 10e-05, meaning that the algorithm will stop if the improvement from one iteration to the next is less than this threshold. The number of particles (n_particles) is set to 50 with the maximum number of iterations (n_itr) set to 50. In this code, the optimum which is [0, 0] is plotted using a star.

```
#Parameter Initialization
#Initializing the velocity calculation parameters
w=0.2
c1=0.5
c2=0.9
seed=12
np.random.seed(seed)
n_itr=50
threshold=10e-05
n_particles=50
```

Next, the particles' positions are randomly initialized by drawing from a uniform distribution. Also, the initial value of the best values and positions are set to infinity since this is a minimization problem and any value will be smaller than infinity. Therefore, the best values will be updated correctly. The velocity values (velocity) are initialized to zero.

```
#Variable Initialization
# (i) Particle position initialization
curr_pos=np.random.uniform(-30,30,(n_particles,2))
part_best_pos=curr_pos
part_best_val=np.array([float('inf') for _ in range(n_particles)])
glob_best_val=float('inf')
glob_best_prev_val=float('inf')
glob_best_pos=np.array([float('inf'), float('inf')])
velocity=([np.array([0, 0]) for _ in range(n_particles)])
iteration=0
markers=['o', 'x', 'c*', 'r+' , 'bo', 'x']
c=0
```

In this for loop, the steps (ii)–(v) are repeated until either the stopping criterion is met or the number of iterations is completed (n_itr). In each iteration, the best position of each particle (part_best_pos) as well as the best position among all particles (glob_best_pos) will be updated. Finally, the velocity of each particle is updated based on its last iteration velocity plus randomly weighted social and local velocity adjustment terms. The new position is calculated based on this new velocity and current position of the particle.

```python
fig, ax0=plt.subplots(1,1)
ax0.set_xlabel('X')
ax0.set_ylabel('Y')
ax0.set_title(r'$-50\cos(x)\cos(y) + x^2 + y ^2$')
ax0.grid(True)
# Mark the optimum Individuals [0,0] as a star
ax0.plot([0], [0], 'g*', markersize=20,
label='Optimum Individual')
ax0.legend(loc='lower left')
plt.tight_layout()
#Steps (ii)-(v)
for iteration in range (n_itr+1):
    print ('Iteration: ', iteration)
    for i in range(n_particles):
# (ii) Evaluate fitness value
        fit_candidate=function(curr_pos[i])
# (iii) Update particle best position and global best position
        if(part_best_val[i] >= fit_candidate):
            part_best_val[i]=fit_candidate
            part_best_pos[i]=curr_pos[i]

    part_vals=np.array([function(x) for x in curr_pos])
    parts_best_val=min(part_vals)
    best_part=curr_pos[np.argmin(part_vals),:]
    if(glob_best_val >= parts_best_val):
        glob_best_prev_val=glob_best_val
        glob_best_val=parts_best_val
        glob_best_pos=best_part
# Plot selected iterations particles positions
    if (iteration in [2,5,7,9,11,14]):
        ax0.plot(part_best_pos[:,0],part_best_pos[:,1],
        markers[c],label='{}'.format(iteration),
        markersize=c+3)
        c=c+1
# (iv)   Update the particle velocities and new positions
    for i in range(n_particles):
        new_velocity=(w*velocity[i]) + (c1*np.random.uniform()) \
        * (part_best_pos[i] - curr_pos \
        [i]) + (c2*np.random.uniform()) * \
        (glob_best_pos-curr_pos[i])
```

```
            new_position=new_velocity + curr_pos[i]
            curr_pos[i]=new_position
            velocity[i]=new_velocity
    # (v) Terminate if the termination condition is met
        if(abs(glob_best_val - glob_best_prev_val) < threshold):
            break
ax0.legend()
print("The best position is ", glob_best_pos, "in iteration number ",
iteration)
```

Multiple snapshots of running the PSO are shown in Fig. 9.15. The positions of the particles at iterations 2, 5, 7, 9, 11, and 14 are plotted. The global best value (glob_best_val) is not improving after iteration 14 which has a value of -49.9984 at a position of [x=-0.00789019 y=-0.00041137]. This value is very close to the optimum which is -50. As observed, at each plotted iteration, the particles are getting closer to the global minima [0, 0]. Note that there are four local minima at locations [-3, -3], [-3, 3], [3, -3], [3, 3] with value of -31 (see Fig. 9.14).

NPV maximization example

In this example, data from a work by Yu et al. (Yu & Sepehrnoori, 2013) are used to optimize the design of hydraulically fractured horizontal wells with the particle swarm optimization algorithm. In Yu's work, a numerical reservoir

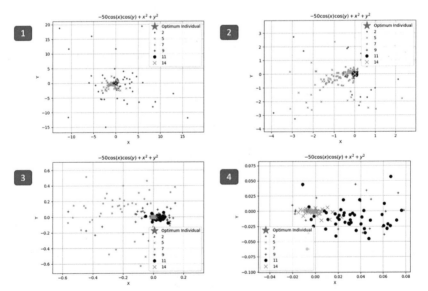

FIGURE 9.15 Swarm optimization at different iterations. The scale of x and y axis gradually decreases from plot 1 to 4 to show how particles move toward the optima.

simulation was built to model the production of Barnett shale gas wells. The simulator was used to find 38 production scenarios by incorporating uncertainty in porosity/permeability and changing design parameters like fracture spacing, fracture half-length, fracture conductivity, and well distance. For each scenario, NPV was calculated with the specific gas price and CAPEX/OPEX assumptions. Finally, a response surface is generated to find a correlation between NPV and design/geological parameters.

For this example, first a polynomial surface function would be generated as objective function. Then PSO would be used to find optimized completions design parameters. Loading the data, importing the libraries, defining input and output, viewing descriptive statistics, and splitting the data into training and testing sets (80% training and 20% testing) can be done in Python with the following codes:

NPV data set can be found in the following location https://www.elsevier.com/books-and-journals/book-companion/9780128219294

```
import numpy as np
import pandas as pd
from sklearn.pipeline import make_pipeline
```

```
       Porosity  Permeability(md)  Fracture-Half Length(ft)  \
count  38.000000        38.000000                 38.000000
mean    0.061579         0.000292                300.000000
std     0.016526         0.000191                 83.827364
min     0.040000         0.000050                200.000000
25%     0.042500         0.000050                200.000000
50%     0.060000         0.000295                300.000000
75%     0.080000         0.000500                400.000000
max     0.080000         0.000500                400.000000

       Fracture Conductivity(md-ft)  Fracture Spacing(ft)  Well Distance(ft)  \
count                     38.000000             38.000000          38.000000
mean                      24.842105             65.526316         757.894737
std                       19.130457             25.542478         198.142153
min                        1.000000             10.000000         500.000000
25%                        4.500000             40.000000         525.000000
50%                       25.000000             60.000000         800.000000
75%                       44.500000             80.000000         975.000000
max                       50.000000            100.000000        1000.000000

       NPV($MM)
count  38.000000
mean    8.905263
std     4.044078
min     2.200000
25%     6.100000
50%     8.500000
75%    12.100000
max    18.900000
```

FIGURE 9.16 NPV data set descriptive statistics.

```
from sklearn.linear_model import LinearRegression
from sklearn.preprocessing import PolynomialFeatures
from sklearn.linear_model import LassoCV
from sklearn.preprocessing import MinMaxScaler
import matplotlib.pyplot as plt
dataset=pd.read_csv('Chapter9_NPV Data Set.csv')
X=dataset.iloc[:,0:6]
y=dataset.iloc[:,6].values
print(dataset.describe())
seed=15
np.random.seed(seed)
from sklearn.model_selection import train_test_split
X_train, X_test, y_train, y_test=train_test_split(X, y,
    test_size=0.2)
```
Python output=Fig. 9.16

For making a correlation between NPV and six design parameters, regularized regression is used to find a fit with polynomial features. Polynomial features with order 3 were generated for the input parameters. "Lasso" (Least Absolute Shrinkage and Selection Operator), also known as L1 regularization, is a technique that adds the absolute values of the weights to the regression objective function as a penalty to prevent overfitting. Alpha is the parameter that regularizes the regression. Epsilon in Lasso is the ratio of alpha-min to alpha-max. Cross-validation (CV) is also used to take the best regression parameters (explained in Chapter 7). Pipeline was applied to make a model including MinMaxScaler for data normalization, generating polynomial features in the input data, implementing Lasso regularization with CV for regression. When the polynomial model is trained, predicted versus actual NPVs with the R^2 score can be plotted with following lines of codes in Python:

```
#Making Polynomial Model with Lasso parameters for the regression,
    #and cross validation with 4 folds
model=make_pipeline(MinMaxScaler(),PolynomialFeatures(3,
    interaction_only=False), LassoCV(eps=0.0005,n_alphas=10,
    max_iter=10000, normalize=True,cv=4))
model.fit(X_train,y_train)
#Predicting the outputs and calculating R2s
y_pred_Test=np.array(model.predict(X_test))
test_scoreTest=model.score(X_test,y_test)
y_Pred_Test_Train=np.array(model.predict(X))
test_scoreTest_Train=model.score(X,y)
#Plotting the Outputs
fig, ax=plt.subplots()
ax.scatter(y_test,y_pred_Test)
ax.plot([0, 18], [0, 18], 'k--', lw=4)
ax.set_xlabel('NPV_Test(MM$)')
```

```
ax.set_ylabel('NPV_Predicted(MM$)')
plt.text(2,17,"R2="+str(test_scoreTest).format("%.2f"))
plt.show()
fig, ax=plt.subplots()
ax.scatter(y,y_Pred_Test_Train)
ax.plot([0, 20], [0, 20], 'k--', lw=4)
ax.set_xlabel('NPV_Test_Train(MM$)')
ax.set_ylabel('NPV_Predicted(MM$)')
plt.text(2,18,"R2="+str(test_scoreTest_Train ).format("%.2f"))
plt.show()
Python output=Fig. 9.17
```

As illustrated in Fig. 9.17, R^2 values for model's prediction versus actual NPVs on the test and the whole data set are 0.63 and 0.91, respectively. This model would be used for the optimization of the NPV. For this example, two

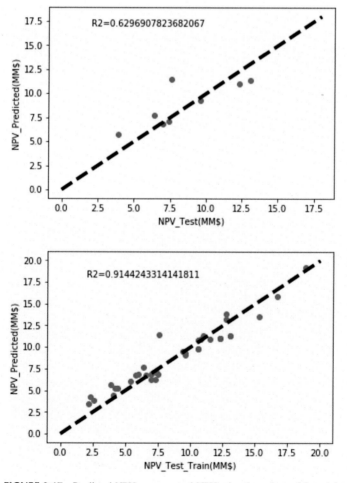

FIGURE 9.17   Predicted NPV versus actual NPV using the polynomial model.

parameters such as fracture half-length and well distance (well spacing) are considered for optimization. Fracture half-length varies from 200 to 400 ft and well distance varies from 500 to 1000 ft. The other parameters are considered constant with porosity=0.06, permeability=1E-4md, fracture conductivity= 26 (md-ft), and fracture spacing=40 ft. Before using PSO, a surface of NPV values versus variable fracture half-length and well distance is plotted to see where the maximum NPV can be found. NPV 3D surface can be plotted in Python with the following codes:

```
# Plot 3D surface for variable Fracture half-length(FH) and #Well
  #Distance(WD)
from mpl_toolkits.mplot3d import axes3d
n=20
m=25
FH_vals=np.linspace(200,400,n+1)
WD_vals=np.linspace(500,1000,m+1)
FH,WD=np.meshgrid(FH_vals, WD_vals)
z=np.zeros_like(FH)
item=np.array([0.0]*6)
item[0]=.06
item[1]= 10e-05
item[3]=26
item[4]=40
for i in range(m+1):
    for j in range(n+1):
        item[2]=FH[i,j]
        item[5]=WD[i,j]
        z[i,j]=model.predict(item.reshape(1,-1))
fig=plt.figure()
#adding 3d for subplot 1,1,1
ax0=fig.add_subplot(111, projection='3d')
surf=ax0.plot_surface(FH, WD, z, rstride=1, cstride=1,
  cmap='viridis', linewidth=0, antialiased=False)
ax0.set_xlabel('Fracture Half Length(ft)')
ax0.set_ylabel('Well Distance(ft)')
ax0.set_title('NPV(MM$)')
Python output=Fig. 9.18
```

It is observed in Fig. 9.18 that high NPV values occur at high fracture half-lengths and long well distances. In Python, "pyswarm" package can be used to perform PSO (pythonhosted). Pyswarm should be installed by "pip install pyswarm" command. Similar to the genetic algorithm, it is necessary to define the range for all the variables in the solution space. This can be done by identifying the lower band (lb) and upper band (ub) for variables in the "lb" and "ub" arrays. All the variables, excluding fracture half-length and well distance, should be constant. Since in pyswarm, upper band values should be higher than the lower band, theta equal to 10E-10 would be added to the lower

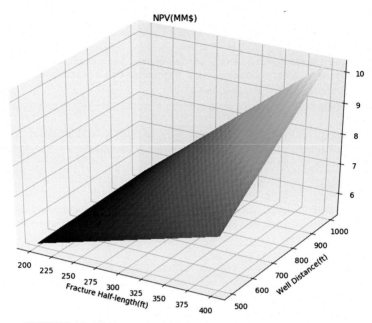

FIGURE 9.18 NPV versus variable fracture half-length and well distance.

band to generate the upper band. The objective of PSO is to maximize the NPV, therefore, the polynomial model should be multiplied by -1 (Pyswarm returns minimum of a function). PSO is used as the particle swarm optimization algorithm. In PSO, objective function (f), lower band (lb), upper band (up), number of particles in the swarm (swarm size, in this example is 200), velocity weight (omega=0.3) , C1 value (phip=0.5), C2 value (phig=0.7), maximum number of iterations (maxiter=1000), and the minimum step size of swarm's best position (minstep=1e-8) should be specified. The solution would be saved into "xopt" and optimum value would be written into "fopt." The Python codes to perform PSO on NPV polynomial function with design variables of fracture half-length and well distance are as follows:

```
from pyswarm import pso
theta=10e-10
lb=np.array([0.06, 10e-05, 200, 26, 40, 500])
ub=np.array([0.06, 10e-05, 400, 26, 40, 1000])
ub += theta
def f(X):
    return -model.predict(X.reshape(1,-1))
xopt, fopt=pso(f, lb, ub,swarmsize=200, omega=0.3, phip=.5,
  phig=0.7, maxiter=1000, minstep=1e-8)
```

```
print(xopt)
print(fopt)
Python output=
Stopping search: Swarm best position change less than 1e-08
[6.00000003e-02 1.00000403e-04 4.00000000e+02 2.60000000e+01
  4.00000000e+01 1.00000000e+03]
[-10.26411093]
```

The results show that a maximum NPV of $10.26 MM is achieved at a fracture half-length of 400 ft and a well distance of 1000 ft.

References

Al-Mudhafar, W. J., Dalton, C. A., & Al Musabeh, M. I. (May 5, 2017). *Metamodeling via hybridized particle swarm with polynomial and splines regression for optimization of CO_2-EOR in unconventional oil reservoirs.* Society of Petroleum Engineers. https://doi.org/10.2118/186045-MS

Al-Nemer, A. A., Issaka, M. B., Awotunde, A. A., & Al-Hashem, H. S. (March 8, 2015). *Global optimization strategies for well tests in dual porosity reservoirs.* Society of Petroleum Engineers. https://doi.org/10.2118/172588-MS

Almeida de, B. S. G., & Leite, V. C. (December 3, 2019). Particle swarm optimization: A powerful technique for solving engineering problems. In J. Del Ser, E. Villar, & E. Osaba (Eds.), *Swarm intelligence - recent advances, new perspectives and applications.* IntechOpen. https://doi.org/10.5772/intechopen.89633. Available from https://www.intechopen.com/books/swarm-intelligence-recent-advances-new-perspectives-and-applications/particle-swarm-optimization-a-powerful-technique-for-solving-engineering-problems.

Bybee, K. (June 1, 2005). *Optimizing cyclic-steam oil production with genetic algorithms.* Society of Petroleum Engineers. https://doi.org/10.2118/0605-0068-JPT

Chaari, M., Ben Hmida, J., Seibi, A. C., & Fekih, A. (August 1, 2020). *An integrated genetic-algorithm/artificial-neural-network approach for steady-state modeling of two-phase pressure drop in pipes.* Society of Petroleum Engineers. https://doi.org/10.2118/201191-PA

Eberhart, R. C., Shi, Y., & Kennedy, J. (April 11, 2001). *Swarm intelligence* (1st ed.). Morgan kaufmann. ASIN: B001IV75CG.

Emerick, A. A., Silva, E., Messer, B., Almeida, L. F., Szwarcman, D., Pacheco, M. A. C., & Vellasco, M. M. B. R. (January 1, 2009). *Well placement optimization using a genetic algorithm with nonlinear constraints.* Society of Petroleum Engineers. https://doi.org/10.2118/118808-MS

Fang, J. H., Karr, C. L., & Stanley, D. A. (January 1, 1992). *Genetic algorithm and its application to petrophysics.* Society of Petroleum Engineers.

Fernandez-Martinez, J. L., García-Gonzalo, M. E., Fernández-Alvarez, J. P., Menéndez Pérez, C. O., & Kuzma, H. A. (January 1, 2008). *Particle swarm optimization (PSO): A simple and powerful algorithm family for geophysical inversion.* Society of Exploration Geophysicists.

Goldberg, D. E. (1983). *Computer aided gas pipeline operation using genetic algorithms and rule learning* (Ph.D. dissertation). Ann Arbor, Michigan: University of Michigan.

https://pypi.org/project/geneticalgorithm/.

https://pythonhosted.org/pyswarm/.

https://www.tutorialspoint.com/genetic_algorithms.

Isebor, O. J., Echeverria Ciaurri, D., & Durlofsky, L. J. (February 18, 2013). *Generalized field development optimization using derivative-free procedures.* Society of Petroleum Engineers. https://doi.org/10.2118/163631-MS

Javadzadeh, E., & Javadzadeh, R. (January 1, 2008). *Bishop's simplified method and particle swarm optimization forlocation the critical failure surface in rock slope stability analysis.* International Society for Rock Mechanics and Rock Engineering.

Jefferys, E. R. (January 1, 1993). *Design applications of genetic algorithms.* Society of Petroleum Engineers. https://doi.org/10.2118/26367-MS

Jesmani, M., Bellout, M. C., Hanea, R., & Foss, B. (September 14, 2015). *Particle swarm optimization algorithm for optimum well placement subject to realistic field development constraints.* Society of Petroleum Engineers. https://doi.org/10.2118/175590-MS

Khan, M. R., Sadeed, A., Asad, A., Tariq, Z., & Tauqeer, M. (January 1, 2018). *Maximizing oil recovery in a naturally fractured carbonate reservoir using computational intelligence based on particle swarm optimization.* Society of Petroleum Engineers. https://doi.org/10.2118/195664-MS

Le Ravalec-Dupin, M., Enchery, G., Baroni, A., & Da-Veiga, S. (January 1, 2010). *Pre-selection of reservoir models from a geostatistics-based petrophysical seismic inversion.* Society of Petroleum Engineers. https://doi.org/10.2118/131310-MS

Martinez, E. R., Moreno, W. J., Moreno, J. A., & Maggiolo, R. (January 1, 1994). *Application of genetic algorithm on the distribution of gas-lift injection.* Society of Petroleum Engineers. https://doi.org/10.2118/26993-MS

Mohaghegh, S. (October 1, 2000). *Virtual-intelligence applications in petroleum engineering: Part 2—evolutionary computing.* Society of Petroleum Engineers. https://doi.org/10.2118/61925-JPT

Mohamed, L., Christie, M. A., & Demyanov, V. (January 1, 2011). *History matching and uncertainty quantification: Multiobjective particle swarm optimisation approach.* Society of Petroleum Engineers. https://doi.org/10.2118/143067-MS

Morales, A. N., Gibbs, T. H., Nasrabadi, H., & Zhu, D. (January 1, 2010). *Using genetic algorithm to optimize well placement in gas condensate reservoirs.* Society of Petroleum Engineers. https://doi.org/10.2118/130999-MS

Ottah, D. G., Ikiensikimama, S. S., & Matemilola, S. A. (August 4, 2015). *Aquifer matching with material balance using particle swarm optimization algorithm − PSO.* Society of Petroleum Engineers. https://doi.org/10.2118/178319-MS

Rajasekaran, S., & Pai, G. A. Vijayalakshmi (June 16, 2013). *Neural networks, fuzzy logic and genetic algorithms: Synthesis and applications.* PHI Learning Private Limited. ISBN-978-81-203-2186-1.

Romero, C. E., Carter, J. N., Gringarten, A. C., & Zimmerman, R. W. (January 1, 2000). *A modified genetic algorithm for reservoir characterisation.* Society of Petroleum Engineers. https://doi.org/10.2118/64765-MS

Salamanca, A., Gutiérrez, E., & Montes, L. (October 23, 2017). *Optimization of a seismic inversion genetic algorithm.* Society of Exploration Geophysicists.

Sarich, M. D. (January 1, 2001). *Using genetic algorithms to improve investment decision making.* Society of Petroleum Engineers. https://doi.org/10.2118/68725-MS

Self, R., Atashnezhad, A., & Hareland, G. (September 14, 2016). *Reducing drilling cost by finding optimal operational parameters using particle swarm algorithm.* Society of Petroleum Engineers. https://doi.org/10.2118/180280-MS

Yan, L., Lu, H., Shen, Q., Wang, H., Fu, X., & Chen, J. (November 30, 2018). *Multilayer-geosteering inversion using particle-swarm optimization.* Society of Exploration Geophysicists.

Yu, W., & Sepehrnoori, K. (2013). Optimization of multiple hydraulically fractured horizontal wells in unconventional gas reservoirs. *Journal of Petroleum Engineering, 2013*, Article 151898, 16 pp.

Zhao, H., Chen, C., Do, S. T., Li, G., & Reynolds, A. C. (January 1, 2011). *Maximization of a dynamic quadratic interpolation model for production optimization.* Society of Petroleum Engineers. https://doi.org/10.2118/141317-MS

Index

'*Note*: Page numbers followed by "f" indicate figures, "t" indicate tables and "b" indicate boxes.'

Printed in the United States
by Baker & Taylor Publisher Services